Yuehua Hu
Wei Sun
Dianzuo Wang

Electrochemistry of Flotation of Sulphide Minerals

Yuehua Hu
Wei Sun
Dianzuo Wang

Electrochemistry of Flotation of Sulphide Minerals

With 287 figures

TSINGHUA UNIVERSITY PRESS

Springer

Authors
Prof. Yuehua Hu
School of Minerals and Bio-engineering
Central South University
410083, Changsha, China
E-mail: hyh@mail.csu.edu.cn

Prof. Wei Sun
School of Minerals and Bio-engineering
Central South University
410083, Changsha, China
E-mail: sunmenghu@mail.csu.edu.cn

Prof. Dianzuo Wang
School of Minerals and Bio-engineering
Central South University
410083, Changsha, China
E-mail: wdz@mail.csu.edu.cn

ISBN 978-7-302-18818-6
Tsinghua University Press, Beijing

ISBN 978-3-540-92178-3 e-ISBN 978-3-540-92179-0
Springer Dordrecht Heidelberg London New York

Library of Congress Control Number: 2008940827

Cover design: Frido Steinen-Broo, EStudio Calamar, Spain

Printed on acid-free paper

Springer is part of Springer Science+Business Media (www.springer.com)

About Authors

Yuehua Hu Born in 1962, graduated from Central South University (CSU, the former Central South Institute of Mining and Metallurgy) in 1982, got the doctor degree in 1989, and elected as professor in 1991. His professional researches are related with the structure-property of flotation reagents and molecular design, the electrochemistry of flotation of sulphide minerals, the solution chemistry of flotation, the interfacial interaction and fine particle flotation. Hu has acquired the better achievement in above fields and has got many top honors, including 1st or 2nd class National Science & Technology Advancement Award, China Book Award, Chinese Youth Award of Science and Technology, National Scientific Award for Outstanding Youth etc. More than 200 papers had been published in China or foreign countries.Hu was honored as Cheung Kong Scholar of the Ministry of Education, elected as vice-chairman of Mineral Processing Committee of China Nonferrous Metals Society, engaged as the adjunct professor of metallurgical department of University of Utah.

Wei Sun Born in 1974, graduated from CSU in 1995, got the doctor degree in 2001. Sun is an associate professor at the School of Minerals Processing & Bioengineering of CSU. He had the post-graduate experience at Shanghai Institute of Microsystems and Information Technology, which belongs to Chinese Academy of Science, and at the Department of Materials and Chemical Engineering, Alberta University in Canada. His research fields include: flotation of minerals, flotation chemistry and fine particle flotation or processing. More than 30 papers had been published in China or foreign countries.

Dianzuo Wang Born in 1934, graduated from CSU in 1961, elected as the member of Chinese Academy of Science in 1991, the initial member of Chinese Academy of Engineering in 1994. He became a foreign associate of the National Academy of Engineering (USA) in 1990, the foreign nationality member of Russian Academy of Engineering in 1994, the foreign nationality member of Russian Academy of Science in 2006. He had the position of president of CSU, the position of director of General Chinese Institute of Nonferrous Metal Research, the position of vice-chairman of Chinese Academy of Engineering.

Preface

Sulphide minerals are the important resources of nonferrous metals, which are widely used in many fields such as electronics, chemical engineering, aviation, transportation, construction and metallurgy etc. Flotation technology is the most important method to beneficiate sulphide ores to obtain concentrate containing nonferrous metals since it has been used in industry at the end of the 19th century. As early as in the 1930s—1950s, considerable amount of fundamental research had been conducted on the flotation of sulphide minerals based on wettability, electrokinetics and chemical interactions. The "chemical reaction" hypothesis based on the solubility product of metal-collector salts involved and competitive adsorption based on the interaction of thio-collectors and hydroxide ions with sulphides were proposed to explain the flotation mechanism of sulphide minerals. A mixed potential model has been established, in which there existed a potential at which an anodic reaction of xanthate to form metal xanthate or dithiolate and the cathodic reduction of oxygen proceed at finite rates, underlying the basis of the researches on the interaction mechanism between thio collectors and sulphide minerals in 1950s. Since then, it was realized that the electrochemistry was very important in flotation of sulphide minerals as it determined the oxidation and dissolution of mineral particles in the pulp, the electrochemical equilibria of reagents, electrochemical interactions among reagents with both soluble and surface species of minerals and the galvanic interactions among grinding media, mineral and reagents. The efficiency of flotation and separation of sulphide minerals and consumption of reagents are thus controlled by the electrochemistry of the pulp. The potential controlled flotation technology has been developed and used in industry recently to obtain higher efficiency of flotation separation in many copper, lead, zinc sulphide mines in China.

In this book, a general review of the fundamental research on flotation electrochemistry of sulphide minerals is made first in Chapter 1. Chapter 2 to Chapter 9 mainly summarize the results of basic research in our group focused on the topics of collectorless floatability of sulphide minerals and hydrophobic entity, interaction mechanisms between thio-collectors and sulphide minerals, role of oxygen in flotation of sulphide minerals and galvanic interactions in potential controlled flotation systems etc. The various electrochemical measurements, electrochemical corrosive method, electrochemical equilibrium calculations, surface analysis, semiconductor energy band and molecular orbital theory, have been used in our studies and introduced in this book. The collectorless and collector-induced flotation behavior of sulphide minerals and mechanism in various flotation systems have been discussed. The electrochemical corrosive mechanism, mechano-electrochemistrical behavior and the molecular orbital approach to flotation of sulphide minerals will provide much new information to the researchers in this area. The examples of potential controlled flotation separation of sulphide ores, especially the examples of industrial application, listed in Chapter 10 demonstrate the future promise of flotation electrochemistry of sulphide minerals for industrial applications.

The authors would like to acknowledge the help of many students and associates in the mineral processing: Dr. Runqing Liu who helped in editing the book, Dr. Guohua Gu, Dr. Shuiyu Sun, Dr. Qin Zhang, Dr. Yunlan Yu and Dr. Daoling Xiong who organized materials for some chapters.

Professor Yuehua Hu
Dr. Wei Sun
Professor Dianzuo Wang
Central South University
Changsha, 410083, China

Contents

Chapter 1 General Review of Electrochemistry of Flotation of Sulphide Minerals

Abstract This chapter reviews the development of froth flotation achieved in the past one hundred years and accounts for the achievements of the theory of flotation of sulphide minerals in four aspects, which are the natural flotability of sulphide minerals, the role of oxygen in the flotation of sulphide minerals, the interaction of collectors with sulphide minerals, the effect of the semi-conductor property of sulphide minerals and electrochemical behaviors in the grinding system. Furthermore, the purpose of this book is revealed in the end.

Keywords sulphide minerals; flotation; electrochemistry

1.1 Three Periods of Flotation of Sulphide Minerals

The one hundred year history of froth flotation may be classified into three periods. The earliest stage is from the end of the 19th century to the early 20th century, i.e. surface flotation or bulk oil flotation. The natural hydrophobic sulphide minerals can be collected by the addition of oil. Froth flotation came into practice in 1909 with the use of pine oil, mechanical flotation machine in 1912, and xanthate and aerofloat as collectors in 1924—1925 (Gaudin, 1932; Sutherland and Wark, 1955).

In the 2nd period ranging from the 1930s to the 1950s, basic research on flotation was conducted widely in order to understand the principles of the flotation process. Taggart and co-workers (1930, 1945) proposed a "chemical reaction" hypothesis, based on which the flotation of sulphide minerals was explained by the solubility product of the metal-collector salts involved. It was plausible at that time that the floatability of copper, lead, and zinc sulphide minerals using xanthate as a collector decreased in the order of increase of the solubility product of their metal xanthate (Karkovsky, 1957). Sutherland and Wark (1955) paid attention to the fact that this model was not always consistent with the established values of the solubility products of the species involved. They believed that the interaction of thio-collectors with sulphides should be considered as adsorption and proposed a mechanism of competitive adsorption between xanthate and hydroxide ions, which explained the Barsky empirical relationship between the upper pH limit of flotation and collector concentration. Gaudin (1957) concurred with Wark's explanation of this phenomenon. Du Rietz

(1957) considered the formation of metal xanthate as the adsorption mechanism of thio-collectors on sulphide minerals and used the Bjerrum diagram to explain the floatability and the upper pH limit of flotation of sulphide minerals. The adsorption of undissociated acid form of the collector was ever noted by Wark and Cox (1933), Cook and Nixon (1950), and Cook and Wadsworth (1957).

In this period, the collectorless floatability of sulphide minerals was referred to as the natural floatability and thought to be typical of sulphides by some investigators (Ravitz and Porter, 1933). Some authors demonstrated that sulphide minerals were not naturally floatable and the so-called natural floatability might have originated from contamination (Hayes et al., 1987). At the same time, the importance of the presence of oxygen for sulphide minerals to float with xanthate collectors was first revealed by Salamy and Nixon (1953, 1954). They proposed that the interaction between the collector and oxygen at the mineral surface takes place by separate electrochemical reactions which proceed simultaneously. An anodic oxidation reaction involving the collector transfers electrons to the mineral and these are returned to the solution phase by the cathodic reduction of oxygen. This model was supported by investigations that showed the presence of oxygen to be necessary for flotation of sulphide minerals with xanthate (Plaksin and Bessonov, 1957). Thus, Nixon (1957) recognized that the theories of flotation could be reconciled by the electrochemical approach.

Since the 1960s', various electrochemical methods such as linear potential sweep voltammetry, cyclic voltammetry etc. and various surface analysis apparatuses such as infrared spectra, X-ray photoelectron spectroscopy etc. have been developed to investigate the electrochemical reaction mechanism involved in the flotation of sulphide minerals (Fuerstenau et al., 1968; Woods, 1976; Ahmed, 1978; Sun, 1990; Feng, 1989; Buckley, 1995; Arce and Gonzalez, 2002; Bulut and Atak, 2002; Costa et al., 2002).

Fuerstenau (1980) found that sulphide minerals are naturally floatable in the absence of oxygen. Yoon (1981) ever attributed the natural floatability of some sulphide minerals to their very low solubility. Finkelstein et al. (1975) considered that the natural floatability of sulphide minerals are due to the formation of elemental sulphur and related to the thickness of formation of elemental sulphur at the surface. Some authors reported that the hydrophobic entity in collectorless flotation of sulphide minerals were the metal-deficient poly sulphide (Buckley et al., 1985). No matter whichever mechanism, investigators increasingly concluded that most sulphide minerals are not naturally floatable and floated only under some suitable redox environment. Some authors considered that the natural floatability of sulphide minerals was restricted to some special sulphide minerals such as molybdenite, stibnite, orpiment etc. owing to the effects of crystal structure and the collectorless floatability of most sulphide minerals could be classified into self-induced and sulphur-induced floatability (Trahar, 1984; Heyes and Trahar, 1984; Hayes et al., 1987; Wang et al., 1991b, c; Hu et al., 2000).

In the aspects of the interaction between thio-collectors and sulphide minerals, a mixed potential model was established, in which there existed a potential at which an anodic reaction of xanthate to form metal xanthate or dithiolate and the cathodic reduction of oxygen proceed at finite rates, underlying the basis of the researches on the interaction mechanism between thio collectors and sulphide minerals (Fuerstenau et al., 1968; Woods, 1971; Rubio and Kitchener, 1964; Allison et al., 1972; Richardson and Maust, 1976). "Potential control flotation theory", either in collector flotation or in collectorless flotation, has been gradually established and accepted, in which the flotation of sulphide minerals is possible only under certain redox potential and the separation of poly-metal sulphide minerals can be accomplished by E_h control. The E_h control flotation of sulphide minerals are investigated based on the knowledge including flotation tests, electrochemical measurements, electrochemical equilibrium calculations, various surface analysis, molecular orbital theory, semiconductor energy band theory, crystal chemistry and has been found to be applicable to mineral industry (Hayes and Ralston, 1988; Wang et al., 1992a; Gu et al., 2002a,c, 2004).

1.2 Natural Floatability and Collectorless Flotation of Sulphide Minerals

The inherent hydrophobicity once thought to be typical of sulphides (Ravitz and Porter, 1933) is now thought to be restricted to sulphides such as molybdenite (Chander et al., 1975) and other minerals or compound with special structural feature (Gaudin et al., 1957b). Common commercial sulphide minerals, which are needed to recover in flotation, are normally composed of anion (S^{2-}) and heavy metal ions such as Cu^+, Cu^{2+}, Pb^{2+}, Zn^{2+}, Hg^+, Sb^{3+}, Bi^{3+}; transitive metal ion such as Fe^{2+}, Co^{2+}, Ni^{2+}; and noble and rare metal ions such as Ag^+, Au^+, Mo^{4+}. On the basis of "structural pattern" or mode of linkage of the atoms or polyhedral units in space, Povarennyk (1972) introduced a crystallochemical classification of sulphide minerals, which have six major patterns as shown in Table 1.1.

The most controversial and contradicting problem is, perhaps, the natural and collectorless floatability of sulphide minerals. Gaudin (1957) classified the natural hydrophobicity of different minerals according to their crystal structure and showed that most sulphide minerals were naturally hydrophobic to some extent, which had been further proved based on van der Waals theory by Chander (1988, 1999). Lepetic (1974) revealed the natural floatability of chalcopyrite in dry grinding. Finklestein (1975, 1977) demonstrated that orpiment, realgar and molybdenite were naturally floatable, and that pyrite and chalcopyrite had natural floatability at certain conditions due to the formation of surface elemental sulphur. Buckley and Woods (1990, 1996) attributed the natural floatability of chalcopyrite

Table 1.1 The sulphide classification scheme (Vanghan and Craig, 1978; Povarennyk, 1972)

Subclass	Division
Coordination	Simple e.g. Troilite (FeS) group Galena (PbS) group
	Complex e.g. Polydymite ((Co, Ni)(Co, Ni)$_2$S$_4$) group Pentlandite ((FeNi)$_9$S$_8$) group Bornite (Cu$_5$FeS$_4$) group
Framework	Simple e.g. Argentite (Ag$_2$S) group
	Complex e.g. Tetrahedrite (Cu$_{12}$Sb$_4$S$_{13}$) group
Insular	e.g. Pyrite-marcasite (FeS$_2$) group Cobaltite-arsenopyrite (CoAsS-FeAsS) group
Ring	e.g. Realgar (As$_4$S$_4$) group
Chain	Simple e.g. Stibnite (Sb$_2$S$_3$) group Millerite (NiS) group
	Complex e.g. Berthierite (FeSb$_2$S$_4$) group Lautite (CuAsS) group
Layer	Simple e.g. Molybdenite (MoS$_2$) group
	Complex e.g. Covellite (Cu$_2$CuS$_2$S) group

to the formation of metal-deficient poly sulphide, which was further tested by other reports (Chander et al., 1988; Woods and Yoon, 1994; Woods et al., 1994). Most sulphide minerals do not exhibit natural floatability. They have, however, self-induced collectorless floatability in specific pulp potential ranges, under the condition of which the sulphide mineral surface has been rendered hydrophobic by the surface self-oxidation. The recovery- size curves of 8 common sulphide minerals in the absence of conventional collectors but at favorable levels of pulp potential (E_{Pt}) and pH ranked collectorless floatability of some sulphides, in descending order are as: chalcopyrite > galena > pyrrhotite > pendlite > covellite > bornite > sphalerite > pyrite > arsenopyrite. The floatability of the first four minerals is strong over a wide size range. Whereas, the floatability of the last four minerals is weak in descending order from bornite to inactivated sphalerite. Covellite floats somewhat like bornite and arsenopyrite is similar to pyrite (Trahar et al., 1983; Guy and Trahar, 1985).

Many authors indicated that self-induced flotation of sulphide minerals can occur only under moderately oxidizing environments. It is obvious that the self-induced collectorless flotation of many sulphide minerals is controlled by pulp potential (Heyes and Trahar, 1977; Gardner and Woods, 1979; Hayes et al., 1987; Hayes and Ralston, 1988; Cheng and Iwasaki, 1992; Sun, 1990; Wang et al., 1991, 1992; Yekeler and Sonmez, 1997; Ekmekci and Demirel, 1997; Zhang Qin et al., 2004c).

The influence of pulp potential on the floatability of galena, arsenopyrite, pyrrhotite and pyrite was reported and given in Fig. 1.1 and Fig. 1.2. Galena, arsenopyrite, and pyrrhotite show strong self-induced collectorless flotation in certain range of potential although these minerals also exhibit different upper and lower potential limit for flotation. In Fig. 1.2, pyrite shows no sign of self-induced flotation, while other investigator claimed self-induced flotation of pyrite, especially in strong acidic pH range (Heyes and Trahar, 1984; Sun et al., 1992, 1993).

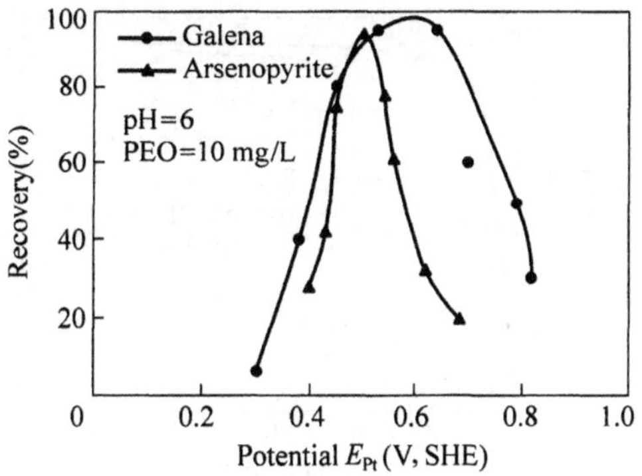

Figure 1.1 Effect of pulp potential on self-induced collectorless flotation behaviors of galena and arsenopyrite at pH = 6 (Sun, 1990)

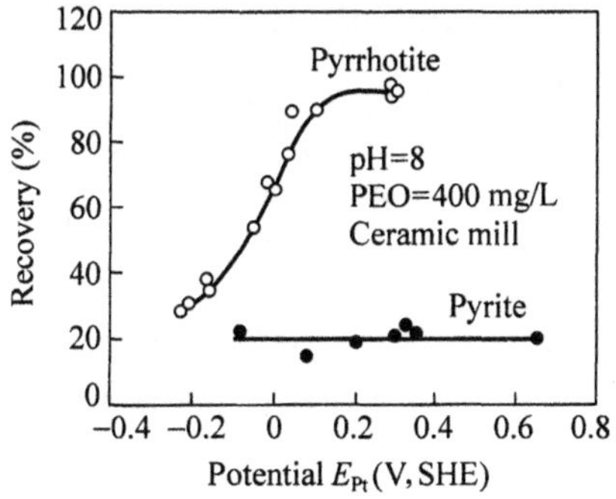

Figure 1.2 Effect of pulp potential on self-induced collectorless flotation behaviors of pyrrhotite and pyrite (Heyes and Trahar, 1984)

It has been reported that although pyrite is only slightly floatable in self-induced flotation, the floatability is pronounced if sodium sulphur is added, which is called sulphur-induced collectorless flotation. As can be seen from

Fig. 1.3, sulphur-induced flotation of pyrite was observed at pH = 8 if the concentration of the sulphur during conditioning was 10^{-3} mol/L or greater, but there was no significant flotation when the concentration of the sulphur was below this level (Trahar, 1984).

Heyes and Trahar (1984) leached pyrite with cyclohexane and compared the extract with a sulphur-containing solution of cyclohexane in a UV spectra photometer as shown in Fig. 1.4, indicating that sulphur was present at the mineral surface. Therefore, the inherent hydrophobicity and natural floatability once thought to be typical of sulphides is now thought to be restricted to sulphides such as molybdenite and other minerals or compound with special structural features. The collectorless floatability that most sulphide minerals showed came from the self-induced or sulphur-induced flotation at certain pulp potential range and certain conditions.

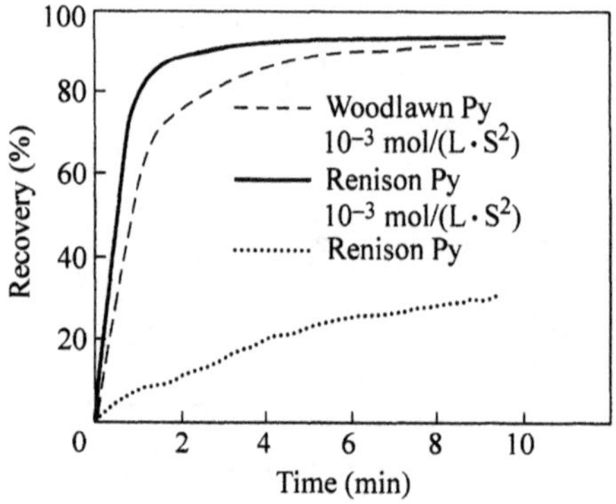

Figure 1.3 Recovery-time data for sulphur-induced flotation of pyrite at pH = 8 (Trahar, 1984)

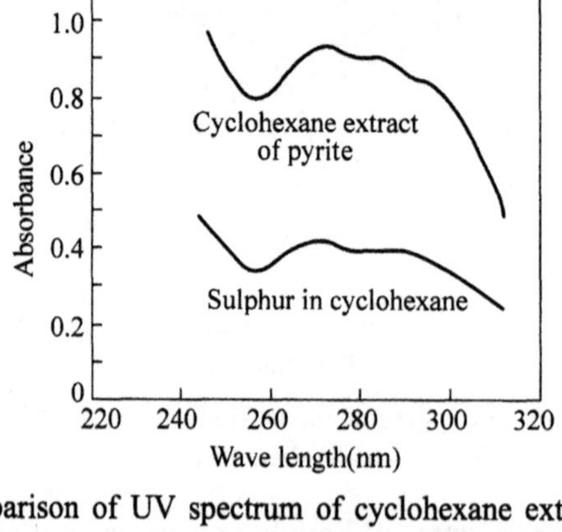

Figure 1.4 Comparison of UV spectrum of cyclohexane extract of sulphidized pyrite after flotation with that of cyclohexane solution of sulphur (Heyes and Trahar, 1984)

1.3 Role of Oxygen and Oxidation of Sulphide Minerals in Flotation

In the early stage of flotation, the presence of oxygen was considered to be harmful to flotation. Gaudin (1957) considered that the oxidation of sulphide surface produced oxygen-sulphur species, which exchanged further with the collector ion to form metal collector salt. Plaksin et al. (1957, 1963) found that the presence of oxygen consumed the free electrons on the galena surface and changed the galena surface from *n*-type to *p*-type, which resulted in the adsorption of xanthate ion due to the presence of hole on *p*-type galena semiconductor.

Most importantly, Tolun and Kitchener (1963, 1964), Fuerstenau et al. (1968) and Majima (1968, 1969) found that the presence of oxygen as a cathodic component was indispensable for the electrochemical reaction on sulphide surface based on electrochemical theory. These works underlay the basis of the later "mixed potential model". In 1975, Biegler et al. (1975) investigated the reduction of oxygen on the surface of different types of pyrite using rotating disc electrode technique and found that the difference of reduction behavior of oxygen was due to the different surface electronic structure. Rand (1975) reported the reduction behavior of oxygen on chalcopyrite, galena, arsenopyrite, covellite, chalcocite, pyrite etc. to show that oxygen was most active at the surface of pyrite and inert at the surface of galena. He considered that the difference of cathodic reduction of oxygen affected the adsorption of collector on mineral surfaces, which was further proved by Neeraj (2000) in the light of their researches using electrochemical impedance spectroscopy. Ahlberg and Broo (1988, 1996, and 1997) studied the oxygen reduction on gold, platinum, pyrite and galena in the presence of xanthate using rotating disc electrode technique. They found that the product of oxygen reduction was dominant as perhydroxide on pyrite surface and as water on galena surface. The current of the cathodic process of oxygen reduction was greater on pyrite than on galena and the formation of dixanthogen on pyrite may relate to the presence of hydrogen peroxide.

Recently, more researches focused on the oxidation products and the rate of sulphide mineral surface. It was reported that the oxidation products of sulphide mineral were affected by pulp potential and pH as well as time. At relative low potential, the surface of galena was oxidized to produce elemental sulphur resulting in collectorless flotation. At high potential, the surface of galena was oxidized to produce lead sulphate at $pH = 6$, lead thiosulphate at $pH = 9$ and lead hydroxide and sulphate at $pH > 12$. Pyrite was oxidized to form ferrous hydroxide and sulphate at the initial stage and further to ferric hydroxide in $pH = 9.2$ buffer solution. It was generally agreed that the oxidation rate of sulphide minerals were faster at high potential than at low potential. The oxidation rate of sulphide minerals were in the order of pyrrhotite > pyrite > chalcopyrite > sphalerite > galena (Feng, 1989; Qin, 1998; Kelsall and Yin, 1999). Brion (1980) reported that the oxidation rate of chalcopyrite and pyrite greater than that of galena and sphalerite in terms of the results of XPS and polarized curve.

Therefore, it has been concluded that the reduction of oxygen as a cathodic process was essential for the electrochemical reaction on sulphide surface and was different for various sulphide minerals. The reduction of oxygen affected the oxidation of sulphide minerals and the interactions with collectors, which had a pronounced influence on flotation behavior of sulphide minerals (Ahmed, 1978; Buckley et al., 1985, 1995; Woods, 1984, 1994; Hu et al., 2004; Yu et al., 2004a; Zhang et al., 2004a, d).

1.4 Interactions between Collector and Sulphide Minerals and Mixed Potential Model

Before the establishment of the "mixed potential mode", there was much controversy on the interactions between thio-collector and sulphide minerals. The chemical reaction, adsorption of ion or undissociated acid had ever been suggested. Since the 1960s' the mode that an anodic oxidation reaction involving the collector transfers electrons to the mineral and these are returned to the solution phase by the cathodic reduction of oxygen was supported by many investigations (Fuerstenau et al., 1968; Woods, 1971, 1976; Allison, 1972; Goold, 1972; Richardson, 1976; Trahar, 1984; Hu et al., 2000; Yu et al., 2004b,c). According to this model, the electrochemical reactions between thio-collector and sulphide mineral included the cathodic reduction of oxygen in Eq. (1-1)

$$\frac{1}{2}O_2 + H_2O + 2e^- \longrightarrow 2OH^- \qquad (1\text{-}1)$$

and the anodic process that the collector transfers electron to the mineral in different ways shown in Eqs. (1-2), (1-3) and (1-4).

The electrochemical adsorption of thio-collector ion

$$X^- \longrightarrow X_{ads} + e^- \qquad (1\text{-}2)$$

The oxidation of the thio-collector to its dithiolate

$$2X^- \longrightarrow X_2 + 2e^- \qquad (1\text{-}3)$$

The chemisorption reaction mechanism

$$MS + 2X^- \longrightarrow MX_2 + S + 2e^- \qquad (1\text{-}4a)$$

$$MS + 2X^- + 4H_2O \longrightarrow MX_2 + SO_4^{2-} + 8H^+ + 8e^- \qquad (1\text{-}4b)$$

$$2MS + 4X^- + 3H_2O \longrightarrow 2MX_2 + S_2O_3^{2-} + 6H^+ + 8e^- \qquad (1\text{-}4c)$$

where X^- represents a thio ion and MS represents a sulphide mineral.

The overall reactions were

$$2X^- + \frac{1}{2}O_2 + H_2O \longrightarrow X_2 + 2OH^- \qquad (1\text{-}5)$$

or

$$MS + 2X^- + \frac{1}{2}O_2 \longrightarrow MX_2 + S + H_2O \qquad (1\text{-}6a)$$

$$MS + 2X^- + 2O_2 \longrightarrow MX_2 + SO_4^{2-} \qquad (1\text{-}6b)$$

$$2MS + 4X^- + \frac{3}{2}O_2 \longrightarrow 2MX_2 + S_2O_3^{2-} \qquad (1\text{-}6c)$$

For overall reactions such as (1-5) or (1-6) to proceed, there must be a potential, termed a mixed potential, at which the anodic reactions, such as (1-3) and (1-4) and the cathodic reaction of oxygen proceed at finite rates.

The mixed-potential model demonstrated the importance of electrode potential in flotation systems. The mixed potential or rest potential of an electrode provides information to determine the identity of the reactions that take place at the mineral surface and the rates of these processes. One approach is to compare the measured rest potential with equilibrium potential for various processes derived from thermodynamic data. Allison et al. (1971, 1972) considered that a necessary condition for the electrochemical formation of dithiolate at the mineral surface is that the measured mixed potential arising from the reduction of oxygen and the oxidation of this collector at the surface must be anodic to the equilibrium potential for the thio ion/dithiolate couple. They correlated the rest potential of a range of sulphide minerals in different thio-collector solutions with the products extracted from the surface as shown in Table 1.2 and 1.3. It can be seen from these Tables that only those minerals exhibiting rest potential in excess of the thio ion/disulphide couple formed dithiolate as a major reaction product. Those minerals which had a rest potential below this value formed the metal collector compounds, except covellite on which dixanthogen was formed even though the measured rest potential was below the reversible potential. Allison et al. (1972) attributed the behavior to the decomposition of cupric xanthate.

The electrochemical studies of pyrite/oxygen/xanthate have been carried out using polarographic method and potential sweep technique by many researchers (Majima and Takeda, 1968; Usul and Tolun, 1974; Janetski et al., 1977; Gardner and Woods, 1977). They found that an anodic current which commenced at a potential close to the reversible value for the xanthate/dixanthogen couple. The typical results investigated by Gardner and Woods (1977) are shown in Fig. 1.5. It shows that the oxidation of each xanthate begins at a potential close to the equilibrium potential of the corresponding xanthate/dixanthogen couple, and that the finite contact angles are formed only at potential where dixanthogen is formed: They also found that a monolayer of dixanthogen was required to float a pyrite particulate bed with ethyl xanthate at $pH = 9.2$.

Table 1.2 Products of the interaction of sulphide minerals with potassium ethyl xanthate and the measured rest potential. Potassium ethyl xanthate (6.25×10^{-4} mol/L at pH = 7) reversible potential for oxidation to dixanthogen is 0.13 V (Allison et al., 1972)

Minerals	Rest potential (V)	Products
pyrite	0.22	dixanthogen
arsenopyrite	0.22	dixanthogen
pyrrhotite	0.21	dixanthogen
molybdenite	0.16	dixanthogen
chalcopyrite	0.14	dixanthogen
alabandite	0.15	dixanthogen
covellite	0.05	dixanthogen
bornite	0.06	metal xanthate
galena	0.06	metal xanthate

Table 1.3 Products of the interaction of sulphide minerals with sodium dimethyl dithio carbamate (5.8×10^{-4} mol/L at pH = 8) and the measured rest potential (reversible potential for oxidation to thiouram disulphide is 0.176 V) (Finkelstein and Goold, 1972)

Minerals	Rest potential (V)	Products
pyrite	0.475	disulphide
covellite	0.115	metal dithiocarbamate
chalcopyrite	0.095	metal dithiocarbamate
galena	−0.035	metal dithiocarbamate
bornite	−0.045	metal dithiocarbamate
chalcocite	−0.155	metal dithiocarbamate

Guy and Trahar (1984) compared the flotation data with the voltammetric results given by Woods (1971), as shown in Fig. 1.6. It can be seen from Fig. 1.6 that if galena is initially reduced, flotation of galena began in the same potential region as an anodic peak occurred on the voltammograms. If galena was ground in an oxidizing environment, the anodic current appeared at about −0.5 V at which flotation occurred, which was attributed to the formation of lead xanthate from lead in the surface formed by the oxidation conditions.

Some instrumental methods have been used for the investigation of sulphide mineral-thio-collector system such as infra-red (IR) spectroscopy (Mielezarski and Yoon, 1989; Leppinen et al., 1989; Persson et al., 1991; Laajalehto et al., 1993; Zhang., et al., 2004a) and X-ray photoelectron spectroscopy (XPS) (Pillai et al., 1983; Page and Hazell, 1989; Grano et al., 1990; Laajalechto et al., 1991). These surface sensitive spectroscopic techniques can be applied for the direct determination of the surface composition at the conditions related to flotation.

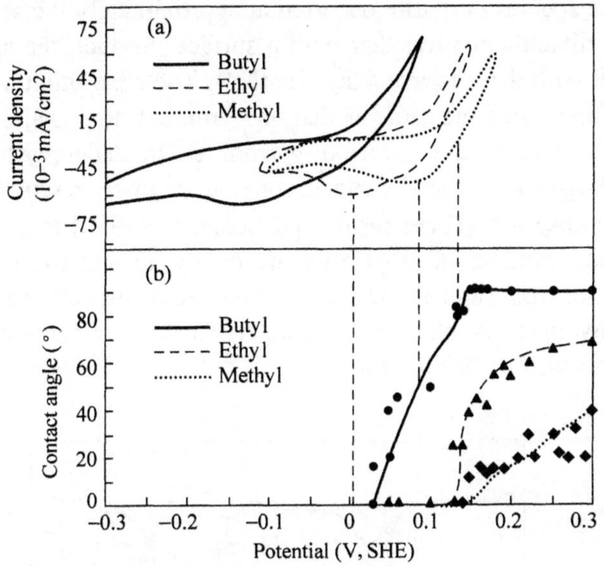

Figure 1.5 Pyrite electrode at pH = 9.2 in 0.05 mol/L sodium tetraborate solution containing 1000 mg/L of three potassium alkylxanthates. (a) Cyclic voltammograms at 4 mV/s; (b) Contact angles measured after holding the electrode at each potential for 30s. Vertical lines are the reversible potential of the xanthate/dixanthogen couples (Gardner and Woods, 1977)

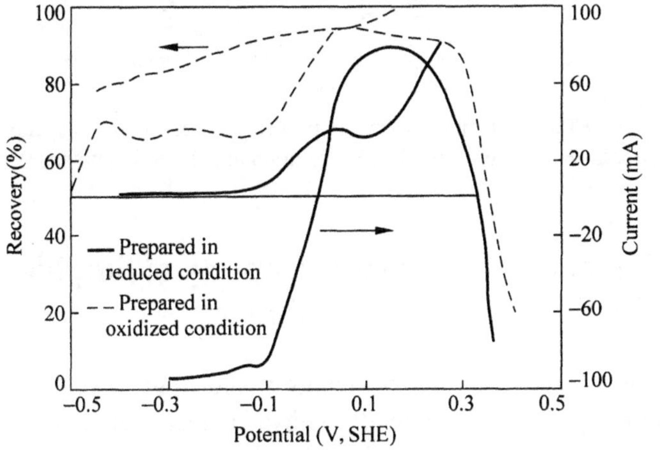

Figure 1.6 Anodic current and floatability of galena as a function of potential in the presence of ethylxanthate at pH = 8 after preparation in oxidizing and reducing environment (Guy and Trahar, 1985)

Leppinen (1990) used FTIR-ATR techniques to study the adsorption of ethyl xanthate on pyrite, pyrrhotite and chalcopyrite. The FTIR spectra of the reaction products on pyrite, pyrrhotite and chalcopyrite after treatment with 1×10^{-4} mol/L KEX solution at pH = 5 is presented in Fig. 1.7. He found that the FTIR signals of the xanthate species on pyrite occurred at approximately the same position as those of bulk dixanthogen together with a surface product, the signal of which coincided well with that of ferric ethyl xanthate. The adsorption product of KEX on pyrrhotite was much the same as that on pyrite but no signal of iron xanthate species was observed on the surface of pyrrhotite. On chalcopyrite the signals of diethyl dixanthogen and a metal xanthate compound closely resembling copper (I) xanthate were observed. These results led Leppinen (1990) to conclude that the main adsorption product of ethyl xanthate on pyrite and pyrrhotite is diethyl dixanthogen. An iron xanthate surface compound is also observed on pyrite at monolayer coverage. A mixture of copper xanthate compound and diethyl dixanthogen occurs on chalcopyrite.

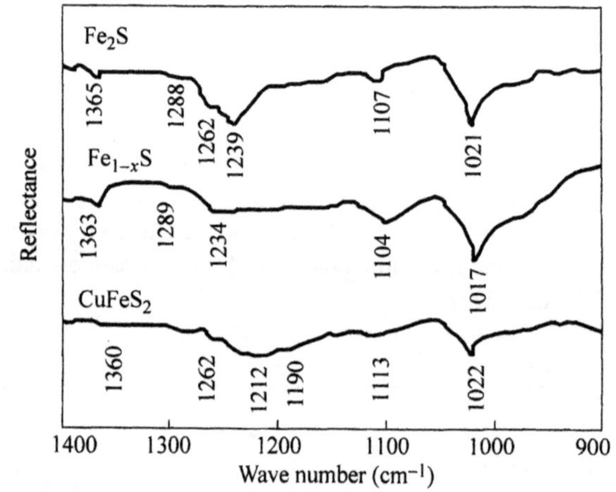

Figure 1.7 FTIR reflection spectra of pyrite, pyrrohotite and chalcopyrite after 15 min treatment with 10^{-4} mol/L KEX solution at pH = 5 (Leppinen, 1990)

1.5 Effect of Semiconductor Property of Sulphide Mineral on Its Electrochemical Behavior

Most sulphide minerals are semiconductors as shown in Table 1.4. The difference of interactions of sulphide minerals with a collector at certain conditions is originated from their different semiconductor properties. Plaksin and Bessonov (1957) reported that *p*-type galena adsorbed xanthate more easily. Later, Eadington and Prosser (1969) investigated the effect of light irradiation on the semiconductor property of galena and its interaction with xanthate showing that the increase of the hole on the galena surface after light irradiation promoted the adsorption of xanthate. Esposito et al., 1987 pointed out that the pyrite in ore is *p*-type and the pyrite in coal is *n*-type giving rise to their different flotation behavior. Salvador and Tafalla (1991) and Mishra (1992) found that *n*-type pyrite-band was helpful for the reduction of oxygen. Richardson and Yoon (1993) revealed the change of the electronic structure (e.g. Femi energy level) of galena and pyrite surface with pH and the concentration of reagent and noted that the interaction between mineral and reagent depended on their energy match. Most importantly, the flotation behavior of sulphide minerals was reported to be able to be controlled by changing the surface electronic structure or energy level through light irradiation and other methods (Cata et al., 1973; Cheng, 1999).

Table 1.4 Semiconductor properties of different sulphide minerals

Minerals	Width of forbidden band (eV) or conductivity	Minerals	Width of forbidden band (eV) or conductivity
PbS	0.41	Sb_2S_3	1.72
ZnS	3.6	As_2S_3	2.44
(Zn,Fe)S*	0.49	HgS	2
Cu FeS_2	0.5	CoS_2	< 0.1
FeS_2	0.9	Cu_2S	Meta conductor
NiS_2	0.27	CoS	Meta conductor
Cu_2S	2.1	FeS	Meta conductor

* The mass fraction of Fe——*w* (Fe) 12.4%.

The authors (Sun et al., 1991, 1993c, d, 1994b; Hu et al., 2000; Wang et al., 1992) used CNDO/2 method of quantum chemistry and band theory to study the flotation mechanism of sulphide minerals in the presence and absence of collectors.

Figure 1.8 shows the energy level diagram of molecular orbital of galena, pyrite, common and activated oxygen and ethyl xanthate ion. The highest occupied molecular orbital (HOMO) of galena surface mainly consists of 3p orbital of sulphur atoms, and the lowest unoccupied molecular orbital (LUMO) mainly consists of the 6p orbital of the lead atom. There is a little overlap between the 6p orbital of the lead atoms and the 3p orbital of the sulphur atoms. It indicates that the Pb—S bond at the galena surface contains a large ionic bond character and a small covalent bond character. In the case of pyrite, the atomic orbital of iron and sulphur have equal contribution to the HOMO and LUMO. There is a great overlap between the atomic orbital of iron and sulphur, showing that Fe—S bond exhibits covalent bond character mostly and the Fe—Fe bond exhibits the metal bond character partly. As above, surface electron structure is much different between the mineral of the type galena and that of the type pyrite. Oxygen possesses the following molecular orbital

$$(\sigma_{1s})^2(\sigma_{1s})^2(\sigma_{2s})^2(\sigma^*_{2s})(\sigma_{2px})^2(\pi_{2py})^2(\pi_{2pz})^2(\pi^*_{2pz})^1(\pi^*_{2py})^1(\sigma^*_{2px})^0$$

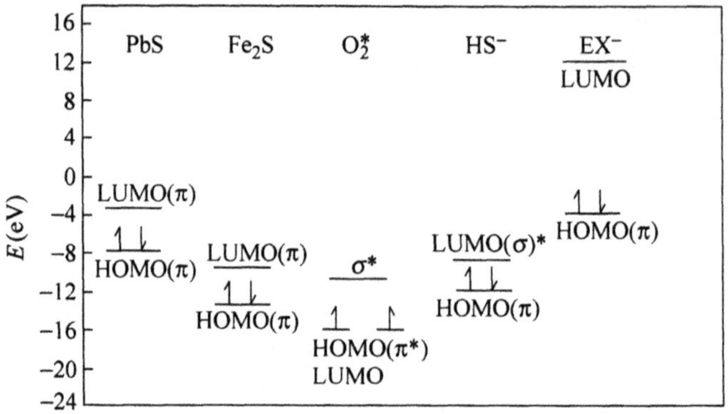

Figure 1.8 Symmetry and energy of frontier molecular orbital (HOMO and LUMO)

In its two anti-π orbital (π^*), there are two single electrons. The orbital can either accept or donate one electron, which can be either HOMO or LUMO. From Fig. 1.8, the energy values and symmetry of the frontier molecular orbital of galena, pyrite, HS^-, EX^-, and oxygen are shown in Table 1.5.

Table 1.5 The energy values and symmetry of the frontier molecular orbital of galena, pyrite, HS^-, EX^- and oxygen

Reactants	Galena	Pyrite	Common oxygen	Activated oxygen	HS^- ion	Ethyl xanthate ion
Energy HOMO	−7.757	−13.27	−15.82	−12.60	−11.70	−3.7
Symmetry	π	π	π	π	π	π
Energy LUMO	−3.133	−9.41	−15.82	−12.6	−8.7	12.1
Symmetry	π	π	π	π	σ	σ

1.6 Electrochemical Behaviors in Grinding System

Grinding of sulphide minerals is essential for liberation in order to achieve effective flotation. Grinding processes, however, may also have a detrimental effect on flotation because grinding system is open to atmosphere. The interactions among the grinding media such as steel, active and noble minerals, and flotation reagents have a pronounced effects on the electrochemical reaction on sulphide surface and their flotation. There were a lot of investigations on the electrochemical behaviors in grinding system (Rao et al., 1976; Hoey et al., 1977; Guy and Trahar, 1984; Dai et al., 2000).

At first, the leakage, dislocation, and the incoming of impurities may be produced in a mineral crystal during grinding resulting in the change of surface properties of the mineral such as electron energy level and electrode potential. Rey and Formanek (1960) first reported the effect of grinding media on flotation. The steel media decreases the activity of sphalerite made the selective flotation separation of galena with sphalerite. Thornton (1973) and Fahlstrom (1974, 1975) observed the improvement of flotation of copper ore in autogenous grinding than in carbon steel media. Rao et al. (1976) considered that the fine iron particles adhered to sulphide surface during grinding decreased the electrode potential and weakened the adsorption of xanthate resulting in the decrease of floatability of sulphide minerals. Harris (1988) found the worsen phenomena of sphalerite flotation due to the replacement of Fe^{2+} for Zn^{2+} during grinding. The compound containing iron on galena and sphalerite surface was detected by XPS. It was obvious that grinding changed the electrode potential and affected the flotation behavior of sulphide minerals.

Table 1.6 summarizes the rest potential of several sulphide, sulfarsenide and arsenide minerals at near neutral pH. This table shows that the rest potential of individual minerals vary depending on the source and on the experimental conditions; nevertheless, those of different minerals are different.

Table 1.6 Rest potential of sulphide, sulfarsenide and arsenide minerals of steel media at near neutral pH

Minerals	Sources	Solutions	Rest potential (mV, SHE)			References
			N_2	Air	O_2	
Pyrrhotite	Sudbury, Ontario	Distilled water	55	160	173	Adam et al. (1984)
	Strathcona Mine, Ontario	Distilled water	125	262	295	Iwasaki et al. (1989)
	Stratheona Mine, Ontario	Distilled water	155	290	335	Cheng (1991)
	Stratheona Mine, Ontario	0.001 mol/L Na_2SO_4	262	277	308	Nakazawa and Iwasaki (1986)
	Stratheona Mine, Ontario	0.05 mol/L Na_2SO_4	58	—	190	Li (1991)
	Sudbury, Ontario	0.5 mol/L NaCl	—	71	—	Adam et al. (1984)
Chalcopyrite	Rouyn, Quebec	Distilled water	190	355	371	Cheng (1991)
	Messina, S.Africa	0.05 mol/L Na_2SO_4	115	—	265	Li (1991)
Pyrite	Huanzala, Peru	Distilled water	405	445	485	Pozzo and Iwasaki (1987)
	Coahuila, New Mexico	0.001 mol/L Na_2SO_4	389	391	393	Nakazawa and Iwasaki (1985)
Galena	Brushy Creek, Missouri	Distilled water	142	172	218	Learmont and Iwasaki (1984)
Arsenopyrite	Unknown	Distilled water	277	303	323	Iwasaki et al. (1989)
Cobaltite	Unknown	Distilled water	200	275	303	Iwasaki et al. (1989)
Nickel Arsenide	Synthetic	Distilled water	−90	5	97	Iwasaki et al. (1989)
	Synthetic	0.001 mol/L Na_2SO_4	152	154	157	Nakazawa and Iwasaki (1986)
Mild steel	AISI 1020 0.2% C	Distilled water	−355	−255	−135	Pozzo and Iwasaki (1987)
	AISI 1020	Distilled water	−515	−335	−175	Adam et al. (1984)
	AISI 1020	0.5 mol/L NaCl	—	−395	—	Adam et al.(1984)

With their rest potential much higher than that of the steel medium listed in Table 1.6, sulphide minerals act as a cathode while the steel medium acts as an anode. During grinding, minerals and grinding media (steel) come in repeated contact with each other, and galvanic current flows between the two surfaces of the sulphide minerals may be altered.

Another important influence of grinding on the floatability of sulphide minerals is the galvanic interactions among the grinding media and different minerals, along with grinding and flotation atmospheres. Galvanic cells are set up by redox reactions taking place due to the difference in their rest potential. One of higher potential acts as a cathode, while that of lower potential as an anode. In grinding-flotation circuit of sulphide minerals, galvanic interactions exist between minerals and grinding media, usually steel, between mineral and mineral, and among multiple mineral and grinding media (steel)system. The galvanic reactions are also governed by the mixed potential principle.

The effect of galvanic interaction between the grinding media and a sulphide mineral on flotation has been studied with respect to pyrrhotite (Adam et al., 1984; Nakazava and Iwasaki, 1985), galena (Learmont and Iwasaki, 1984), chalcopyrite (Yelloji Rao and Natarajan, 1989a and 1989b), and sphalerite (Yelloji Rao and Natarajan, 1989c).The balls undergo anodic dissolution whereas oxygen reduction could be the cathodic reaction at the surfaces of the sulphide minerals. The most prominent anodic reaction of steel ball would be the dissolution of iron:

$$Fe \longrightarrow Fe^{2+} + 2e^{-} \tag{1-7}$$

Thus the dissolved ferrous species would further react with the hydroxyl ions generated at the cathodic sulphide mineral sits forming iron oxy-hydroxide species:

$$Fe^{2+} \longrightarrow Fe^{3+} + e^{-} \tag{1-8}$$

$$\downarrow \qquad\qquad \downarrow$$

$$Fe(OH)_2 \longrightarrow FeOOH,\ Fe(OH)_3 \tag{1-9}$$

The cathodic reaction is oxygen reduction in Eq. (1-1). Because $Fe(OH)_3$ and metal oxy-hydroxide species of iron, lead and zinc formed will coat the cathodic mineral surface, affecting its floatability.

In the case of mineral-mineral interactions, a mineral with higher potential acts as a cathode, while a mineral with lower potential acts as an anode. For a multiple mineral/grinding media(steel)system. The galvanic interactions become more complex than the two-electrode systems. The galvanic reactions among multielectrode systems are also governed by the mixed potential principle as shown in an example of polarization curves involving pyrite, pyrrhotite and mild steel in Fig. 1.9 (Pozzo and Iwasaki, 1987).

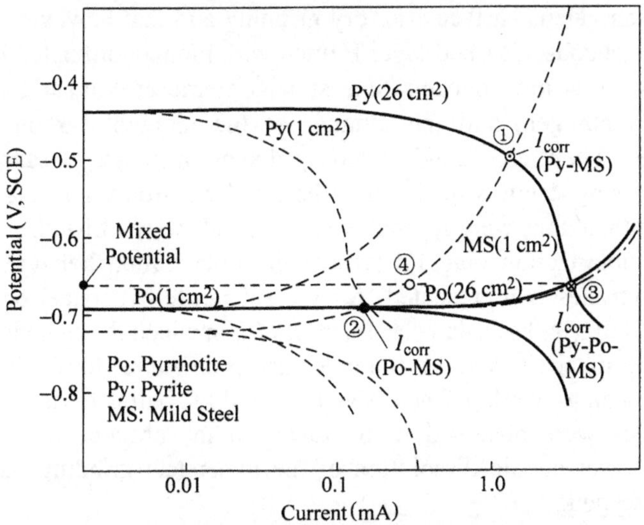

Figure 1.9 Determination of corrosion currents for pyrite-mild steel (Py-MS) pyrrhotite-mild steel (Po-MS) and pyrite-pyrrhotite-mild steel (Py-Po-MS)systems under abrasion in a quartzite slurry by adjusting to the surface area ratios of ground minerals and steel balls (Pozzo and Iwasaki, 1987)

The effect of galvanic interactions among sulphide minerals, such as chalcopyrite, galena and sphalerite in the presence and absence of a grinding ball material in the flotation of chalcopyrite, was studied by Rao and Natarajan (1989, 1990). They found that the floatability of chalcopyrite was affected significantly when the chalcopyrite contacted with active minerals (galena and sphalerite), either alone or together. Such an effect became more pronounced in the added presence of a grinding medium as well as oxygen.

Nakazawa and Iwasaki (1985) and Pozzo, Malicsi and Iwasaki (1988) investigated a pyrite-pyrrhotite contact and a pyrite-pyrrhotite-grinding media contact on flotation, respectively. They found that the floatability of pyrrhotite increased in the presence of pyrite, whereas it decreased in the presence of both pyrite and grinding media (mild steel). Similarly, a galvanic contact between nickel arsenide and pyrrhotite decreased the floatability of pyrrhotite (Nakazawa and Iwasaki, 1986).

Wang and Xie (1990), reviewing the influence of grinding media and environment on flotation responses of sulphide minerals, concluded that grinding media and environment affected the surface properties and floatability of sulphide minerals mainly by electrochemical interaction, surface morphology and mechano-chemical reaction. Other important factors include surface area, solution composition, etc. To minimize galvanic interaction, it is necessary to choose an inert grinding atmosphere, such as ceramic or stainless steel mill balls and nitrogen aeration.

Cases et al. (1990, 1991) studied the influence of grinding media on the adsorption of xanthate on galena and pyrite using XPS and FTIR. They found that only lead xanthate, either monocoordinated or in the three dimensional form was formed on galena surface after dry grinding and that however, wet grinding led to heterogeneous adsorbed layer formed with monocoordinated lead xanthate and dixanthogen at low concentration; stoichiometric or non-stoichiometric lead xanthate, dixanthogen, and the dimmer of the monothiocarbonate at higher concentration. They also found that only dixanthogen was observed on pyrite surface after using stainless steel rods. Whereas, ferric iron xanthate and physically absorbed xanthate ions were formed when iron rods were utilized.

A mixed polarization diagram (where the polarization behavior of the two different electrodes is represented) for the sphalerite-hypersteel combination is given in Fig. 1.10 (Vathsala and Natarajan, 1989), in which the cathodic polarization curves for the sphalerite and the anodic polarization curves for the hypersteel ball material are seen to overlap. The active nature of the ball material is evident. The current values were observed to be lower in the absence of oxygen which indicated a lower anodic dissolution of the hypersteel grinding medium in the absence of oxygen.

The presence of oxygen enhances galvanic interactions. Pozzo et al. (1990) found that the floatability of pyrite and pyrrhotite in the flotation of their mixtures were affected very little at neutral or acidic pH when the mixture was ground in a nitrogen atmosphere, whereas the recovery of pyrrhotite was greatly improved when the mixture was ground in an oxygen atmosphere. Therefore, the amount of oxygen used in flotation must be carefully controlled in order to alleviate the interactions among minerals. Trahar (1984) pointed out that the interactions among different minerals could be diminished if the minerals were ground separately and combined immediately before flotation, thereby demonstrating the effect of multi-electrode galvanic contacts.

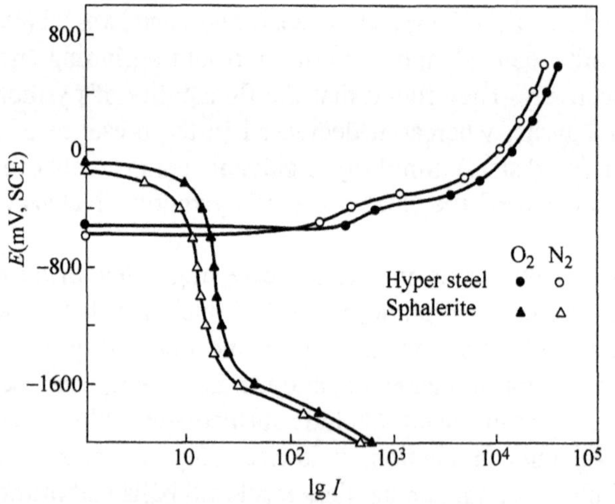

Figure 1.10 Mixed polarization diagram involving hypersteel ball material in contact with sphalerite (unit of I: μA; Vathsala and Natarajan, 1989)

The separation of chalcopyrite and pyrrhotite at natural pH was reported to be possible when mild steel filings addition combined with magnetizing treatment were used under oxygen aeration (Cheng and Iwasaki, 1992). Mild steel filings adhere to the pyrrhotite surface by magnetic interaction. They form a localized galvanic cell involving oxygen reduction and the hydroxyl ion thereby generated reacts with ferrous ion from the pyrrhotite as well as from the mild steel filings and results in an iron hydroxide layer.

$$FeS = Fe^{2+} + S \quad (1\text{-}10)$$

$$\searrow \qquad \searrow SO_4^{2-}$$

$$\nearrow Fe(OH)_2 \longrightarrow Fe(OH)_3 \text{ or } Fe(OH)SO_4$$

$$Fe = Fe^{2+} + 2e^- \quad (1\text{-}11)$$

The presence of oxygen enhances the formation of the surface coating and depresses the flotation of pyrrhotite. It appears, therefore, that although the floatability of individual mineral may be controlled by pulp potential, the presence of several sulphide minerals, particularly when they are ground with steel media, leads to galvanic interaction among them and the alteration of certain mineral surfaces may be accelerated. Then the pulp potential dependence of their floatability may not follow those of individual mineral.

1.7 The Purpose of This Book

This book systematically summarizes the researches on electrochemistry of sulphide flotation in our group. The various electrochemical measurements, especially electrochemical corrosive method, electrochemical equilibrium calculations, surface analysis and semiconductor energy band theory, practically, molecular orbital theory, have been used in our studies and introduced in this book. The collectorless and collector-induced flotation behavior of sulphide minerals and the mechanism in various flotation systems have been discussed. The electrochemical corrosive mechanism, mechano-electrochemical behavior and the molecular orbital approach of flotation of sulphide minerals will provide much new information to the researchers in this area. The example of electrochemical flotation separation of sulphide ores listed in this book will demonstrate the good future of flotation electrochemistry of sulphide minerals in industrial applications.

Chapter 2 Natural Floatability and Collectorless Flotation of Sulphide Minerals

Abstract This chapter first explains the natural flotability of some minerals in the aspect of the crystal structure and demonstates the collectorless flotaiton of some minerals and its dependence on the E_h and pH of pulp. And then the surface oxidation is analysed eletrochemically and the relations of E_h to the composition of the solutions are calculated in accordance with Nernst Equation. The E_h-pH diagrams of several minerals are obtained. Thereafter, electrochemical determination such as linear potential sweep voltammetry (LPSV) and cyclic voltammetry (CV) and surface analysis of surface oxidation applied to the sulphide minerals are introduced. And recent researches have proved that elemental sulfur is the main hydrophobic entity which causes the collectorless flotability and also revealed the relation of the amount of sulfur formed on the mineral surfaces to the recoveries of minerals, which is always that the higher the concentration of surface sulphur, the quicker the collectorless flotation rate and thus the higher the recovery.

Keywords natural flotability; collectorless flotation; E_h-pH diagram; linear potential sweep voltammetry; cyclic voltammetry; XPS; UV

2.1 Crystal Structure and Natural Floatability

The structure of galena is shown in Fig. 2.1. Galena is a common and popular mineral among rock hounds. The structure of galena is identical to that of halite, NaCl, which has face-centered cubic structure in the crystal, each Pb atom is bonded to 6 S atoms and each S atom is bonded to 6 Pb atoms. It have perfect cleavage in [001], [010] and [100] direction. Pyrite is cubic with the octahedrally coordinated metal atoms at the corners and face centers of the cube unit cell (see Fig. 2.2). "Dumb-bell" shaped disulphide atoms S_2^{2-} lie at the center of the cube and at the midpoints of the cube edges. The mid-point of the S_2 group occupies the Cl sites of the NaCl structure, while the Fe atoms occupy the Na positions. The S_2^{2-} pairs are oriented such that their axes are parallel to four non-intersecting bodies' diagonals of the cubic space lattice and give pyrite hemihedral symmetry. Each S atom is coordinated to three Fe atoms and one S atom in a distorted tetrahedral configuration. Therefore, the cleavage planes for both galena and pyrite exhibit hydrophilic characteristics due to their structural feature.

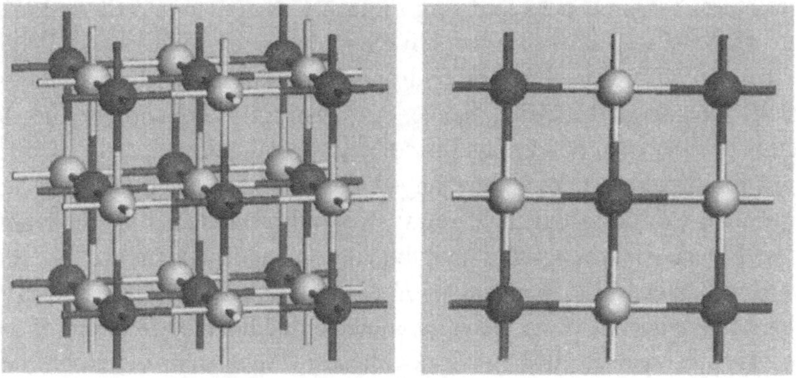

(a) (b)

Figure 2.1 Crystal structure of galena
(a) Side view of galena crystal; (b) Top view of galena crystal

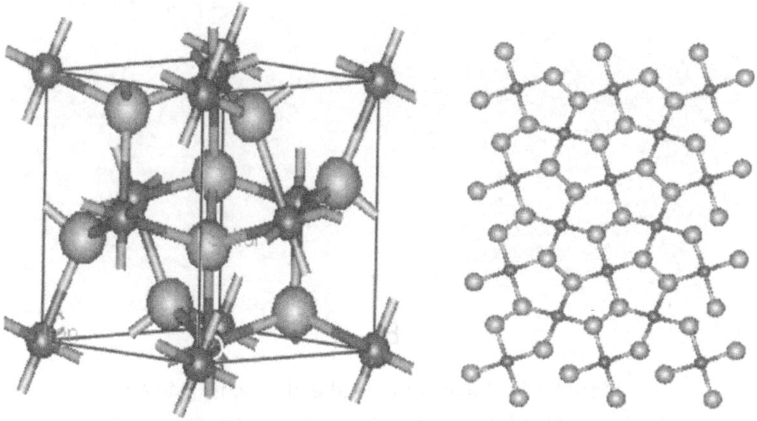

(a) (b)

Figure 2.2 Crystal structure of pyrite
(a) Pyrite (100) plane; (b) Showing the coordination of ions

In the layered structure of molybdenite (MoS_2), each atom is surrounded by six sulfurs at the vertices of a trigonal prism. The prism is linked by their edges into a continuous layer within which the centers of half of the prisms are occupied by cations as shown in Fig. 2.3. That is, a molybdenite crystal consists of S—Mo—S sandwich-like layers held together by van der Waals bonds. During comminution, fracture occurs by the breakage of these weak residual bonds. Thus the breakage faces are naturally hydrophobic. The stibnite has a chain structure (Vanghan and Craig, 1978). It appears that the sulphur-antimony bonds are "weak", at least weaker than most metal-sulphur bonds and the surface fracture faces are somewhat non-polar in nature. Orpiment has a layer structure consisting of corrugated pyramidal AsS_3 groups linked via common S atoms. Realgar has a ring structure, having a puckered ring molecule (Vanghan and Craig,1978). The fracture of these two minerals occurs by the rupture of weaker residual bonds and hence they will also display natural hydrophobicity. Minerals such as molybdenite, orpiment, realgar and stibnite, hence, are easily floated without the need of collectors or of special conditions for collectorless flotation, i.e. no specific E_h is required for flotation, nor is there any display of variable floatability with E_h. Such mineral exhibit is called "natural floatability" induced by their structural features. Figure 2.4 shows that molybdenite is highly floatable. The flotation recovery is basically independent of E_h, i.e. no specific E_h is required for the flotation of molybdenite. It has been reported that the flotation recovery of orpiment and realgar can reach to 100% without collector (Finkelstein and Goold, 1972).

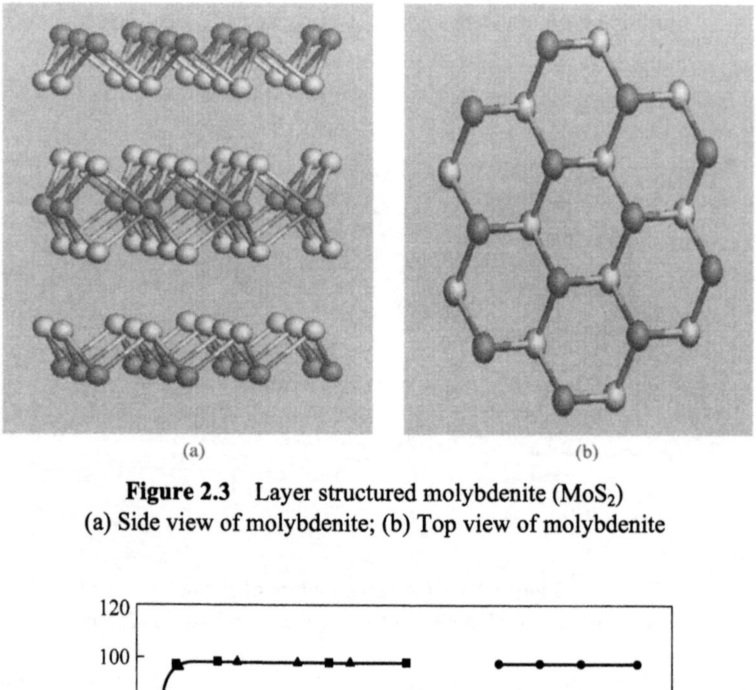

Figure 2.3 Layer structured molybdenite (MoS_2)
(a) Side view of molybdenite; (b) Top view of molybdenite

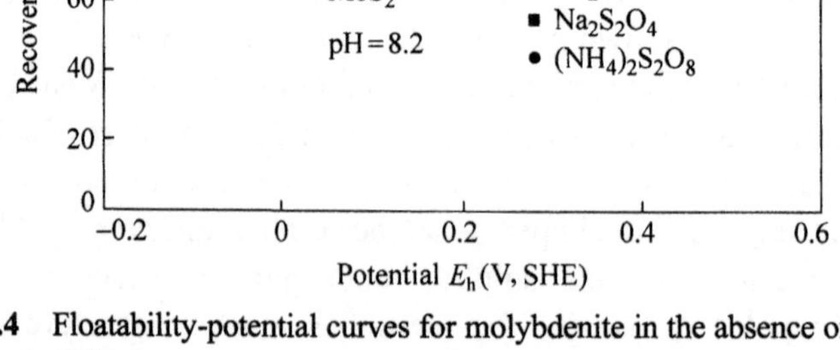

Figure 2.4 Floatability-potential curves for molybdenite in the absence of collector (Wang et al., 1993a)

2.2 Collectorless Flotation

2.2.1 Effect of Pulp Potential on Flotation at Certain pH

Most sulphide minerals do not exhibit natural floatability. They have, however, collectorless floatability in specific pulp potential ranges, under the condition of which the sulphide mineral surface has been rendered hydrophobic by surface self-oxidation. Many other authors indicated that collectorless flotation of sulphide minerals can occur only under moderately oxidizing environments and be controlled by pulp potential (Gardner and Woods, 1979; Hayes et al., 1987, 1988; Sun, 1990; Wang et al., 1991a,b,c,d; Cheng and Iwasaki, 1992; Zhang et al., 2004a,b,c,d,e,f).

The collectorless floatability of chalcopyrite has been studied in some detail and some results are shown in Fig. 2.5 (Guy and Trahar, 1985; Wang, 1992). It has been found that there is a clear distinction between flotation and non-flotation, which appear to be pH and E_h dependent. The upper limit and lower limit of pulp potential for collectorless flotation of chalcopyrite change with pH.

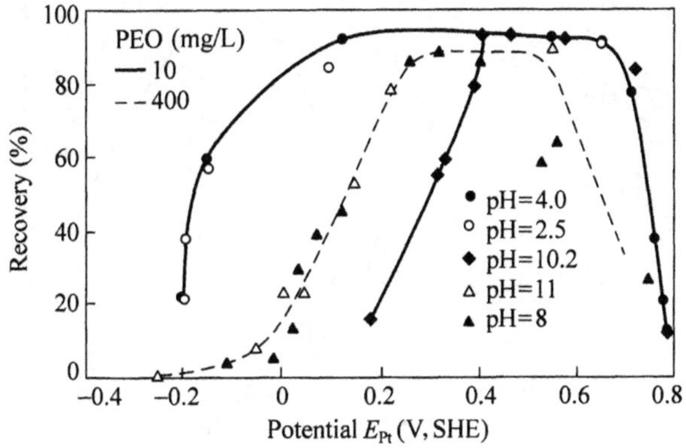

Figure 2.5 Floatability-potential curve for chalcopyrite in collectorless flotation (Guy and Trahar, 1985; Wang, 1992)

The influences of pulp potential on the floatability of galena, jamesonite, pyrite, pyrrhotite, sphalerite and marmatite are, respectively given in Fig. 2.6, Fig. 2.7 and Fig. 2.8. It can be seen that these minerals show collectorless flotation in certain range of potential and exhibit different upper and lower potential limit for flotation. In Fig. 2.6, the potential range of collectorless flotation is 0.15 – 0.3 V for galena at pH = 6 and 0.45 – 0.6 V for jamesonite at pH = 4.7. Pyrrhotite exhibits good collectorless floatability in the potential range of 0.25 – 0.35 V at pH = 8.8, but pyrite shows weak collectorless flotation, the range of potential is around 0.17 V at pH = 6 as seen in Fig. 2.7. The potential range of collectorless flotation is 0.45 – 0.55 V for marmatite and 0.15 – 0.25 V for sphalerite at pH = 4.7.

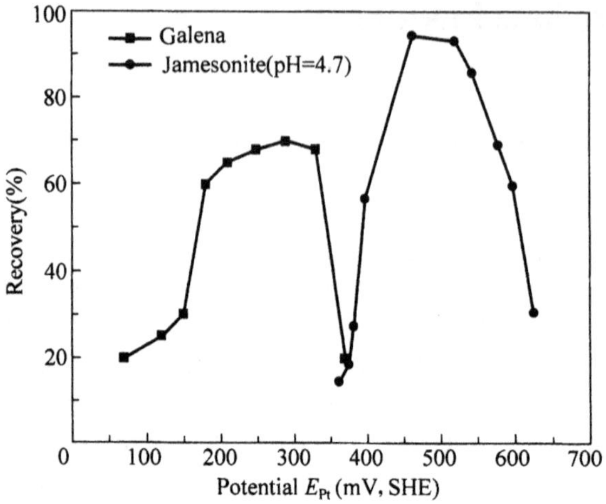

Figure 2.6 Effect of pulp potential on collectorless flotation behaviors of galena and jamesonite

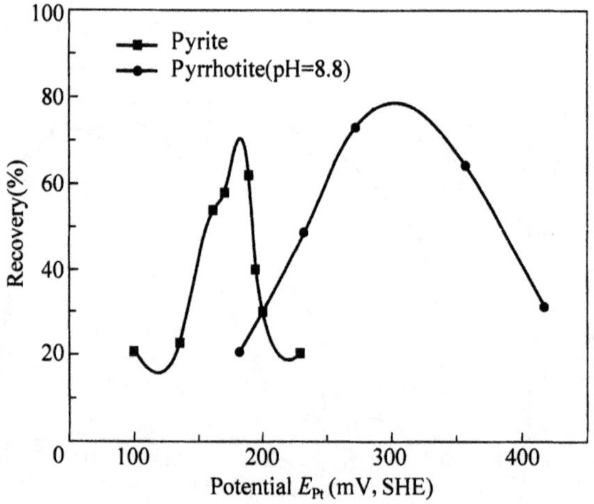

Figure 2.7 Effect of pulp potential on collectorless flotation behaviors of pyrite and pyrrhotite

2.2.2 Pulp Potential and pH Dependence of Collectorless Floatability

Figures 2.6, 2.7 and 2.8 provided the evidence that there exists the critical upper and lower limit of pulp potential for collectorless flotation at certain pH. Figure 2.9 further demonstrated the flotation recovery of jamesonite as a function of potential at different pH. It is obvious that jamesonite has very good collectorless floatability in different potential range, which much depended on different pH. The

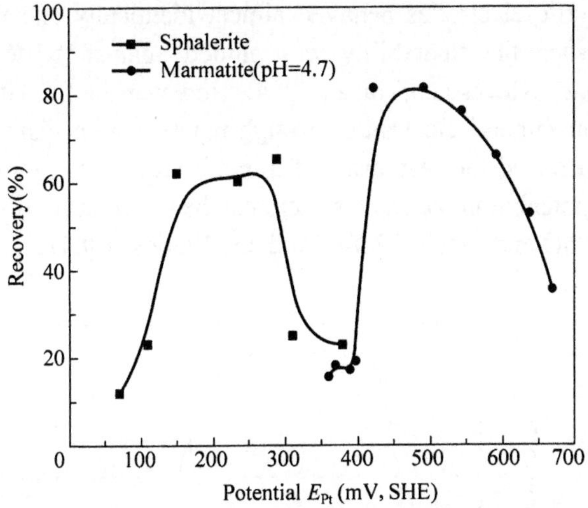

Figure 2.8 Effect of pulp potential on collectorless flotation behaviors of sphalerite and marmatite

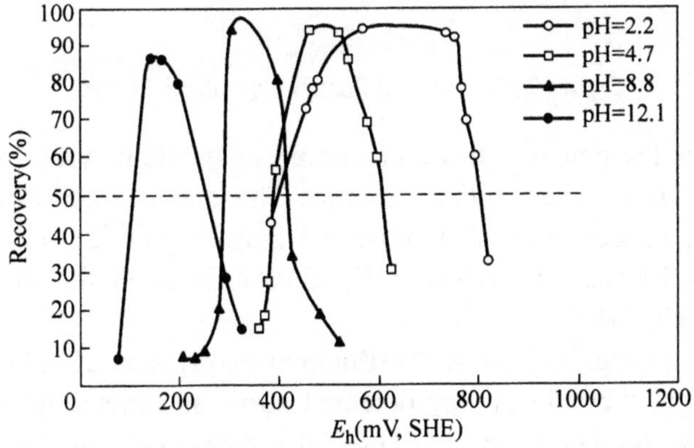

Figure 2.9 Effect of pulp potential on collectorless flotation behaviors of jamesonite at different pH

floatable potential region located more oxidized atmosphere in acid medium and toward lower potential region in alkaline medium. If we take recovery to be 50% as the floatable and non-floatable criterion, Floatability- potential curve at certain pH for sulphide minerals can be generally plotted as Fig. 2.10. Define the upper limit potential (E^U), below which recovery is greater than 50% and lower limit potential (E^L), above which recovery is greater than 50%, The upper limit potential and lower limit potential of some minerals at different pH can be determined through the curves obtained from flotation tests. The E^U_{Pt}, E^L_{Pt} and E^U_m, E^L_m, respectively, expressed the upper and lower limit of potential determined using platinum and mineral electrode. It has been reported that platinum and pyrite electrodes behaved almost identically and the same curve was obtained when the floatability was plotted against pyrite electrode and platinum electrodes (Heyes and Trahar, 1984). However the formation of oxidized surface layers on mineral electrodes through reaction with dissolved species in flotation pulps affected the responses of mineral electrodes. Some investigators found that pulp potential in the same system can be different for different indicator electrodes (Labonte and Finch, 1988; Rand and Woods, 1984).

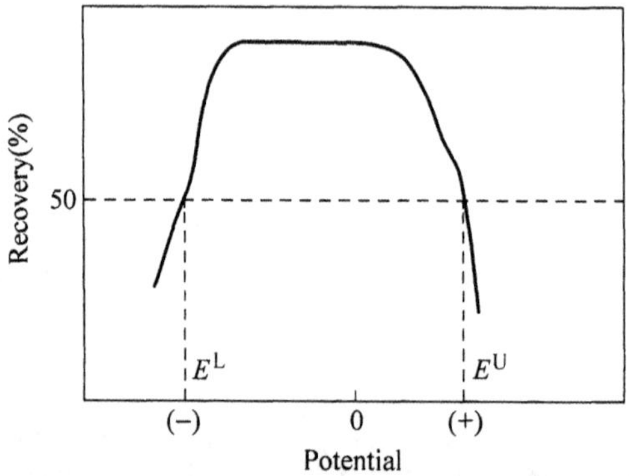

Figure 2.10 General floatability-potential curve

Nevertheless the potential to be monitored in flotation system would be the mineral potential, not the solution potential. The electrode constructed from the mineral being concentrated should be the most appropriate electrode for E_h measurements because the relevant E_h is established at the mineral/solution interface (Woods, 1991).

Figure 2.11 to Fig. 2.13 show the flotation E_h-pH area of sulphide minerals determined from their floatability-potential curve at various pH using mineral electrode. It can be seen from the Figures that there exists an E_h-pH area where the collectorless flotation of sulphide minerals is possible. Figure 2.11 and Fig. 2.12 show that jamesonite and pyrrhotite exhibit wider potential and pH range for collectorless flotation, especially in strong acidic media they exhibit a wider potential range for collectorless flotation. The upper limit potential E_m^U and the lower limit potential E_m^L of pyrrhotite collectorless flotation are, respectively, 750 mV and 450 mV in strong acidic medium, and 300 mV and 50 mV in strong alkaline medium. The upper limit potential E_m^U and lower limit potential E_m^L of jamesonite collectorless flotation are, respectively, 820 mV and 400 mV in strong acidic medium, and 300 mV and 100 mV in strong alkaline medium. With the increase of pH, the potential of collectorless flotation of pyrrhotite and jamesonite decreases. Figure 2.13 shows that marmatite covers a wide range of potential for collectorless flotation only in strong acidic pH region.

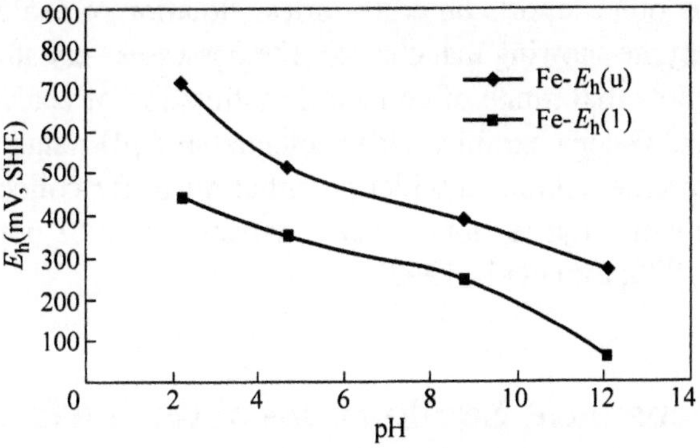

Figure 2.11 E_h-pH area of collectorless flotation of pyrrhotite

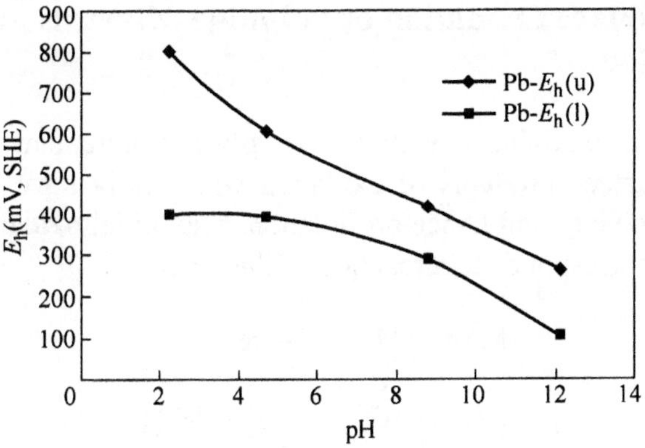

Figure 2.12 E_h-pH area of collectorless flotation of jamesonite

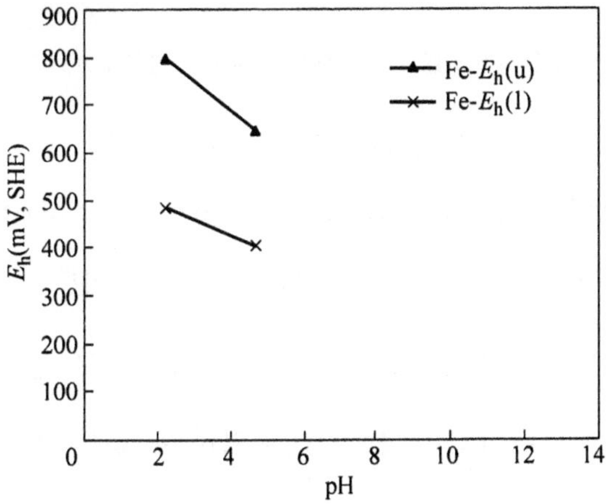

Figure 2.13 E_h-pH area of collectorless flotation of marmatite

There are some other reports on collectorless flotation of chalcopyrite, galena, pyrite and arsenopyrite showing that chalcopyrite possesses very strong collectorless floatability. The potential range of collectorless flotation of chalcopyrite is wider in wider pH range. Galena exhibits wider potential and pH range for collectorless flotation. Arsenopyrite exhibits a wider potential range for collectorless flotation in weak acidic media. Pyrite floats over a narrow range of potential in strong acidic pH (Sun, 1990; Sun et al., 1992).

2.3 Electrochemical Equilibriums of the Surface Oxidation and Flotation of Sulphide Minerals

2.3.1 The Surface Oxidation of Sulphide Minerals and Nernst Equation

It appears that an oxidizing potential is usually a requirement for collectorless flotation. The surface products of oxidation will clearly have an influence on surface hydrophobicity and hence on flotation. The initial oxidation of sulphides in acid solution corresponds to a reaction of the type

$$\mathrm{MS{=\!=}M^{\mathit{n}+} + S + \mathit{n}e^-} \tag{2-1a}$$

$$\Delta G^0 = \Delta G^0_{\mathrm{M}^{n+}} + \Delta G^0_{\mathrm{S}} - \Delta G^0_{\mathrm{MS}} \tag{2-1b}$$

where ΔG^0 is the standard free energy above reaction.

The oxidation of sulphides corresponds to a reaction of the type in neutral or alkaline solution

$$\mathrm{MS + \mathit{n}H_2O{=\!=}M(OH)_\mathit{n} + S + \mathit{n}H^+ + \mathit{n}e^-} \tag{2-2a}$$

$$\Delta G^0 = \Delta G^0_{\mathrm{M(OH)}_n} + \Delta G^0_{\mathrm{S}} - \Delta G^0_{\mathrm{MS}} - n\Delta G^0_{\mathrm{H_2O}} \tag{2-2b}$$

Alternative oxidation reactions may lead to the production of oxy-sulphur species

$$\mathrm{2MS + 7H_2O{=\!=}2M(OH)_2 + S_2O_3^{2-} + 10H^+ + 8e^-} \tag{2-3a}$$

$$\Delta G^0 = 2\Delta G^0_{\mathrm{M(OH)}_n} + \Delta G^0_{\mathrm{S_2O_3^{2-}}} - \Delta G^0_{\mathrm{MS}} - 7\Delta G^0_{\mathrm{H_2O}} \tag{2-3b}$$

$$\mathrm{MS + 6H_2O{=\!=}M(OH)_2 + SO_4^{2-} + 10H^+ + 8e^-} \tag{2-4a}$$

$$\Delta G^0 = 2\Delta G^0_{\mathrm{M(OH)}_n} + \Delta G^0_{\mathrm{SO_4^{2-}}} - \Delta G^0_{\mathrm{MS}} - 6\Delta G^0_{\mathrm{H_2O}} \tag{2-4b}$$

The nature of the oxidation products will in general determine the subsequent flotation behavior. It seems reasonable to assume then that the hydrophobic entity is sulphur itself produced by the reactions similar to (2-1) or (2-2), the potential of which defines the lower limit of potential at which the collectorless flotation of sulphide minerals begins. However, if oxidation is by reactions (2-3) or (2-4), collectorless floatability would not be expected. The potential corresponding to reactions (2-3) or (2-4) define the upper limit of potential at which the collectorless flotation of sulphide minerals ceased. Therefore, the potential relations of surface oxidation of sulphide minerals to produce elemental sulphur on flotation and the potential dependence of oxidation to produce thiosulphate on depression may be obtained.

The oxidation-reduction potential or redox potential (E_h) is a measure of the tendency of a solution to be oxidizing or reducing. Oxidation and reduction are basically electrical processes that are readily measured by an electrode potential. All measurements are referred to the standard hydrogen electrode, the potential of which is taken as 0.00 V at 298 K, the H_2 pressure as 101325 N/m^2 (1 atm) and activities of H_2 and H^+ as unity. When the half-cell reaction is written as an oxidation reaction:

$$\text{Reduced state} = \text{Oxidized state} + n\text{Electrons}$$

The relationship between any redox potential and the standard electrode potential is given by the Nernst equation:

$$E_h = E^0 + \frac{RT}{nF}\ln\frac{[\text{Oxidized state}]}{[\text{Redued state}]} = \frac{2.303RT}{nF}\lg\frac{[\text{Oxidized state}]}{[\text{Reduced state}]} \tag{2-5}$$

where R is the gas constant (8.314 $J \cdot K^{-1} \cdot mol^{-1}$), T is absolute temperature (K), F is the Faraday constant (96490 $C \cdot mol^{-1}$), and n is the number of electrons involved in the redox reaction. E^0 is the electrode potential that is related to standard free energy as:

$$E^0 = \frac{\Delta G^0}{nF} \tag{2-6}$$

The [oxidized state] and [reduced state] are, respectively, the various items in the right and left of the reactions like Eqs. (2-1a) to (2-4a). ΔG^0 can be calculated using various standard free energy dates reported from literature (Vanghan and Craig, 1978; Garrels and Christ, 1965; Wang and Hu, 1989).

For reactions at 298 K, $(2.303RT/F) = 0.059$, so that

$$E_h = \frac{\Delta G^0_{298}}{nF} + \frac{0.059}{n}\lg\frac{[\text{Oxidized state}]}{[\text{Reduced state}]} \tag{2-7}$$

In the light of Eq. (2-7), the potential of surface oxidation of sulphide minerals to form elemental sulphur or oxy-sulphur species could be calculated.

2.3.2 Electrochemical Equilibriums in Collectorless Flotation

Although the nature of the hydrophobic entity responsible for the self-induced flotation of sulphide minerals remains somewhat obscure, most reported results clearly show that it is only when the environment becomes slightly oxidizing that flotation is observed. The elemental sulphur and polysulphide-intermediates in the oxidation of sulphide to sulphur have ever been suggested to be of the hydrophobic species. Whatever it is, there is no doubt that sulphur can generate hydrophobicity and floatability.

For chalcopyrite, the sulphur-producing reactions include:
in alkaline medium

$$CuFeS_2 + 3H_2O = CuS + Fe(OH)_3 + S + 3H^+ + 3e^- \quad (2\text{-}8a)$$

$$\Delta G^0 = \Delta G^0_{Fe(OH)_3} + \Delta G^0_{CuS} - \Delta G^0_{CuFeS_2} - 3\Delta G^0_{H_2O} \quad (2\text{-}8b)$$

where the date of $\Delta G^0_{Fe(OH)_3}$, ΔG^0_{CuS}, $\Delta G^0_{H_2O}$ can be taken from literature (Vanghan and Craig, 1978; Garrels and Christ, 1965; Wang and Hu, 1989), and $\Delta G^0_{CuFeS_2}$ can be calculated according to the reaction of Eq. (2-9):

$$CuFeS_2 = Cu^{2+} + Fe^{2+} + 2S^{2-} \quad (2\text{-}9a)$$

$$K_{sp} = 10^{-61.5} \quad (2\text{-}9b)$$

$$\begin{aligned}\Delta G^{0'} &= \Delta G^0_{Fe^{2+}} + \Delta G^0_{Cu^{2+}} + \Delta G^0_{S^{2-}} - \Delta G^0_{CuFeS_2} \\ &= -RT\ln K_{sp} = 350.01(\text{kJ/mol})\end{aligned} \quad (2\text{-}9c)$$

$$\begin{aligned}\Delta G^0_{CuFeS_2} &= \Delta G^0_{Fe^{2+}} + \Delta G^0_{Cu^{2+}} + 2\Delta G^0_{S^{2-}} - \Delta G^{0'} \\ &= -84.935 + 64.978 + 2\times 91.881 - 350.01 \\ &= -187.106(\text{kJ/mol})\end{aligned} \quad (2\text{-}9d)$$

thus

$$\Delta G^0 = -694.544 - 48.953 + 187.106 + 3\times 237.19 = 155.179(\text{kJ/mol}) \quad (2\text{-}10a)$$

$$E^0 = \frac{\Delta G^0}{nF} = \frac{155.179\times 1000}{3\times 96490} = 0.536(\text{V}) \quad (2\text{-}10b)$$

The potential-pH relationship of Eq. (2-8a) is

$$E_h = 0.536 - 0.059\text{pH} \quad (2\text{-}10c)$$

Similarly, in weak acidic medium

$$CuFeS_2 = CuS + Fe^{2+} + S + 2e^- \quad (2\text{-}11a)$$

$$\Delta G^0 = 53.2(\text{kJ/mol}) \quad (2\text{-}11b)$$

$$E^0 = 0.276(\text{V}) \tag{2-11c}$$

$$E_{\text{h}} = 0.276 + 0.0295\lg[\text{Fe}^{2+}] \tag{2-11d}$$

The formation of sulphur on chalcopyrite surface would be expected to occur at 0.15 V at pH = 6 by assuming $[Fe^{2+}] = 10^{-4}$ mol/L according to reaction (2-8), 0.08 V at pH = 8 and –0.1 V at pH = 11 according to reaction (2-9). While flotation does begin very near these potential for the experimental conditions reported by some authors as shown in Table 2.1.

Table 2.1 Relationship between initial potential producing elemental sulphur and flotation

Minerals	Galena PbS		Chalcopyrite $CuFeS_2$	
	E_{hit} (V)	E_{hie} (V)	E_{hit} (V)	E_{hie} (V)
pH = 6	0.24 (Eq. (2-12b))	0.26 (Sun, 1990)	0.16 (Eq. (2-11b))	0.18 (Feng, 1989)
pH = 8	0.28 (Eq. (2-13b))	0.27 (Guy and Trahar, 1984)	0.08 (Eq. (2-10b))	0.08 (Trahar, 1984)
pH = 11	0.1 (Eq. (2-14b))	0.16 (Sun, 1990)	−0.11 (Eq. (2-10b))	−0.1 (Trahar, 1984)

E_{hit} is the theoretical potential calculated on the basis of electrochemical equilibrium;
E_{hie} is the experimental values reported by some authors.

For galena, the reactions producing elemental sulphur may include in acidic media

$$\text{PbS} = \text{Pb}^{2+} + \text{S} + 2\text{e}^- \tag{2-12a}$$

$$\begin{cases} \Delta G^0 = 68.315(\text{kJ/mol}) \\ E^0 = 0.354(\text{V}) \\ E_{\text{h}} = 0.354 + 0.0295\lg[\text{Pb}^{2+}] = 0.236(\text{V}) \end{cases} \tag{2-12b}$$

in weak alkaline media

$$\text{PbS} + 2\text{H}_2\text{O} = \text{Pb(OH)}_2 + \text{S} + 2\text{H}^+ + 2\text{e}^- \tag{2-13a}$$

$$\begin{cases} \Delta G^0 = 146.085(\text{kJ/mol}) \\ E^0 = 0.757(\text{V}) \\ E_{\text{h}} = 0.757 - 0.059\text{pH} \end{cases} \tag{2-13b}$$

in alkaline media

$$\text{PbS} + 2\text{H}_2\text{O} = \text{HPbO}_2^- + \text{S} + 3\text{H}^+ + 2\text{e}^- \tag{2-14a}$$

$$\begin{cases} \Delta G^0 = 228.102(\text{kJ/mol}) \\ E^0 = 1.182(\text{V}) \\ E_{\text{h}} = 1.064 - 0.0885\text{pH} \end{cases} \tag{2-14b}$$

The reaction potential producing elemental sulphur are 0.24 V at pH = 6 based on the reaction (2-10), 0.28 V at pH = 8 on the reaction (2-11), and 0.1 V at pH = 11 on the reaction (2-12) with 10^{-4} mol/L concentration of dissolved species. The reported flotation initial potential (see Table 2.1) are very close to these theoretical calculation values. The theoretical and experimental values in Table 2.1 indicate that the elemental sulphur might be responsible for the hydrophobicity of sulphide surfaces. At different pH media the formation of elemental sulphur occurs and hence the flotation behavior undergoes different processes.

2.3.3 E_h-pH Diagrams of Potential and pH Dependence of Flotation

E_h-pH diagrams have become a standard method for illustrating equilibrium relationships between dissolved and solid species and have been applied successfully to investigate flotation systems (Hepel and Pomianowski, 1977; Chander et al., 1979; Pritzker and Yoon, 1984a,b; Johnson and Munro, 1988). In the following calculations all the thermodynamic data are from Garrels and Christ (1965) and Vanghan and Craig (1978). The E_h-pH diagrams of surface oxidation of sulphide minerals are constructed based on the reactions below, assuming the concentration of all dissolved species to be 10^{-4} mol/L and considering elemental sulphur as a metastable phase because thiosulphate and even sulphate may be formed but frequently at potential higher than the calculated equilibrium potential, i.e. at over potential (η) varying magnitudes (take $\eta = 0.5$ V in subsequent calculation, Peters, 1977, 1986).

1. E_h-pH Diagram of Chalcopyrite

Besides reactions (2-8) and (2-11), the reactions, which could produce sulphur on chalcopyrite, may involve the progressive oxidation of the products of these two reactions CuS as following:

$$CuS = Cu^{2+} + S + 2e^- \qquad (2\text{-}15a)$$

$$\begin{cases} \Delta G^0 = 113.858(\text{kJ/mol}) \\ E^0 = 0.59(\text{V}) \\ E_h = 0.59 + 0.0295\lg[Cu^{2+}] \end{cases} \qquad (2\text{-}15b)$$

$$CuS + 2H_2O = Cu(OH)_2 + S + 2H^+ + 2e^- \qquad (2\text{-}16a)$$

$$\begin{cases} \Delta G^0 = 166.44(\text{kJ/mol}) \\ E^0 = 0.862(\text{V}) \\ E_h = 0.862 - 0.059\text{pH} \end{cases} \qquad (2\text{-}16b)$$

At certain high potential, chalcopyrite surface is further oxidized, and thus the reactions producing hydrophilic species are:

$$2CuFeS_2 + 3H_2O = 2CuS + 2Fe^{2+} + S_2O_3^{2-} + 6H^+ + 8e^- \quad (2\text{-}17a)$$

$$\begin{cases} \Delta G^0 = 285.768(\text{kJ/mol}) \\ E^0 = 0.37(\text{V}) \\ E_h = 0.331 - 0.044\text{pH} \end{cases} \quad (2\text{-}17b)$$

$$2CuS + 3H_2O = 2Cu^{2+} + S_2O_3^{2-} + 6H^+ + 8e^- \quad (2\text{-}18a)$$

$$\begin{cases} \Delta G^0 = 407.229(\text{kJ/mol}) \\ E^0 = 0.528(\text{V}) \\ E_h = 0.44 - 0.044\text{pH} \end{cases} \quad (2\text{-}18b)$$

$$2CuFeS_2 + 9H_2O = 2CuS + 2Fe(OH)_3 + S_2O_3^{2-} + 12H^+ + 10e^- \quad (2\text{-}19a)$$

$$\begin{cases} \Delta G^0 = 489.278(\text{kJ/mol}) \\ E^0 = 0.507(\text{V}) \\ E_h = 0.48 - 0.071\text{pH} \end{cases} \quad (2\text{-}19b)$$

$$2CuS + 7H_2O = 2Cu(OH)_2 + S_2O_3^{2-} + 10H^+ + 8e^- \quad (2\text{-}20a)$$

$$\begin{cases} \Delta G^0 = 512.247(\text{kJ/mol}) \\ E^0 = 0.664(\text{V}) \\ E_h = 0.635 - 0.0738\text{pH} \end{cases} \quad (2\text{-}20b)$$

The E_h-pH diagram of chalcopyrite is constructed combining the Eqs. (2-8), (2-11), and (2-15) to (2-20) as given in Fig. 2.14, where the E_h limits of water stability and other unimportant species are not considered for the sake of simplification. It can be seen from Fig. 2.14 that the upper limit and the lower limit potential of collectorless flotation of chalcopyrite at various pH values well agree with the potential defined respectively by reactions producing thiosulphate and elemental sulphur. The E_h-pH area where chalcopyrite possesses self-induced floatability is just the region where metastable elemental sulphur exists, indicating that elemental sulphur is a hydrophobic entity. When above the upper limit potential, hydrophilic species such as thiosulphate and metal hydroxides are formed, and the flotation ceases. Figure 2.14 also demonstrates that chalcopyrite exhibits the best self-induced floatability, which has wider upper and low limit potential of flotation at almost all pH regions.

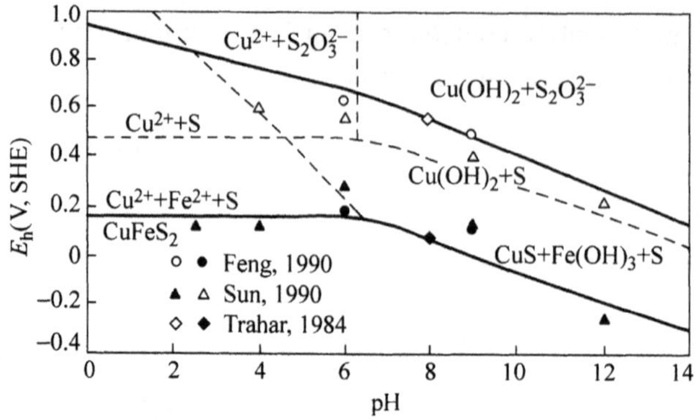

Figure 2.14 Electrochemical phase diagram for chalcopyrite with elemental sulphur as metastable phase. Equilibrium lines (solid lines) correspond to dissolved species at 10^{-4} mol/L. Plotted points show the upper and lower limit potential of collectorless flotation of chalcopyrite reported from Sun (1990), Feng (1989) and Trahar (1984)

2. E_h-pH Diagram of Galena

For galena, the reactions not producing hydrophobic species are

$$2PbS + 3H_2O = 2Pb^{2+} + S_2O_3^{2-} + 6H^+ + 8e^- \quad (2\text{-}21a)$$

$$\begin{cases} \Delta G^0 = 316.101(\text{kJ/mol}) \\ E^0 = 0.409(\text{V}) \\ E_h = 0.32 - 0.044\text{pH} \end{cases} \quad (2\text{-}21b)$$

$$2PbS + 7H_2O = 2Pb(OH)_2 + S_2O_3^{2-} + 10H^+ + 8e^- \quad (2\text{-}22a)$$

$$\begin{cases} \Delta G^0 = 471.643(\text{kJ/mol}) \\ E^0 = 0.611(\text{V}) \\ E_h = 0.581 - 0.0737\text{pH} \end{cases} \quad (2\text{-}22b)$$

$$2PbS + 7H_2O = 2HPbO_2^- + S_2O_3^{2-} + 12H^+ + 8e^- \quad (2\text{-}23a)$$

$$\begin{cases} \Delta G^0 = 635.675(\text{kJ/mol}) \\ E^0 = 0.823(\text{V}) \\ E_h = 0.734 - 0.0885\text{pH} \end{cases} \quad (2\text{-}23b)$$

The E_h-pH diagram of galena is constructed through the reactions (2-12) to (2-14) and (2-21) to (2-23) and presented in Fig. 2.15. It may be seen from Fig. 2.15 that the lower limit potential of collectorless flotation of galena at various pH are well defined by the conditions producing elemental sulphur. The upper limit potential of self-included collectorless flotation of galena at various pH is

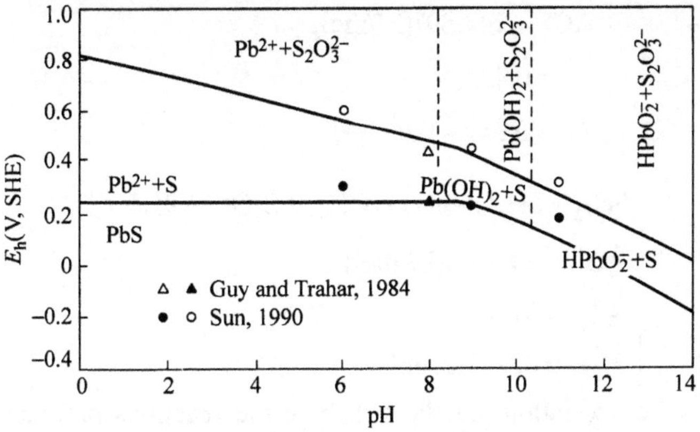

Figure 2.15 E_h-pH diagram for galena in aqueous solutions with elemental sulphur as metastable phase. Equilibrium lines correspond to dissolved species at 10^{-4} mol/L. Plotted points show the upper and lower limit potential of collectorless flotation of galena reported from Sun (1990, 1992) and from Guy and Trahar (1984)

determined by the reactions producing thiosulphate, lead hydroxide and plumbate. The E_h-pH area where galena has self-induced collectorless floatability is just the region where metastable elemental sulphur exists, indicating that metastable elemental sulphur is a hydrophobic entity. Galena appears to have better collectorless floatability at wider pH and potential range although it shows less floatablive than chalcopyrite.

3. E_h-pH Diagram of Pyrite and Arsenopyrite

Pyrite oxidation reactions include the reactions producing hydrophobic species:

$$FeS_2 = Fe^{2+} + 2S + 2e^- \quad (2\text{-}24a)$$

$$\begin{cases} \Delta G^0 = 65.689(\text{kJ/mol}) \\ E^0 = 0.34(\text{V}) \\ E_h = 0.34 + 0.059\lg[Fe^{2+}] = 0.24(\text{V}) \end{cases} \quad (2\text{-}24b)$$

$$FeS_2 + 3H_2O = Fe(OH)_3 + 2S + 3H^+ + 3e^- \quad (2\text{-}25a)$$

$$\begin{cases} \Delta G^0 = 167.603(\text{kJ/mol}) \\ E^0 = 0.579(\text{V}) \\ E_h = 0.579 - 0.059\text{pH} \end{cases} \quad (2\text{-}25b)$$

and the reactions producing hydrophilic species:

$$FeS_2 + 3H_2O = Fe^{2+} + S_2O_3^{2-} + 6H^+ + 6e^- \quad (2\text{-}26a)$$

$$\begin{cases} \Delta G^0 = 244.891(\text{kJ/mol}) \\ E^0 = 0.423(\text{V}) \\ E_h = 0.344 - 0.059\text{pH} \end{cases} \tag{2-26b}$$

$$FeS_2 + 6H_2O = Fe(OH)_3 + S_2O_3^{2-} + 9H^+ + 7e^- \tag{2-27a}$$

$$\begin{cases} \Delta G^0 = 347.021(\text{kJ/mol}) \\ E^0 = 0.514(\text{V}) \\ E_h = 0.48 - 0.076\text{pH} \end{cases} \tag{2-27b}$$

For arsenopyrite, oxidation reactions include the reactions producing elemental sulphur:

$$FeAsS + 4H_2O = Fe^{2+} + H_3AsO_4 + S + 5H^+ + 7e^- \tag{2-28a}$$

$$\begin{cases} E^0 = 0.303(\text{V}) \\ E_h = 0.236 - 0.042\text{pH} \end{cases} \tag{2-28b}$$

$$FeAsS + 4H_2O = Fe^{2+} + H_2AsO_4^- + S + 6H^+ + 7e^- \tag{2-29a}$$

$$\begin{cases} E^0 = 0.333(\text{V}) \\ E_h = 0.266 - 0.0506\text{pH} \end{cases} \tag{2-29b}$$

$$FeAsS + 7H_2O = Fe(OH)_3 + HAsO_4^{2-} + S + 10H^+ + 8e^- \tag{2-30a}$$

$$\begin{cases} E^0 = 0.4771(\text{V}) \\ E_h = 0.448 - 0.0738\text{pH} \end{cases} \tag{2-30b}$$

and the reaction producing thiosulphate:

$$2FeAsS + 11H_2O = 2Fe^{2+} + 2H_3AsO_4 + S_2O_3^{2-} + 16H^+ + 18e^- \tag{2-31a}$$

$$\begin{cases} E^0 = 0.339(\text{V}) \\ E_h = 0.273 - 0.052\text{pH} \end{cases} \tag{2-31b}$$

$$2FeAsS + 11H_2O = 2Fe^{2+} + 2H_2AsO_4^- + S_2O_3^{2-} + 18H^+ + 18e^- \tag{2-32a}$$

$$\begin{cases} E^0 = 0.362(\text{V}) \\ E_h = 0.296 - 0.059\text{pH} \end{cases} \tag{2-32b}$$

$$2FeAsS + 17H_2O = 2Fe(OH)_3 + 2HAsO_4^{2-} + S_2O_3^{2-} + 26H^+ + 20e^- \tag{2-33a}$$

$$\begin{cases} E^0 = 0.475(\text{V}) \\ E_h = 0.416 - 0.0767\text{pH} \end{cases} \tag{2-33b}$$

The E_h-pH diagrams of surface oxidation of arsenopyrite and pyrite are shown in Fig. 2.16 and Fig. 2.17, respectively. Figure 2.16 shows that E_h-pH area of self-induced collectorless flotation of arsenopyrite is close to the area forming sulphur. The reactions producing elemental sulphur determine the lower limit potential of flotation. The reactions producing thiosulphate and other hydrophilic species define the upper limit of potential. In acid solutions arsenopyrite demonstrates wider potential region for collectorless flotation, but almost non-floatable in alkaline environment. It suggests that the hydrophobic entity is metastable elemental sulphur. However, in alkaline solutions, the presence of

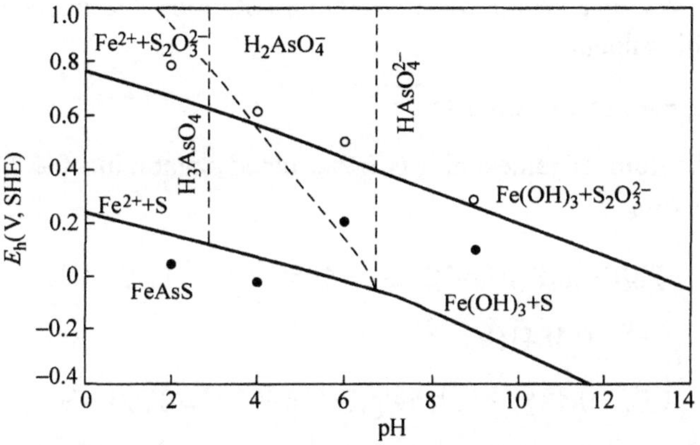

Figure 2.16 E_h-pH diagram for arsenopyrite in aqueous solutions with elemental sulphur as metastable phase. Equilibrium lines correspond to dissolved species at 10^{-4} mol/L. Plotted points show the upper and lower limit potential of collectorless flotation of arsenopyrite reported from Sun (1990, 1992)

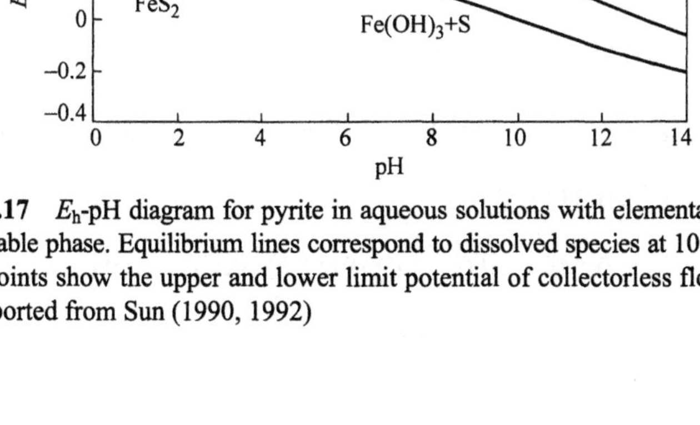

Figure 2.17 E_h-pH diagram for pyrite in aqueous solutions with elemental sulphur as metastable phase. Equilibrium lines correspond to dissolved species at 10^{-4} mol/L. Plotted points show the upper and lower limit potential of collectorless flotation of pyrite reported from Sun (1990, 1992)

metastable elemental sulphur does not render arsenopyrite hydrophobic and exhibits collectorless floatability. It may be attributed to the presence of $Fe(OH)_3$ and $HAsO_4^{2-}$ at the same time on the arsenopyrite surface. The hydrophobicity of elemental sulphur is not sufficient to overcome the hydrophilicity by $Fe(OH)_3$ and $HAsO_4^{2-}$. Therefore, it might be supposed that the relative amounts of $Fe(OH)_3$, $HAsO_4^{2-}$ and sulphur and accordingly the rations of hydrophobic to hydrophilic species perhaps determine the floatability of arsenopyrite.

Figure 2.17 shows that although the metastable elemental sulfur may be present at pyrite surface, pyrite does not exhibit self-induced collectorless floatability except in very strong acidic media and a narrow oxidized Eh range. Such behavior may be similar to that of arsenopyrite in alkaline solutions due to the formation of hydroxides, thiosulphate.

4. E_h-pH Diagram of Jamesonite

The E_h-pH diagram of jamesonite is constructed combining the Eqs. (2-34) to (2-49) as following:

$$Pb_4FeSb_6S_{14} = 3Sb_2S_3 + 4Pb^{2+} + Fe^{2+} + 5S + 10e^- \quad (2\text{-}34a)$$

$$\begin{cases} E^0 = 0.4841(V) \\ E_h = 0.4841 + 0.0059\lg[Fe^{2+}][Pb^{2+}]^4 = 0.3213(V) \end{cases} \quad (2\text{-}34b)$$

$$Pb_4FeSb_6S_{14} + 12H_2O = 6HSbO_2 + 4Pb^{2+} + Fe^{2+} + 14S + 18H^+ + 28e^- \quad (2\text{-}35a)$$

$$\begin{cases} E^0 = 0.4690(V) \\ E_h = 0.4690 + 0.00211\lg[Fe^{2+}][HSbO_2]^6[Pb^{2+}]^4 - 0.0379pH \\ \quad = 0.3447 - 0.0379pH \end{cases} \quad (2\text{-}35b)$$

$$Pb_4FeSb_6S_{14} + 18H_2O = 6SbO_3^- + 4Pb^{2+} + Fe^{2+} + 14S + 36H^+ + 40e^- \quad (2\text{-}36a)$$

$$\begin{cases} E^0 = 0.5317(V) \\ E_h = 0.5317 + 0.00148\lg[SbO_3^-]^6[Pb^{2+}]^4[Fe^{2+}] - 0.0531pH \\ \quad = 0.4445 - 0.0531pH \end{cases} \quad (2\text{-}36b)$$

$$Pb_4FeSb_6S_{14} + 21H_2O = 6SbO_3^- + 4Pb^{2+} + Fe(OH)_3 + 14S + 39H^+ + 41e^- \quad (2\text{-}37a)$$

$$\begin{cases} E^0 = 0.5445(V) \\ E_h = 0.5445 + 0.00144\lg[SbO_3^-]^6[Pb^{2+}]^4 - 0.0561pH \\ \quad = 0.4683 - 0.0561pH \end{cases} \quad (2\text{-}37b)$$

$$Pb_4FeSb_6S_{14}+42H_2O = 6SbO_3^-+4Pb^{2+}+Fe(OH)_3+7S_2O_3^{2-}+81H^++69e^- \quad (2\text{-}38a)$$

$$\begin{cases} E^0 = 0.5120(V) \\ E_h = 0.5120+0.000855\lg[Pb^{2+}]^4[SbO_3^-]^6[S_2O_3^{2-}]^7-0.0693pH \\ \quad = 0.4359\text{-}0.0693pH \end{cases} \quad (2\text{-}38b)$$

$$Pb_4FeSb_6S_{14}+77H_2O = 6SbO_3^-+4Pb^{2+}+Fe(OH)_3+14SO_4^{2-}+151H^++125e^- \quad (2\text{-}39a)$$

$$\begin{cases} E^0 = 0.4165(V) \\ E_h = 0.4165+0.000472\lg[Pb^{2+}]^4[SbO_3^-]^6[SO_4^{2-}]^{14}-0.0713pH \\ \quad = 0.3594-0.0713pH \end{cases} \quad (2\text{-}39b)$$

$$Pb_4FeSb_6S_{14}+12H_2O = 6HSbO_2+4PbS+Fe^{2+}+10S+18H^++20e^- \quad (2\text{-}40a)$$

$$\begin{cases} E^0 = 0.5150(V) \\ E_h = 0.5150+0.00295\lg[Fe^{2+}][HSbO_2]^6-0.0531pH = 0.4049-0.0531pH \end{cases} \quad (2\text{-}40b)$$

$$Pb_4FeSb_6S_{14}+15H_2O = 6HSbO_2+4PbS+Fe(OH)_3+10S+21H^++21e^- \quad (2\text{-}41a)$$

$$\begin{cases} E^0 = 0.5405(V) \\ E_h = 0.5405+0.00282\lg[HSbO_2]^6-0.0592pH = 0.4521-0.0592pH \end{cases} \quad (2\text{-}41b)$$

$$Pb_4FeSb_6S_{14}+15H_2O = 6HSbO_2+4Pb^{2+}+Fe(OH)_3+14S+21H^++29e^- \quad (2\text{-}42a)$$

$$\begin{cases} E^0 = 0.4893(V) \\ E_h = 0.4893+0.00204\lg[HSbO_2]^6[Pb^{2+}]^4-0.0427pH = 0.3813-0.0427pH \end{cases} \quad (2\text{-}42b)$$

$$Pb_4FeSb_6S_{14}+21H_2O = 6SbO_3^-+4PbS+Fe(OH)_3+10S+39H^++33e^- \quad (2\text{-}43a)$$

$$\begin{cases} E^0 = 0.5907(V) \\ E_h = 0.5907+0.00179\lg[SbO_3^-]^6-0.0697pH = 0.5346-0.0697pH \end{cases} \quad (2\text{-}43b)$$

$$Pb_4FeSb_6S_{14}+29H_2O = 6SbO_3^-+4Pb(OH)_2+Fe(OH)_3+14S+47H^++41e^- \quad (2\text{-}44a)$$

$$\begin{cases} E^0 = 0.6231(V) \\ E_h = 0.6231+0.00144\lg[SbO_3^-]^6-0.0676pH = 0.5780-0.0676pH \end{cases} \quad (2\text{-}44b)$$

$$Pb_4FeSb_6S_{14}+50H_2O = 6SbO_3^-+4Pb(OH)_2+Fe(OH)_3+7S_2O_3^{2-}+89H^++69e^- \quad (2\text{-}45a)$$

$$\begin{cases} E^0 = 0.5588(\text{V}) \\ E_h = 0.5588 + 0.000855\lg[SbO_3^-]^6[S_2O_3^{2-}]^7 - 0.0761\text{pH} = 0.5012 - 0.0761\text{pH} \end{cases} \quad (2\text{-}45b)$$

$$Pb_4FeSb_6S_{14} + 85H_2O = 6SbO_3^- + 4Pb(OH)_2 + Fe(OH)_3 + 14SO_4^{2-} + 159H^+ + 125e^- \quad (2\text{-}46a)$$

$$\begin{cases} E^0 = 0.4443(\text{V}) \\ E_h = 0.4443 + 0.000472\lg[SbO_3^-]^6[SO_4^{2-}]^{14} - 0.0750\text{pH} = 0.3974 - 0.0750\text{pH} \end{cases} \quad (2\text{-}46b)$$

$$Pb_4FeSb_6S_{14} + 29H_2O = 6SbO_3^- + 4HPbO_2^- + Fe(OH)_3 + 14S + 51H^+ + 41e^- \quad (2\text{-}47a)$$

$$\begin{cases} E^0 = 0.7059(\text{V}) \\ E_h = 0.7059 + 0.00144\lg[HPbO_2^-]^4[SbO_3^-]^6 - 0.0734\text{pH} = 0.6297 - 0.0734\text{pH} \end{cases} \quad (2\text{-}47b)$$

$$Pb_4FeSb_6S_{14} + 50H_2O = 6SbO_3^- + 4HPbO_2^- + Fe(OH)_3 + 7S_2O_3^{2-} + 93H^+ + 69e^- \quad (2\text{-}48a)$$

$$\begin{cases} E^0 = 0.6080(\text{V}) \\ E_h = 0.6080 + 0.000855\lg[HPbO_2^-]^4[SbO_3^-]^6[S_2O_3^{2-}]^7 - 0.0795\text{pH} \\ \quad = 0.5319 - 0.0795\text{pH} \end{cases} \quad (2\text{-}48b)$$

$$Pb_4FeSb_6S_{14} + 85H_2O = 6SbO_3^- + 4HPbO_2^- + Fe(OH)_3 + 14SO_4^{2-} + 163H^+ + 125e^- \quad (2\text{-}49a)$$

$$\begin{cases} E^0 = 0.4715(\text{V}) \\ E_h = 0.4715 + 0.000472\lg[HPbO_2^-]^4[SbO_3^-]^6[SO_4^{2-}]^{14} - 0.0769\text{pH} \\ \quad = 0.4144 - 0.0769\text{pH} \end{cases} \quad (2\text{-}49b)$$

The E_h-pH diagram for jamesonite is given in Fig. 2.18. It may be seen from Fig. 2.18 that in acid and neutral solutions, the reactions producing elemental sulphur may render jamesonite surface hydrophobic and jamesonite shows good collectorless flotation. In alkaline solution, the hydrophilic species Fe^{2+}, Pb^{2+}, $HSbO_2$, SbO_3^-, are produced, the collectorless floatability of jamesonite becomes weaker, it may be attributed to the presence of $Fe(OH)_3$ and $Pb(OH)_2$ at the same time on the jamesonite surface. The relative amounts of hydrophobic sulphur and hydrophilic $Fe(OH)_3$ and $Pb(OH)_2$ perhaps determine the collectorless floatability of jamesonite.

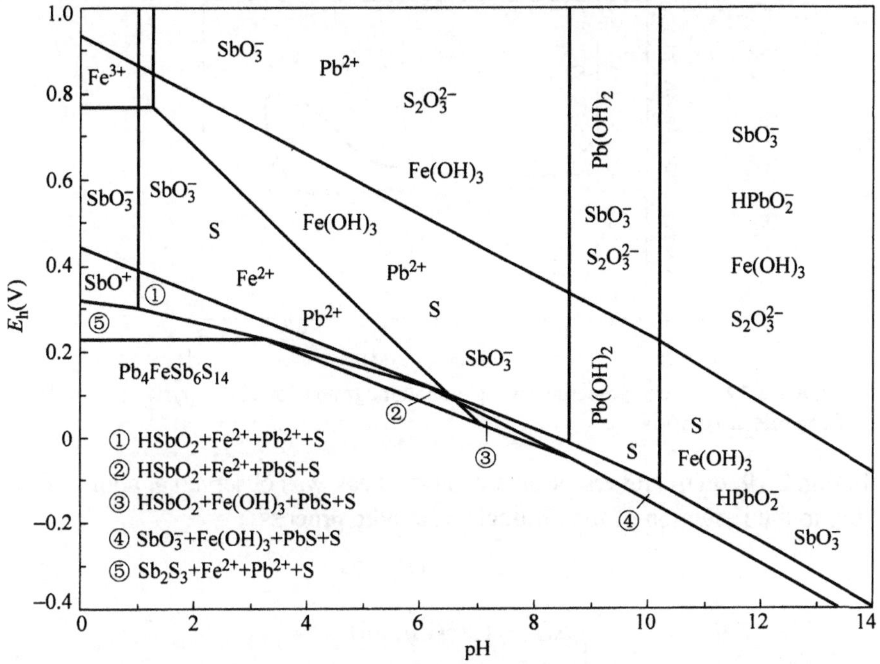

Figure 2.18 E_h-pH diagram for jamesonite in aqueous solutions with elemental sulphur as metastable phase. Equilibrium lines correspond to dissolved species at 10^{-4} mol/L

2.4 Electrochemical Determination of Surface Oxidation Products of Sulphide Minerals

Many investigators have used different techniques to study the electrochemical behavior of different sulphide mineral electrodes in solutions of different compositions. Linear potential sweep voltammetry (LPSV), and cyclic voltammetry (CV) have been perhaps, used most extensively and applied successfully to the investigation of reactions of sulphide minerals with aqueous systems. These techniques have provided valuable information on the extent of oxidation as a function of potential for various solution conditions and have allowed the identity of the surface products to be deduced.

The oxidation of chalcopyrite has been studied in some detail (Gardner and Woods, 1979; Biegler and Horne,1985; Guy and Trahar, 1985; Kelsall and Page, 1985; Pang and Chander, 1990). Figure 2.19 presents the voltammogram of chalcopyrite electrode at pH = 11. It is evident that the anodic peaks occur due to the anodic oxidation of chalcopyrite. The potential at which the anodic current rises are in reasonable agreement with those calculated based on reaction (2-8) at pH = 11. It suggests that the formation of sulphur would be expected to occur at −0.12 V at pH = 11 by reaction (2-8).

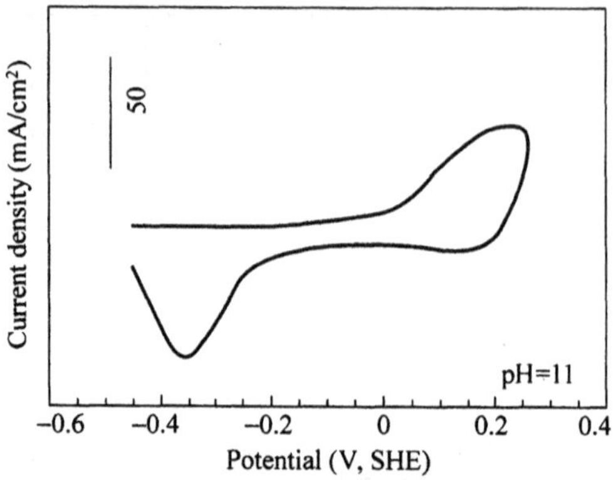

Figure 2.19 Linear potential sweep voltammograms for chalcopyrite electrode. Scan rate at 10 mV/s

In Fig. 2.19, on the reverse scan a cathodic peak was observed at about −0.27 V owing to the reduction of the products of anodic process.

$$HS^- = S + H^+ + 2e^- \tag{2-50a}$$

$$\begin{cases} \Delta G^0 = 12.6(\text{kJ/mol}) \\ E^0 = -0.065(\text{V}) \\ E_h = 0.053 - 0.0295\text{pH} \end{cases} \tag{2-50b}$$

The potential calculated using above reaction is −0.27 V at pH = 11 at dissolved species 10^{-4} mol/L corresponding to the potential of the cathodic current rise, further indicating the elemental sulphur as the oxidation product.

Voltammograms obtained with galena (see Fig. 2.20) show that the initial potential of anodic oxidation of galena is around 0.15 V, which is very close to the potential of formation of elemental sulphur in Eqs. (2-12) and (2-13) and corresponding to the initial potential of flotation beginning of galena in Fig. 2.6 suggested that the oxidation product of galena surface is sulphur. The potential at which the anodic current and anodic oxidation of galena increased evidently is around 0.4 V, which is close to the potential according to reactions (2-21) and (2-22) considering elemental sulphur as a metastable phase. It indicates that the further oxidation of galena formed hydrophilic oxy-sulphur species or hydroxides and galena collectorless flotation ceased as shown in Fig. 2.6. It has also been reported that the initial product of oxidation was elemental sulphur at all pH with Pb^{2+} being formed in acid and $Pb(OH)_2$ and $HPbO_2^{2-}$ in alkaline media (Hamilton and Woods, 1984; Lamacha et al., 1984; Pritzker and Yoon, 1984a,b; Sun et al., 1993a,b,c,d).

The voltammogram for jamesonite is given in Fig. 2.21. It can be seen that there are four obvious anodic peaks at the first cycle voltammetry. The peak current of ap_1, ap_3 and cp_2 constantly increase with the number of cycle adding, but the peak height of ap_2 and ap_4 decreases at the second cycle and then keeps constant even by multi-cycle according to the experimental appearances. It was

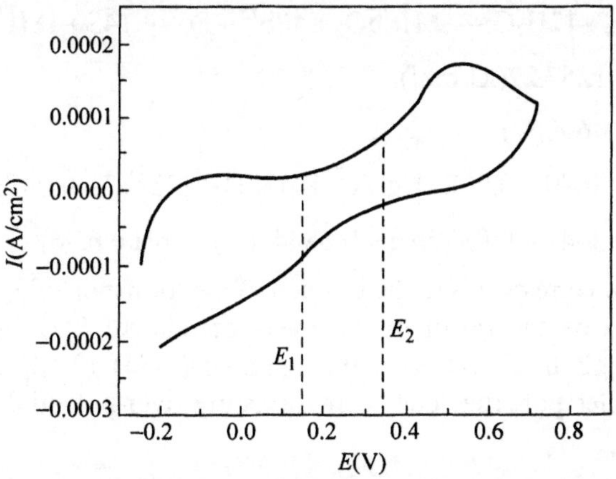

Figure 2.20 Voltammograms for galena at natural pH (sweep rate: 30 mV/s)

assumed that ap_1 peak corresponds to the reaction of forming the deficient-metal and sulfur-rich compound, whereas ap_2 is related to the reaction in which the deficient-metal sulphide is further oxidized into elemental sulfur. As described in the following, the E_h values of the reactions (2-51), (2-52) are unanimous with the peak potential in Fig. 2.21.

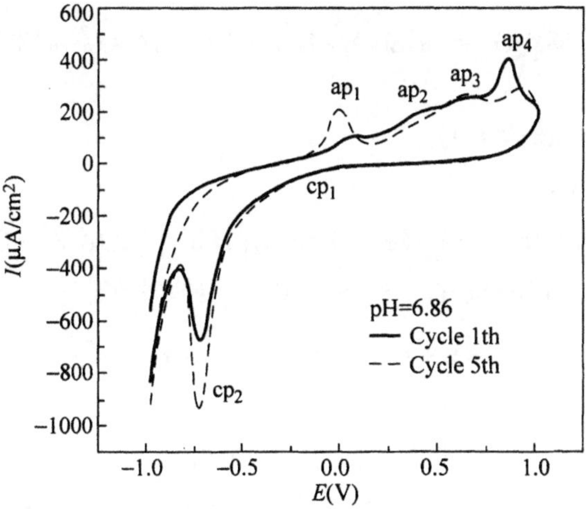

Figure 2.21 Multi-voltammograms for jamesonite electrode in 0.1 mol/L KNO_3 mixed phosphate buffer solution at pH = 6.86 and scanning rate of 50 mV/s

$$ap_1:\ Pb_4FeSb_6S_{14}+12H_2O = 6HSbO_2+4PbS+Fe^{2+}+10S+18H^++20e^- \quad (2\text{-}51a)$$

$$\begin{cases} \Delta G^0 = 994.02(\text{kJ/mol}) \\ E^0 = 0.5150(\text{V}) \\ E_h = 0.5150 + 0.00295\lg[Fe^{2+}][HSbO_2]^6 - 0.0531\text{pH} \\ \quad = 0.4049 - 0.0531\text{pH} = 0.04063(\text{V}) \quad (\text{pH} = 6.86) \end{cases} \quad (2\text{-}51b)$$

ap$_2$: $Pb_4FeSb_6S_{14}+12H_2O = 6HSbO_2+4Pb^{2+}+Fe^{2+}+14S+18H^++28e^-$ (2-52a)

$$\begin{cases} \Delta G^0 = 1267.22(\text{kJ/mol}) \\ E^0 = 0.4690(\text{V}) \\ E_h = 0.4690 + 0.00211\lg[Fe^{2+}][HSbO_2]^6[Pb^{2+}]^4 - 0.0379\text{pH} \\ \quad = 0.3447 - 0.0379\text{pH} = 0.08447(\text{V}) \quad (\text{pH} = 6.86) \end{cases} \quad (2\text{-}52b)$$

The peak of ap$_3$ corresponds to the reaction of the formation of $S_2O_3^{2-}$. The change of ap$_4$ peak may be the result of the over-potential of SO_4^{2-} by the following reactions. Although the E_h values of the reactions (2-53), (2-54) are smaller, they occur at the higher potential due to the existing over-potential of $S_2O_3^{2-}$, SO_4^{2-} ions, respectively.

ap$_3$: $Pb_4FeSb_6S_{14}+33H_2O = 6HSbO_2+4Pb^{2+}+Fe^{2+}+7S_2O_3^{2-}+60H^++56e^-$

(2-53a)

$$\begin{cases} \Delta G^0 = 2522.48(\text{kJ/mol}) \\ E^0 = 0.4668(\text{V}) \\ E_h = 0.4668 + 0.00106\lg[Fe^{2+}][HSbO_2]^6[Pb^{2+}]^4[S_2O_3^{2-}]^7 - 0.0634\text{pH} \\ \quad = 0.3661 - 0.0634\text{pH} = -0.06903(\text{V}) \quad (\text{pH}=6.86) \end{cases} \quad (2\text{-}53b)$$

ap$_4$: $Pb_4FeSb_6S_{14}+68H_2O = 6HSbO_2+4Pb^{2+}+Fe^{2+}+14SO_4^{2-}+130H^++112e^-$

(2-54a)

$$\begin{cases} \Delta G^0 = 4161.62(\text{kJ/mol}) \\ E^0 = 0.3850(\text{V}) \\ E_h = 0.3850 + 0.000529\lg[Fe^{2+}][HSbO_2]^6[Pb^{2+}]^4[SO_4^{2-}]^{14} - 0.0685\text{pH} \\ \quad = 0.3179 - 0.0685\text{pH} = -0.1519(\text{V}) \quad (\text{pH} = 6.86) \end{cases} \quad (2\text{-}54b)$$

cp$_1$: $Fe(OH)_2 + H_2O = Fe(OH)_3 + H^+ + e^-$ (2-55a)

$$\begin{cases} E^0 = 0.271(\text{V}) \\ E_h = 0.271 - 0.059\text{pH} \end{cases} \quad (2\text{-}55b)$$

The cyclic voltammetry for sphalerite electrode is presented in Fig. 2.22. It follows from Fig. 2.22 that the potential range of collectorless flotation of sphalerite is 155 – 270 mV. For the collectorless flotation of sphalerite, the lower limit of flotation corresponds to the following reactions:

$$ZnS \longrightarrow S + Zn^{2+} + 2e^- \quad (2\text{-}56a)$$

$$E^0 = 0.311(\text{V}) \quad (2\text{-}56b)$$

$$ZnS + 2H_2O \longrightarrow S + Zn(OH)_2 + 2H^+ + 2e^- \quad (2\text{-}56c)$$

$$E^0 = 0.645(\text{V}) \quad (2\text{-}56d)$$

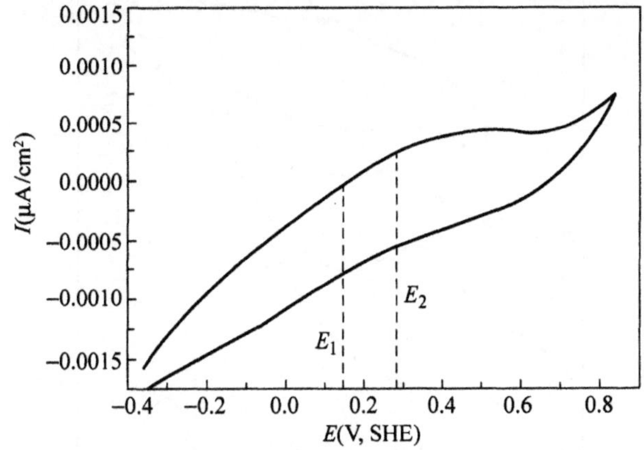

Figure 2.22 Voltammograms for sphalerite electrode in natural pH value at scanning rate of 0.50 mV/s (0.1 mol/L KNO_3)

The potential is 155 mV based on reaction (2-56a) and 280 mV based on reaction (2-56c), with 10^{-6} mol/L concentration of dissolved species, the initial reaction of sphalerite correspond to reaction (2-56a).

The further oxidation of sphalerite surface are:

$$2ZnS + 3H_2O \longrightarrow S_2O_3^{2-} + 2Zn^{2+} + 6H^+ + 8e^- \quad (2\text{-}57a)$$

$$E^0 = 0.174(V) \quad (2\text{-}57b)$$

$$2ZnS + 7H_2O \longrightarrow S_2O_3^{2-} + 2Zn(OH)_2 + 10H^+ + 8e^- \quad (2\text{-}58a)$$

$$E^0 = 0.345(V) \quad (2\text{-}58b)$$

Considering the formation of $S_2O_3^{2-}$ needing 0.5 V over-potential, assuming the concentration of all dissolved species to be 10^{-6} mol/L, the reaction potential are 0.28 V and 0.37 V respectively, corresponding to reactions (2-57) and (2-58), the upper limit potential of flotation of sphalerite depends on reaction (2-57).

Pyrrhotite is one of many sulphides which display collectorless flotation resulting from the formation of sulphur on the mineral surface (Hamilton and Woods, 1981; Heyes and Trahar, 1984; Hodgson and Agar, 1984). The anodic scan section of cyclic voltammetry for pyrrhotite electrode in pH = 2.2, 4.7, 7.0, 8.8, 11, 12.1, 12.7 buffer solutions respectively, is presented in Fig. 2.23. The cyclic voltammograms curve at pH = 8.8 is also presented in Fig. 2.23. It can be seen from Fig. 2.23 that anodic current peak emerges at about −0.1—0 V when $pH < 11$. As pH increases, the peak moves to the left. This peak may correspond to the following reactions:

$$FeS \longrightarrow Fe^{2+} + S + 2e^- \quad (2\text{-}59a)$$

$$FeS + 3H_2O \longrightarrow Fe(OH)_3 + S + 3H^+ + 3e^- \quad (2\text{-}59b)$$

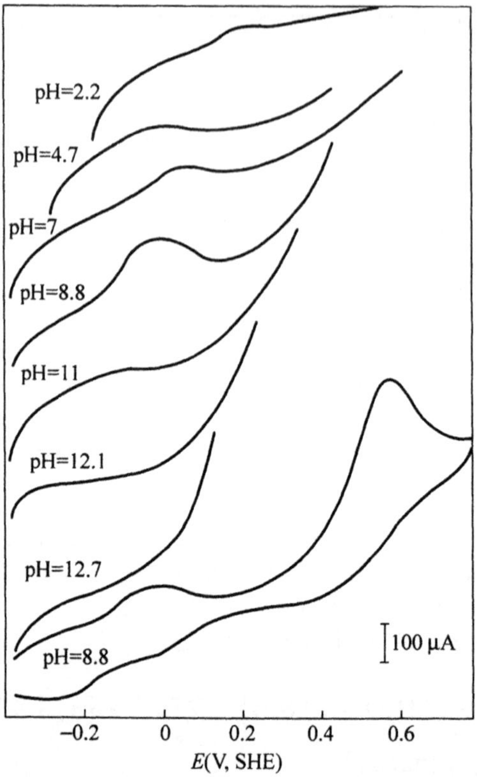

Figure 2.23 Voltammograms for pyrrhotite electrode in 0.1 mol/L KNO_3 mixed buffer solution of different pH at scanning rate of 10 mV/s

Due to the formation of S on the surface of pyrrhotite, the collectorless flotation of pyrrhotite can be carried out. When pH = 12.7, the anodic peak disappeares, and the anodic current increases rapidly. It indicates that only the oxidation of pyrrhotite takes place on the surface of pyrrhotite. So the flotation of pyrrhotite could not occur. These phenomena may correspond to the following reactions.

$$2FeS + 3H_2O = 2Fe^{2+} + S_2O_3^{2-} + 6H^+ + 8e^- \tag{2-60a}$$

$$2FeS + 9H_2O = 2Fe(OH)_3 + S_2O_3^{2-} + 12H^+ + 8e^- \tag{2-60b}$$

Other reports also demonstrate the potential of pyrrhotite flotation commences is corresponding to the initial potential of anodic oxidation process explained in terms of the following reaction by Hamilton and Woods (1981, 1984), Heyes and Trahar (1984).

$$FeS_{1.13} \longrightarrow Fe^{2+} + 1.13S + 2e^- \tag{2-61a}$$

$$FeS_{1.13} + 3H_2O \longrightarrow Fe(OH)_3 + 1.13S + 3H^+ + 3e^- \tag{2-61b}$$

Pyrite and arsenopyrite have similar oxidation and self-induced collectorless flotation behavior. It is generally suggested that anodic oxidation of pyrite occurs according to reactions (2-24) in acidic solutions (Lowson, 1982; Heyes and Trahar, 1984; Trahar, 1984; Sun et al., 1991; Chander et al., 1993). The oxidation of pyrite in basic solutions takes place according to reactions (2-25). Since pyrite is flotable only in strong acidic solutions, it seems reasonable to assume that reaction (2-24) is the dominant oxidation at acidic solutions. Whereas pyrite oxidizes to oxy-sulfur species with minor sulphur in basic solutions.

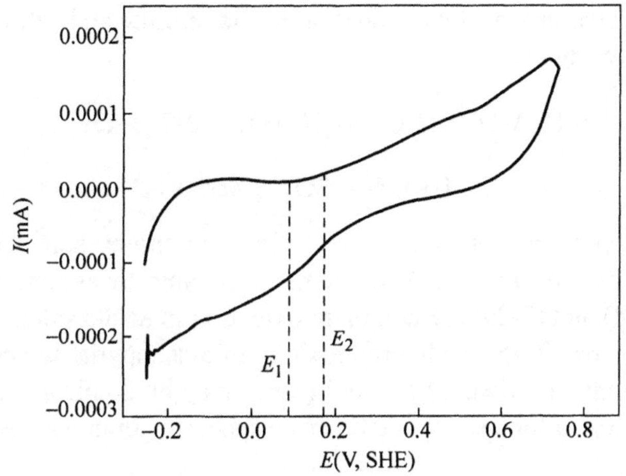

Figure 2.24 Voltammograms for pyrite electrode in natural pH at scanning rate of 0.50 mV/s (0.1 mol/L KNO_3)

Figure 2.24 shows the voltammograms of pyrite electrode in neutral media. It follows that the potential range of flotation relates to reactions (2-24) and (2-25). The potential calculated in reaction (2-24) is 161 – 180 mV, pyrite exhibit collectorless floatability due to the formation of elemental sulphur.

The oxidation of arsenopyrite has some more arguments. Kostina and Chernyak (1979) concluded that the reaction was significant only in caustic soda solution and that the adsorption of hydroxide ion was a key initiating step in the oxidation reaction. Beattie and Poling (1987), Sanchez and Hiskey (1988) and Feng (1989) considered that the oxidation of arsenopyrite was observed only under basic conditions by a reaction of the type:

$$FeAsS + 11H_2O \longrightarrow Fe(OH)_3 + HAsO_4^{2-} + SO_4^{2-} + 18H^+ + 14e^- \quad (2\text{-}62)$$

Dunn et al. (1989) attributed the oxidation of arsenopyrite electrode in acid to the reaction of type:

$$FeAsS + 4H_2O = Fe^{3+} + H_2AsO_4^- + S + 6H^+ + 8e^- \quad (2\text{-}63a)$$

$$\begin{cases} E^0 = 0.388(\text{V}) \\ E_\text{h} = 0.329 - 0.0443\text{pH} \end{cases} \tag{2-63b}$$

The reaction product was identified as α-sulphur using XRD and SEM analysis.

Sanchez and Hiskey (1991) reinvestigated the oxidation reaction of arsenopyrite and a two-step reaction sequence was suggested by the electrochemical measurements. The initial step was described by

$$\text{FeAsS} + 6\text{H}_2\text{O} = \text{Fe(OH)}_3 + \text{H}_2\text{AsO}_3^- + \text{S} + 7\text{H}^+ + 6\text{e}^- \tag{2-64}$$

The second step involved the oxidation of the arsenite and sulfur according to the following reactions:

$$\text{H}_2\text{AsO}_3^- + \text{H}_2\text{O} = \text{HAsO}_4^{2-} + 3\text{H}^+ + 2\text{e}^- \tag{2-65}$$

$$\text{S} + 4\text{H}_2\text{O} = \text{SO}_4^{2-} + 8\text{H}^+ + 6\text{e}^- \tag{2-66}$$

Because arsenopyrite is floatable in acidic conditions and non-floatable in basic conditions (see Fig. 2.16), it seems reasonable to assume that reactions (2-63) or (2-28) and (2-29) are dominant oxidation in acidic solutions. Elemental sulphur is responsible for the hydrophobicity of arsenopyrite in acidic media. In alkaline solutions, reactions (2-64) and (2-65) may be dominant resulting in the formation of oxy-sulfur species and arsenate species with minor sulphur.

2.5 Surface Analysis of Oxidation of Sulphide Minerals

Although the results of flotation, E_h-pH diagram and voltammetry studies show the evidence that the hydrophobic entity responsible for the collectorless flotation of most sulphide minerals is sulphur. The attempts to identify sulphur on the surface of sulphide minerals and to relate the amount produced to their floatability have not produced unequivocal results. Luttrell and Yoon (1984a,b) were unable to establish a relation between surface coverage and floatability. They suggested that the hydrophobic entity might be copper polysulphide rather than sulphur. Buckley et al. (1985) maintained that the absence of any detectable elemental sulphur in the X-ray photoelectron spectrum (XPS) of sulphide minerals indicated that the surface oxidation of sulphide minerals involves progressive removal of metal atoms from the sulphur lattice leaving metal-deficient sulphides with sulphur lattices little altered from the original structure. They also found that when the potential was made more positive in the oxidant solutions than in air-saturated solutions, the metal-deficient layer decomposed to form elemental sulphur. Chen et al. (1986) detected the presence of elemental sulphur on galena surface in the XPS at pH = 6 and E_h = 0.45 V. It can be seen from Fig. 2.25 that S^{2-}, S and SO_2^{2-} are present with binding energy of 161 eV, 163 eV and 167 eV.

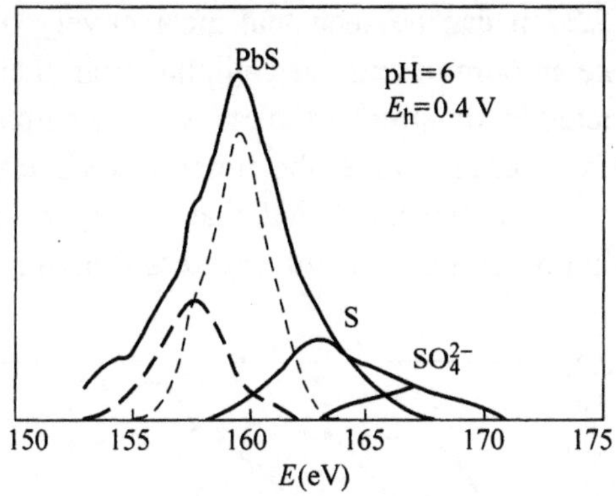

Figure 2.25 S (2p) XPS of floated concentrate of galena in the absence of collector pH = 6, E_h = 0.4 V (Chen et al., 1986)

Despite the conflicting evidence, Heyes and Trahar (1984) believe there is sufficient evidence to confirm the presence of sulphur on mineral surface. They leached the surface of floated pyrrhotite from a typical test with cyclohexane and have examined the leach solution in a UV spectrophotometer. They found that sulphur could be extracted from the surface of pyrrhotite, which had been floated in the absence of collector. As can be seen from Fig. 2.26, the spectrum from the leached pyrrhotite was compared with the spectrum of sulphur dissolved in cyclohexane indicating that sulphur was present at the surface. Kelebek and Smith (1989) used UV spectrophotometer to determine sulphur in the ethanol extract from the surface of floated galena and chalcopyrite showing that the amount of sulphur on the minerals can be correlated with their flotation rate which was found to be first order within the critical surface tension range.

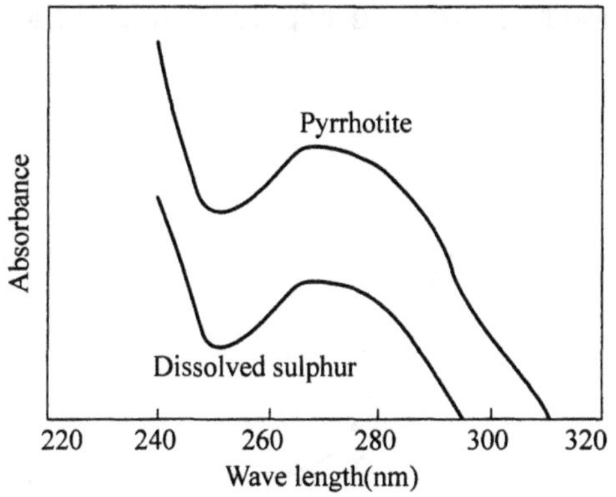

Figure 2.26 Comparison of spectra of cyclohexane extracts of floated pyrrhotite with that of sulphur dissolved in cyclohexane (Heyes and Trahar, 1984)

The correlation between the amount of extracted sulphur and floatability was further investigated. Figure 2.27 represents the relationship between the recovery of marmatite, pyrrhotite and jamesonite and the amount of extracted sulphur at pH = 4.7. From Fig. 2.27 it can be seen that the recovery of sulphide minerals corresponds well to the amount of extracted sulphur from their surface. The more the amount of S extracted is, the greater the recovery of sulphide minerals. When the amount of S attain a certain value, the flotation recovery of minerals is the maximum. Therefore it can be concluded that S may be responsible for the hydrophobicity of sulphide minerals in collectorless flotation.

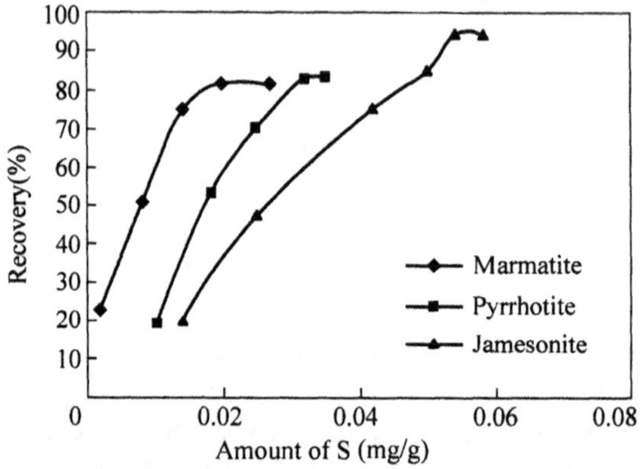

Figure 2.27 The relationship between the recovery of marmatite, pyrrhotite and jamesonite and the amount of extracted sulphur

The collectorless flotation recovery of marmatite, pyrrhotite and jamesonite and the amount of extracted sulphur as a function of pH are shown in Figs. 2.28, 2.29 and 2.30. Figure 2.28 shows that the trend of recovery of marmatite is consistent with change of the amount of extracted sulphur. They all decrease with the increase of pH. In acidic pH media, both collectorless flotation recovery of marmatite and the amount of extracted sulphur from its surface are the maximum.

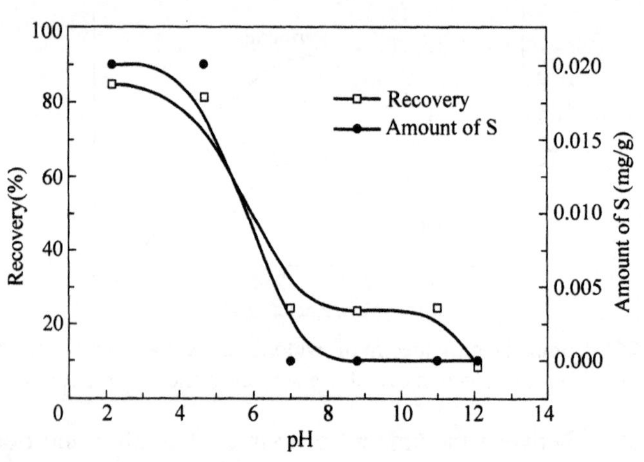

Figure 2.28 Collectorless flotation recovery of marmatite and the amount of extracted sulphur as a function of pH

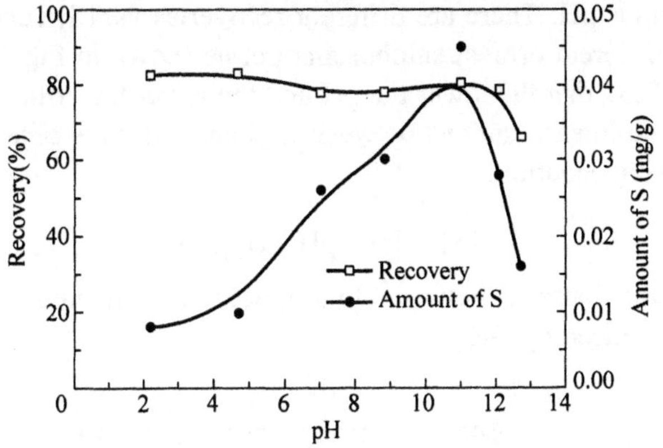

Figure 2.29 Collectorless flotation recovery of pyrrhotite and the amount of extracted sulphur as a function of pH

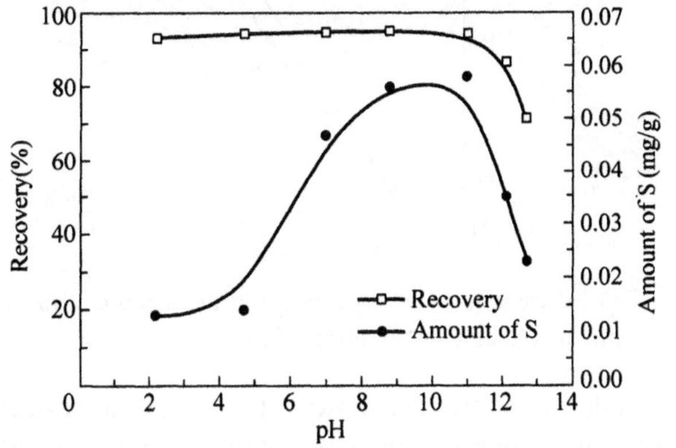

Figure 2.30 Collectorless flotation recovery of jamesonite and the amount of extracted sulphur as a function of pH

However, Fig. 2.29 and Fig. 2.30 show that the trend of recovery of pyrrhotite and jamesonite are not so well consistent with that of the amount of extracted sulphur. At lower pH region, the lower value of extracted S is required for collectorless flotation of jamesonite and pyrrhotite. At higher pH region, the greater amount of extracted S is required for good flotation of jamesonite and pyrrhotite. In strong alkaline medium (pH > 11), although there is enough amount of extracted S, the recovery of jamesonite and pyrrhotite still decline, which may be due to the hydrophilicity of OH^-.

Sun et al. (1993c; 1994a,b) reported that the potential range where galena is able to be induced flotation is in excellent agreement with the range where sulphur could be extracted. The flotation recovery of galena and the amount of extracted sulphur occurs at a plateau in the range of potential 0.45 to 0.65 V. The self-induced flotation recovery of galena is increased with the increase of the amount of extracted sulphur. If we take the amount of extracted sulphur [S] at recovery 50% as the critical value of [S] required for self-induced collectorless flotation at various pH. There are different recoveries vs. [S] curves in different pH and hence different critical sulphur amount as shown in Fig. 2.31. It may be seen from Fig. 2.31 that the lower the pH and the lower the critical sulphur value are required. The linear relations between $[S]_c$ and pH were obtained for galena as in the following equation.

$$[S]_c = a + b\text{pH} \quad (\text{mg/g}) \tag{2-67}$$

which shows the balance between hydrophobicity rendered by sulphur and hydrophilicity induced by OH^-.

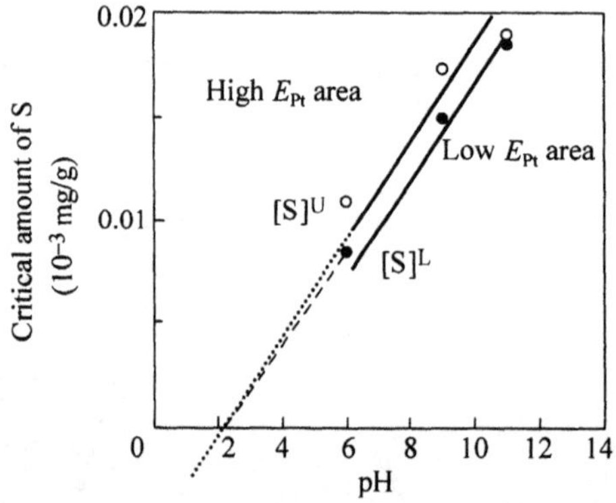

Figure 2.31 Critical value (at recovery > 50%) of the amount of extracted sulphur as a function of pH in collectorless flotation of galena

It may be concluded from the discussion above that the collectorless flotation was observed for many sulphides in moderately oxidizing potential region, but not in strongly reducing potential region. The available evidence suggests that the hydrophobic entity could be sulphur produced by superficial oxidation of the mineral. In general, flotation is observed in the potential-pH areas where elemental sulphur is metastable.

There is significant agreement between the lower potential boundary of the flotation region and the potential at which the anodic current begins in a potential sweep. The amount of extracted sulphur on the sulphide minerals can be correlated with their collectorless flotation behaviors. The higher the concentration of surface sulphur, the faster the collectorless flotation rate and thus the higher the recovery.

Chapter 3 Collectorless Flotation in the Presence of Sodium Sulphide

Abstract The sodium sulphide-induced collectorless flotation of several minerals are first introduced in this chapter. The results obtained are that sodium sulphide-induced collectorless flotation of sulphide minerals is strong for pyrite while galena, jamesonite and chalcopyrite have no sodium sulphide-induced collectorless flotability. And the nature of hydrophobic entity is then determined through E_h-pH diagram and cyclic voltammogram, which is element sulphur. It is further proved with the results of surface analysis and sulphur-extract. In the end, the self-induced and sodium sulphide-induced collectorless flotations are compared. And it is found that the order is just reverse in sodium sulphide-induced flotation to the one in self-induced collectorless flotation.

Keywords sodium sulphide-induced collectorless flotation; E_h-pH diagram

3.1 Description of Behavior

The flotation results of three sulphide minerals in the presence of sodium sulphide are presented in Fig. 3.1 as recovery-concentration curves. In contrast to self-induced collectorless flotation, pyrite and arsenopyrite are strongly floatable while chalcopyrite and galena are weakly floatable in the presence of sodium sulphide.

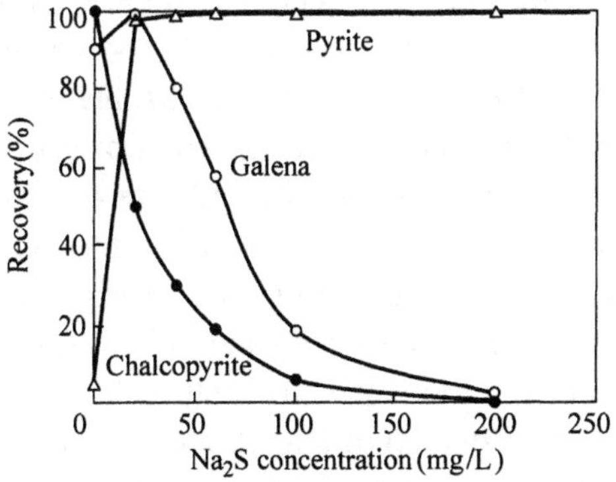

Figure 3.1 Flotation recovery of sulphide minerals as a function of Na_2S concentration at pH = 11 (Wang and Hu, 1992; Wang and Long, 1991)

It can be seen from Fig. 3.1 that pyrite is strongly floatable with recovery above 90%. Chalcopyrite and galena are not floatable. At pH = 11, the recovery of chalcopyrite is sharply decreased with the increase of the concentration of Na_2S and exhibits no floatability. The recovery of galena is also sharply decreased when the concentration of Na_2S is greater than 40 mg/L and also exhibits poor sulphur-induced collectorless flotation. In other words, the self-induced collectorless flotation of chalcopyrite and galena is depressed by the addition of sulphide. Alternatively, the recovery of pyrite is evidently increased when the concentration of Na_2S added is above 2×10^{-4} at pH = 11, showing strong Na_2S induced collectorless floatability.

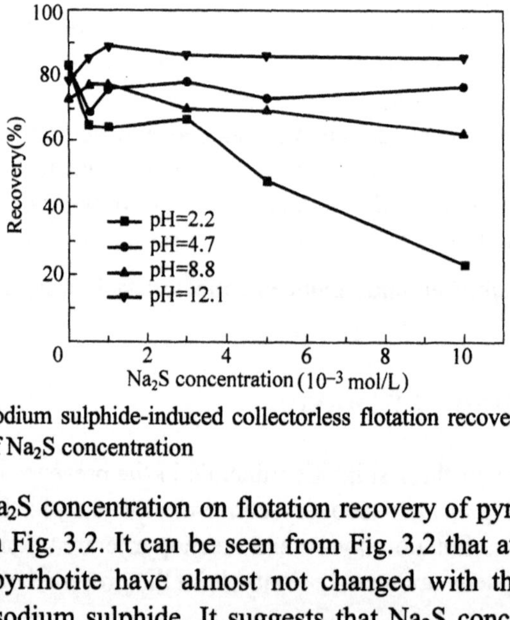

Figure 3.2 Sodium sulphide-induced collectorless flotation recovery of pyrrhotite as a function of Na_2S concentration

The effect of Na_2S concentration on flotation recovery of pyrrhotite at different pH was present in Fig. 3.2. It can be seen from Fig. 3.2 that at pH = 4.7 and 8.8 the recovery of pyrrhotite have almost not changed with the increase of the concentration of sodium sulphide. It suggests that Na_2S concentration may not affect the self-induced collectorless flotation of pyrrhotite. At pH = 2.2, the recovery of pyrrhotite decreases with the increase of concentration and when the concentration of Na_2S is greaten than 3×10^{-3} mol/L the recovery of pyrrhotite is sharply decreased, indicating the depressing action of Na_2S on the self-induced collectorless flotation of pyrrhotite. At pH = 12, the addition of Na_2S improves slightly the self-induced collectorless flotation of pyrrhotite.

Flotation recovery of jamesonite as a function of Na_2S concentration is shown in Fig. 3.3. It follows that the recovery of jamesonite is decreased by the addition of low dosage of Na_2S at various pH. In contrast to self-induced collectorless flotation, the collectorless flotation of jamesonite is depressed in the presence of sodium sulphide. In alkaline pH range, the self-induced collectorless flotation of jamesonite is sharply depressed by the addition of Na_2S. In acidic pH range, the influence of Na_2S on the self-induced collectorless flotation of jamesonite is much less than in alkaline media.

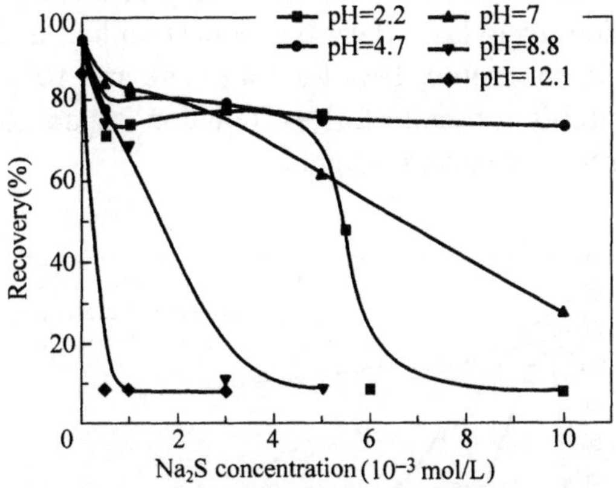

Figure 3.3 Flotation recovery of jamesonite as a function of Na_2S concentration

The effect of Na_2S concentration on flotation recovery of marmatite at different pH value is presented in Fig. 3.4. It can be seen from Fig. 3.4 that the collectorless flotation of marmatite is depressed by the addition of Na_2S at acidic condition. The lower pH, the stronger the depress effect. The addition of Na_2S has no effect on the flotation of marmatite in strong alkali. In neutral and weak alkali media, the addition of low dosage of Na_2S can improve the flotation of marmatite, but when the concentration of Na_2S added is above 1×10^{-3} mol/L, the flotation of marmatite is also depressed.

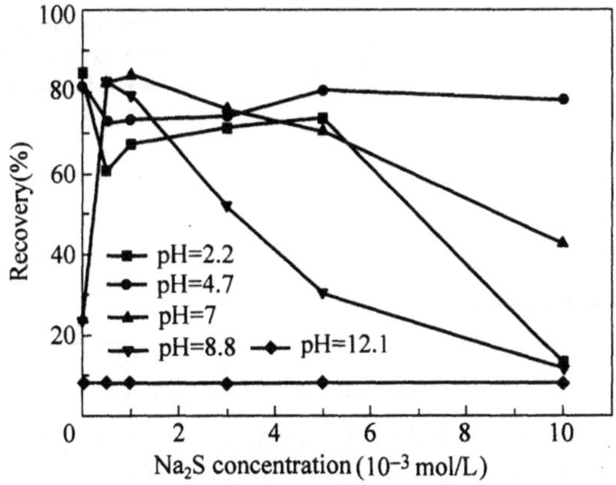

Figure 3.4 Sulphur-induced collectorless flotation recovery of marmatite as a function of Na_2S concentration

Replot the date points on three minerals at pH = 8.8 and pH = 12.1 from Fig. 3.2 to Fig. 3.4 is shown in Figs. 3.5 and 3.6. The collectorless floatability difference of three minerals in the presence of sodium sulphide is clearly seen in descending order of pyrrhotite, marmatite and jamesonite. At pH = 8.8, the collectorless flotation recovery when the concentration is 2×10^{-3} mol/L is, respectively, 72% for pyrrhotite, 60% for marmatite and 30% for jamesonite. At pH = 12.1, the collectorless flotation recovery when the concentration is 2×10^{-3} mol/L is, respectively, 90% for pyrrhotite, 10% for marmatite and 10% for jamesonite. It indicates the possibility of collectorless flotation separation of these three minerals in the presence of sodium sulphide.

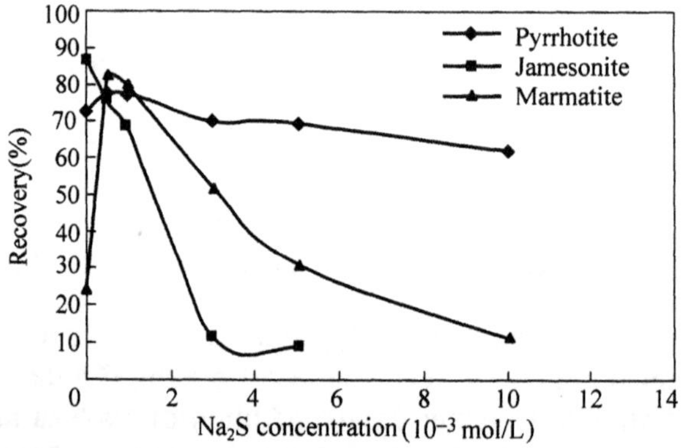

Figure 3.5 Sulphur-induced collectorless flotation recovery of pyrrhotites, jamesonite and marmatite as a function of Na_2S concentration at pH = 8.8

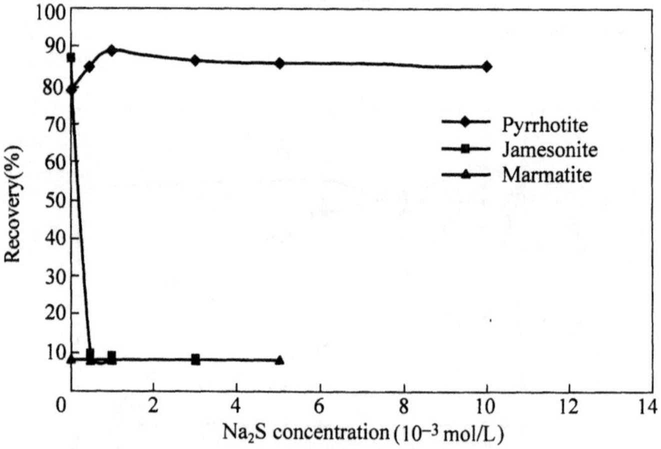

Figure 3.6 Sulphur-induced collectorless flotation recovery of pyrrhotite, jamesonite and marmatite as a function of Na_2S concentration at pH = 12.1

The influences of pulp potential on the Na_2S -induced flotation of pyrite at pH = 8 was examined by Heyes and Trahar (1984) for an initial sodium concentration of 10^{-3} mol/L. The region of fairly strong Na_2S -induced floatability over a wider potential range between 0 and 0.5 V was reported. They noted that the floatability of pyrite was much reduced compared with self-induced collectorless flotation for essentially similar conditions. They believe that this result was related to the rates of oxidation of the pyrite samples which were from different sources. Guy and Trahar (1984) found that as with chalcopyrite there was no flotation in the presence of free hydrosulphide ion and the galena ground in sodium sulphide solution exhibited some floatability over a similar potential range to that in self-induced flotation, but began at more reducing potential and recovery was relatively lower.

The results above show that the sodium sulphide-induced collectorless floatability of sulphide minerals is strong for pyrite. Galena, jamesonite and chalcopyrite have no sodium sulphide-induced collectorless floatability. Marmatite and pyrrhotite showed some sodium sulphide-induced collectorless floatability in certain conditions.

3.2 Nature of Hydrophobic Entity

It was established by Tolun and Kitchener (1963—1964) that a platinum electrode became hydrophobic when hydrosulphide ion (HS^-) was oxidized at the electrode surface. A contribution to the hydrophobicity may be the production of sulphur. According to thermodynamics, the hydrolysis reaction of sodium sulphide is

$$Na_2S + 2H_2O = H_2S + 2NaOH \quad (3\text{-}1)$$

$$H_2S = HS^- + H^+ \quad (3\text{-}2a)$$

$$K_{a_1} = 10^{-7.02} \quad (3\text{-}2b)$$

$$HS^- = H^+ + S^{2-} \quad (3\text{-}3a)$$

$$K_{a_2} = 10^{-13.9} \quad (3\text{-}3b)$$

The redox reactions of species S^{2-}, HS^- and H_2S in solution and E_h-pH relations are (assuming the total concentration of dissolved sulphur species to be 10^{-3} mol/L):

$$S^{2-} + 4H_2O = SO_4^{2-} + 8H^+ + 8e^- \quad (3\text{-}4a)$$

$$\begin{cases} E^0 = 0.157(V) \\ E_h = 0.157 - 0.059pH \end{cases} \quad (3\text{-}4b)$$

$$HS^- + 4H_2O = SO_4^{2-} + 9H^+ + 8e^- \quad (3\text{-}5a)$$

$$\begin{cases} E^0 = 0.252(V) \\ E_h = 0.252 - 0.066pH \end{cases} \quad (3\text{-}5b)$$

$$H_2S + 4H_2O = SO_4^{2-} + 10H^+ + 8e^- \quad (3\text{-}6a)$$

$$\begin{cases} E^0 = 0.303(\mathrm{V}) \\ E_h = 0.303 - 0.074\mathrm{pH} \end{cases} \tag{3-6b}$$

$$H_2S + 4H_2O = HSO_4^- + 9H^+ + 8e^- \tag{3-7a}$$

$$\begin{cases} E^0 = 0.289(\mathrm{V}) \\ E_h = 0.289 - 0.066\mathrm{pH} \end{cases} \tag{3-7b}$$

$$H_2S = S + 2H^+ + 2e^- \tag{3-8a}$$

$$\begin{cases} E^0 = 0.1418(\mathrm{V}) \\ E_h = 0.2303 - 0.059\mathrm{pH} \end{cases} \tag{3-8b}$$

$$HS^- = S + H^+ + 2e^- \tag{3-9a}$$

$$\begin{cases} E^0 = -0.065(\mathrm{V}) \\ E_h = 0.0235 - 0.0295\mathrm{pH} \end{cases} \tag{3-9b}$$

$$S^{2-} = S + 2e^- \tag{3-10a}$$

$$\begin{cases} E^0 = -0.4761(\mathrm{V}) \\ E_h = -0.3876(\mathrm{V}) \end{cases} \tag{3-10b}$$

$$S + 4H_2O = SO_4^{2-} + 8H^+ + 6e^- \tag{3-11a}$$

$$\begin{cases} E^0 = 0.3572(\mathrm{V}) \\ E_h = 0.3276 - 0.07867\mathrm{pH} \end{cases} \tag{3-11b}$$

$$S + 4H_2O = HSO_4^- + 7H^+ + 6e^- \tag{3-12a}$$

$$\begin{cases} E^0 = 0.3384(\mathrm{V}) \\ E_h = 0.3088 - 0.0688\mathrm{pH} \end{cases} \tag{3-12b}$$

$$SO_4^{2-} + H^+ = HSO_4^- \tag{3-13a}$$

$$K = 10^{-1.91} \tag{3-13b}$$

From the Eqs. (3-1) to (3-13), the E_h-pH diagram of sodium sulphide solution is constructed with element sulphur as metastable phase considering the presence of barrier (about 300 kJ/mol) or overpotential (about 3.114 mV) of sulphide oxidation to sulphate and shown in Fig. 3.7. It is obvious that the lower limit of potential of sodium sulphide-induced collectorless flotation of pyrite, pyrrhotite and arsenopyrite at various pH agree well with the potential defined respectively by reactions of Eq. (3-9) producing elemental sulphur. The initial potential

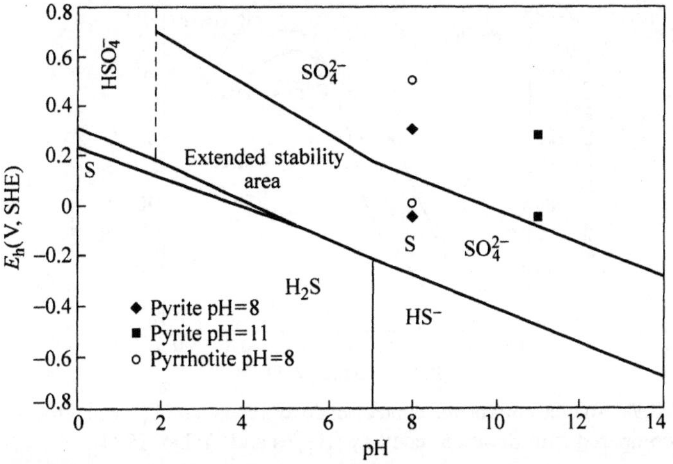

Figure 3.7 Electrochemical phase diagram for sodium sulphide with elemental sulphur as metastable phase. Equilibrium lines (solid lines) correspond to dissolved species at 10^{-3} mol/L. Plotted points show the upper and lower limit potential of sulphur-included flotation of pyrite, pyrrhotite and arsenopyrite from literature

where these minerals begin to possess sodium sulphide-induced floatability is corresponding to the metastable elemental sulphur, indicating that elemental sulphur is hydrophobic entity. When above the upper limit potential, hydrophilic species such as sulphate is dominant, and the flotation ceases.

The voltammetric behavior of pyrite at pH = 8.8 (see Fig. 3.8) shows that an anodic current commenced at about −0.25 V to give an anodic peak at about 0 V. On the reverse scan a cathodic current that appeared at the same potential could be presumed to represent the reduction of the initial oxidation products. According to the reaction (3-9), the formation of sulphur would be expected to occur at −0.26 V for the HS^- concentration of 10^{-2} mol/L at pH = 8.8 which is consistent with an anodic current that begins to occur.

Heyes and Trahar (1984) further compared the floatability of pyrite with the electrochemical and contact angle result reported by Walker and Richardson. Their results are listed in Fig. 3.9. It indicates that the onset of anodic current during a

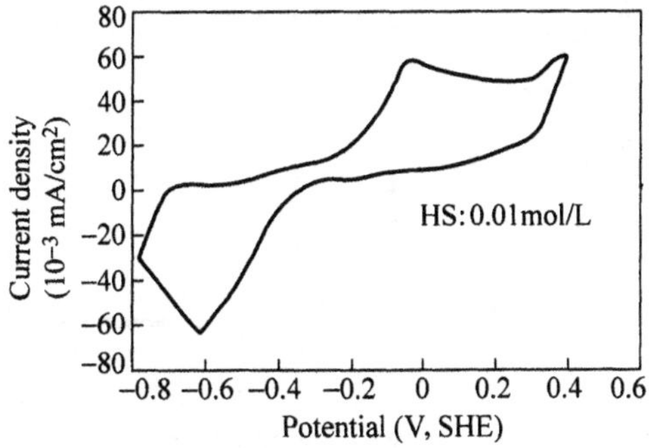

Figure 3.8 Cyclic voltammogram for pyrite in buffer solution at pH = 8.8

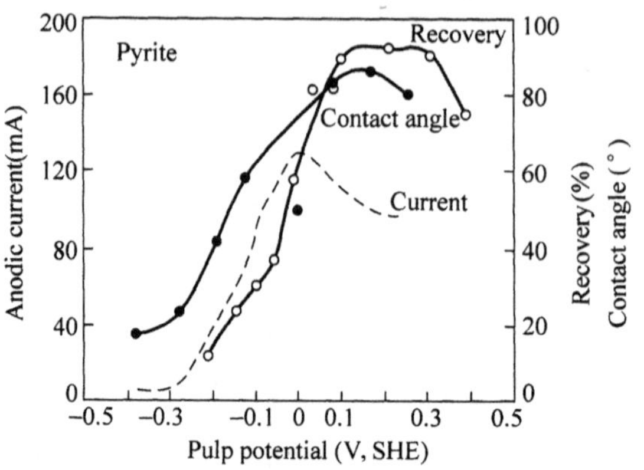

Figure 3.9 Anodic current and contact angle as a function of potential for sulphidized pyrite compared with flotation recovery (Heyes and Trahar, 1984)

potential sweep of a pyrite electrode and an increase in the contact angle with potential of sulphidized pyrite is in reasonable agreement with flotation edge. They also attributed the results to the production of sulphur in terms of reaction (3-9) by which the oxidation of hydrosulphide ion to sulphur should be expected at -0.213 V for the hydrosulphide ion concentration of 10^{-3} mol/L at pH = 8.

3.3 Surface Analysis and Sulphur-Extract

The X-ray diffraction diagrams of pyrite before and after treatment with sodium sulphide are shown in Fig. 3.10, which appears to confirm the presence of sulphur because the diffraction band of elemental sulphur occurs after pyrite treated with Na_2S.

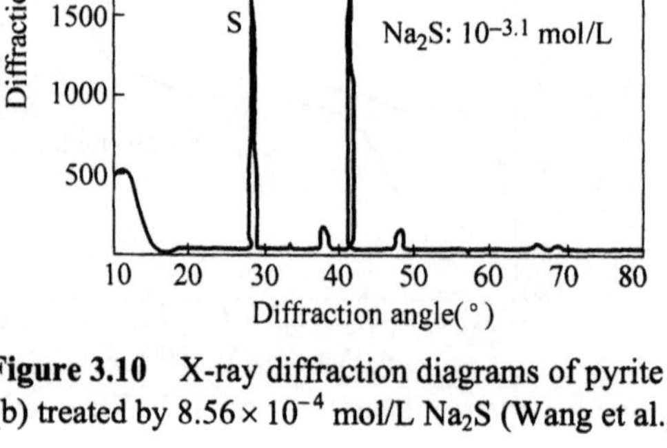

Figure 3.10 X-ray diffraction diagrams of pyrite
(a) untreated; (b) treated by 8.56×10^{-4} mol/L Na_2S (Wang et al., 1991a,b,c,d)

Sun et al. (1993a) reported the effects HS^- ion concentration on the adsorption of HS^-, the amount of extracted sulphur and sulphur-induced flotation of pyrite as shown in Fig. 3.11. The results show that during sodium sulphide-induced collectorless flotation, it involves the adsorption of HS^- ion on the mineral and the HS^- adsorbed can be oxidized into sulphur to render pyrite and arsenopyrite surface hydrophobic due to the fact that the adsorption density of HS^- ion increases with the HS^- ion concentration and the amount of extracted sulphur and hence the flotation rate increases with the increase of adsorption density. It suggests that the mechanism of sodium sulphide-induced collectorless flotation of pyrite takes place by reactions:

$$HS^-_{(aq)} = HS^-_{(ad)} \tag{3-14}$$

$$HS^-_{(ad)} = S^-_{(ad)} + H^+ + 2e^- \tag{3-15}$$

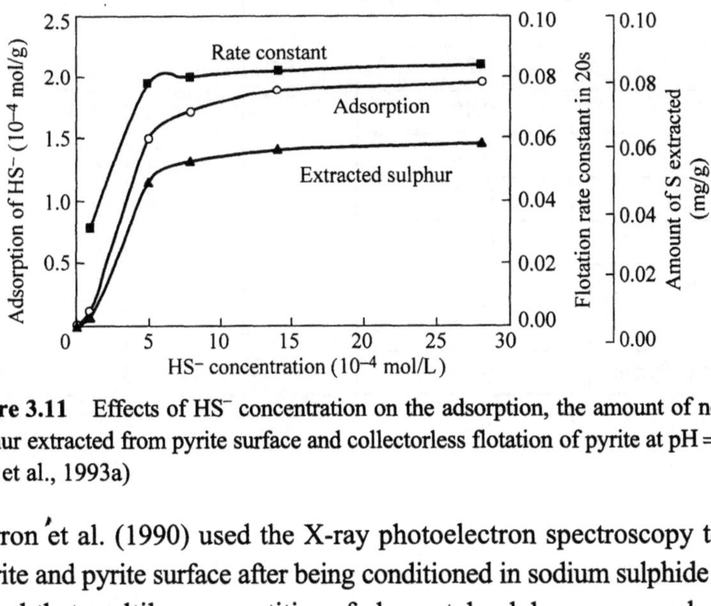

Figure 3.11 Effects of HS^- concentration on the adsorption, the amount of neutral sulphur extracted from pyrite surface and collectorless flotation of pyrite at pH = 11.0 (Sun et al., 1993a)

McCarron et al. (1990) used the X-ray photoelectron spectroscopy to analyze chalcopyrite and pyrite surface after being conditioned in sodium sulphide solutions. They found that multilayer quantities of elemental sulphur were produced at the surface of both minerals in 3×10^{-3} and 3×10^{-4} mol/L sulphide solutions although for a given sulphide concentration, the surface coverage of elemental sulphur for pyrite was greater than that for chalcopyrite under open circuit conditions. Eliseev et al. (1982) concluded that elemental sulphur was responsible for the hydrophobicity of pyrite and chalcopyrite treated with sodium sulphide. Luttrell and Yoon (1984a, b) observed a shoulder due to elemental sulfur near 164 eV in the S (2p) spectra from relatively pure chalcopyrite floated after being conditioned at different pulp potential established by different hydrosulphide concentration.

3.4 Comparison between Self-Induced and Sodium Sulphide-Induced Collectorless Flotation

It may be seen from the results in Chapter 2 that the floatability descends in the order of chalcopyrite galena, pyrrhotite, bornite, arsenopyrite and pyrite in self-induced collectorless flotation but the order is just reversed in sodium sulphide-induced flotation, the phenomena of which may be explained in the light of their rest potential as shown in Table 3.1.

Table 3.1 Rest potential for sulphide minerals in water at pH = 4 (Hayes et al., 1987) and the order of their self-induced and sulphur-induced collectorless floatability

Minerals	Rest potential (V, SHE)	Self-induced floatability	Sulphur-induced floatability
Molybdenite	0.11	High	Low
Stibnite	0.12	|	|
Argentite	0.28	|	|
Galena	0.40	|	|
Bornite	0.42	|	|
Covellite	0.45	|	|
Sphalerite	0.46	|	|
Chalcopyrite*	0.56*	|	|
Marcasite	0.63	↓	↓
Arsenopyrite Pyrite	0.66	Low	High

* anomalous.

Chapter 4 Collector Flotation of Sulphide Minerals

Abstract In the beginning, the mixed potential model, which is generally used to explain the adsorption of collectors on the sulphide minerals, is illustrated. And the collector flotation of several kinds of minerals such as copper sulphide minerals, lead sulphide minerals, zinc sulphide minerals and iron sulphide minerals is discussed in the aspect of pulp potential and the nature of hydrophobic entity is concluded from the dependence of flotation on pulp potential. In the following section, the electrochemical phase diagrams for butyl xanthate/water system and chalcocite/oxygen/xanthate system are all demonstrated from which some useful information about the hydrophobic species are obtained. And some instrumental methods including UV analysis, FTIR analysis and XPS analysis can also be used to investigated sulphide mineral-thio-collector sytem. And some examples about that are listed in the last part of this chapter.

Keywords collector flotation; electrochemical phase diagram; UV; FTIR; XPS

Ever since the mixed potential model has been proposed, the interaction mechanism between thio-collector and sulphide minerals has been usually explained on the basis of this model. The principle of the mixed potential model can be schematically shown in Fig. 4.1. Here $E_{rev}^{O_2}$, E_{rev}^{Xads}, $E_{rev}^{X_2}$, $E_{rev}^{MX_2}$ respectively represent the reversible potential of reactions (1-1), (1-2), (1-3), (1-4), $E_{restpotential}^{s}$ represents the rest potential of a mineral. Figure 4.1 demonstrates that the reduction of oxygen is the only cathodic process. When $E_{rev}^{Xads} < E_{restpotential}^{s} < E_{rev}^{MX_2}$, the interaction of thio collector with a mineral is dominant by electrochemical adsorption in Eq. (1-2). When $E_{rev}^{MX_2} < E_{restpotential}^{s} < E_{rev}^{X_2}$, the electrochemical reaction of thio collector on a mineral produces collector-metal salts in Eq. (1-4). When $E_{rev}^{X_2} < E_{restpotential}^{s} < E_{rev}^{O_2}$, the thio collector is oxidized to its dithiolate like Eq. (1-3).

Table 4.1 shows the measured rest potential of sulphide electrode in thio collector solutions at pH = 6.86 and the equilibrium potential calculated for possible processes. In terms of the mixed potential model, the reaction products should be metal collector salts between four thio collectors and galena and jamesonite; and should be disulphide between four thio collectors and pyrite and arsenopyrite. For chalcopyrite the reaction products may be dixanthogen in xanthate solution and metal collector compound in other collector solution.

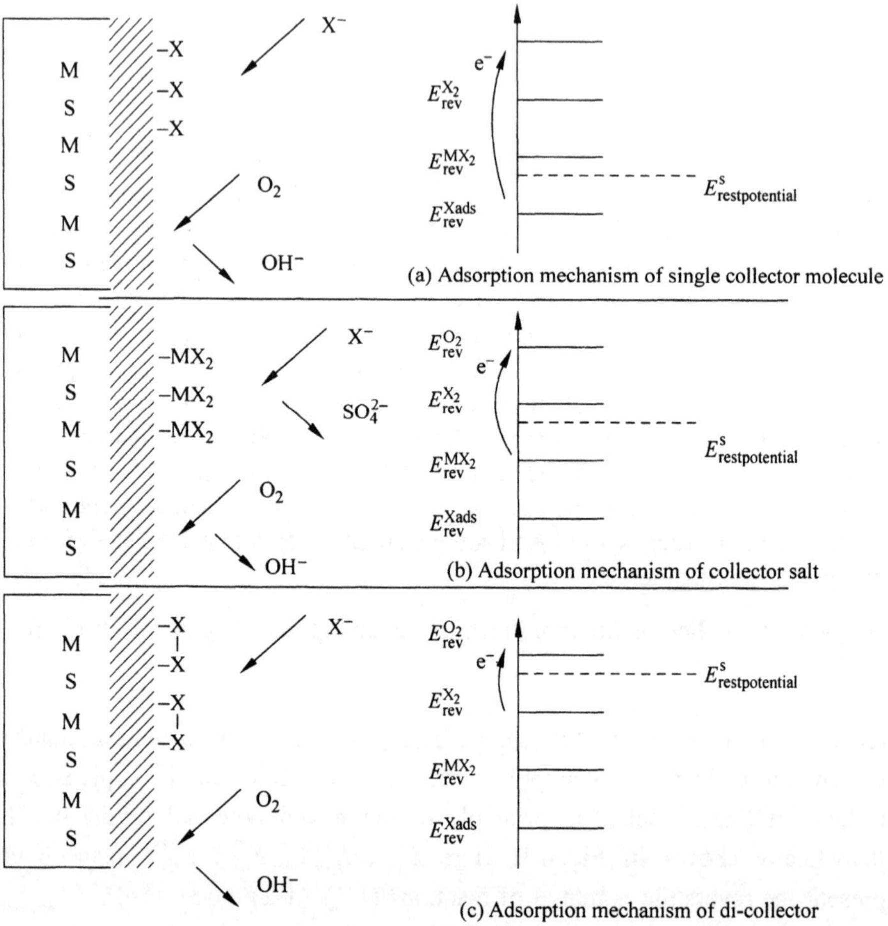

Figure 4.1 The mixed potential model of redox reaction on sulphide surface

Table 4.1 The measured rest potential in 10^{-4} mol/L thio collector solution at pH = 6.86 (Feng, 1989; Yu et al., 2004a,b,c,d,e,f,g; Zhang et al., 2004a,b,c,d,e,f)

Collectors	Rest potential (V, SHE)					Reversible potential (V, SHE)
	PbS	$CuFeS_2$	FeS_2	FeAsS	PbSbS	
Ethyl xanthate	0.08	0.19	0.295	0.26	0.15	0.176
Butyl xanthate	0.04	0.11	0.30	0.245	0.09	0.107
Dithiocarbamate	0.09	0.145	0.32	0.28	0.13	0.168
Dithiophosphate	0.13	0.20	0.48	0.37	0.21	0.340

Although there have been a lot of investigations on the interactions of sulphide minerals with thio-collectors in terms of the mixed potential principle, there are still much controversy about the products formed on a sulphide mineral in the presence of a collector in different conditions. In the following sections, the effects of potential on the flotation and formation of surface products of many kinds of sulphide minerals will be introduced based on the results of flotation, electrochemical measurement, surface analyses and thermodynamic calculations.

4.1 Pulp Potential Dependence of Collector Flotation and Hydrophobic Entity

4.1.1 Copper Sulphide Minerals

1. Chalcocite

The collector flotation of chalcocite has been studied in some detail (Basiollio et al., 1985; Heyes and Trahar, 1979; O'Dell et al., 1984; Richardson et al., 1984, 1985; Walker et al., 1984). Figure 4.2 compares the results reported by different authors. It can be seen that the lower potential limit of flotation found in the different experimental conditions agrees favorably and is independent of pH; but the upper limit is pH dependent.

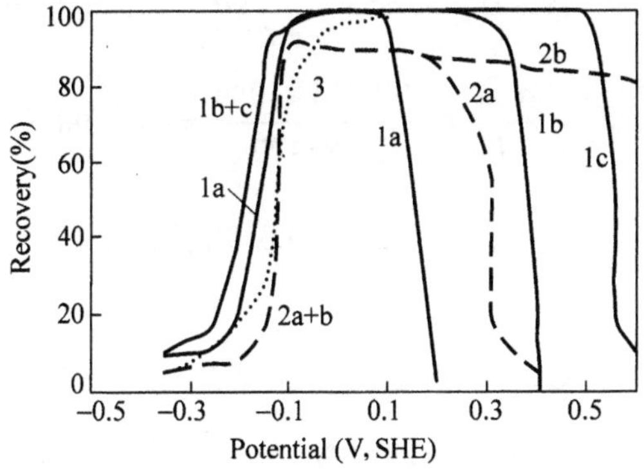

Figure 4.2 Flotation recovery of chalcocite with ethyl xanthate as a function of potential: curve 1−from Basiollio et al. (1985), solution pH = 11 (1a),8(1b),5(1c); curve 2−from Heyes and Trahar (1979), solution pH = 11(2a) and 8(2b); curve 3−from Richardson et al. (1984), at pH = 9.2

Gaudin and Schuhmann (1936) and Harris and Finkelstein (1977) have shown that the most likely adsorbed hydrophobic entity is cuprous xanthate which could be formed by a combination of the oxidation of chalcocite and the reaction of the oxidation product (cupric-ion) with xanthate (Hepel and Pomianowski, 1977). It was assumed therefore that the reactions to account for the formation of cuprous xanthate may be:

$$Cu_2S + EX^- = CuS + CuEX + e^- \tag{4-1}$$

According to the date in literature (Du Tietz, 1975)

$$CuEX = Cu^+ + EX^- \tag{4-2a}$$

$$K_{sp} = [Cu^+][EX^-] = 10^{-19.28} \tag{4-2b}$$

The change of standard free energy of reaction (4-2) is

$$\Delta G^{0'} = -RT\ln K_{sp} = -2.303RT\lg K_{sp} = 110.009(\text{kJ/mol}) \tag{4-3a}$$

Since

$$\Delta G^{0'} = \Delta G^0_{EX^-} + \Delta G^0_{Cu^+} - \Delta G^0_{CuEX} \tag{4-3b}$$

Thus

$$\begin{aligned}\Delta G^0_{CuEX} - \Delta G^0_{EX^-} &= \Delta G^0_{Cu^+} - \Delta G^{0'} \\ &= 50.208 - 110.009 = -59.801(\text{kJ/mol})\end{aligned} \tag{4-3c}$$

The change of standard free energy of reaction (4-1) is

$$\begin{aligned}\Delta G^0 &= \Delta G^0_{CuEX} + \Delta G^0_{CuS} - \Delta G^0_{EX^-} - \Delta G^0_{Cu_2S} \\ &= -59.801 - 48.953 + 86.190 = -22.564(\text{kJ/mol})\end{aligned} \tag{4-4}$$

$$E^0 = \frac{\Delta G^0}{nF} = \frac{-22.564 \times 1000}{2 \times 96490} = -0.234(\text{V}) \tag{4-5}$$

$$E_h = -0.234 - 0.059\lg[EX^-] \tag{4-6}$$

Similarly,

$$Cu_2S + 2EX^- = 2CuEX + S + 2e^- \tag{4-7a}$$

$$\begin{cases} E^0 = -0.173(\text{V}) \\ E_h = -0.173 - 0.059\lg[EX^-] \end{cases} \tag{4-7b}$$

$$Cu_2S + 2EX^- + 4H_2O = 2CuEX + SO_4^{2-} + 8H^+ + 8e^- \tag{4-8a}$$

$$\begin{cases} E^0 = 0.225(\text{V}) \\ E_h = 0.225 - 0.01481\lg[X^-] + 0.007375\lg[SO_4^{2-}] - 0.059\text{pH} \end{cases} \tag{4-8b}$$

$$2Cu_2S + 4EX^- + 3H_2O \Longrightarrow 4CuEX + S_2O_3^{2-} + 6H^+ + 8e^- \quad (4\text{-}9a)$$

$$\begin{cases} E^0 = 0.147(\mathrm{V}) \\ E_h = 0.147 - 0.0295\lg[X^-] + 0.007375\lg[S_2O_3^{2-}] - 0.04425\mathrm{pH} \end{cases} \quad (4\text{-}9b)$$

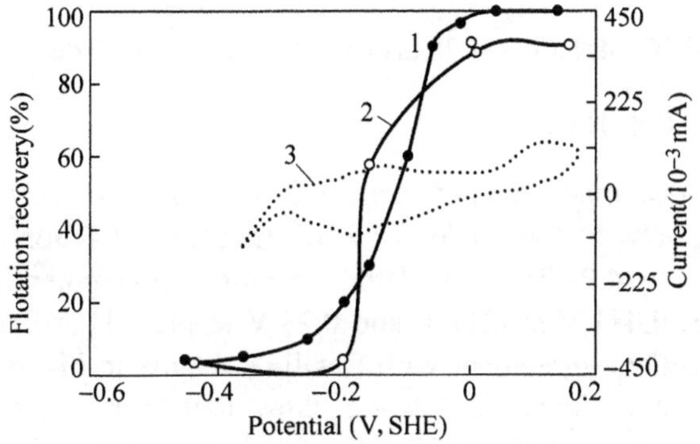

Figure 4.3 Linear voltammogram and floatability of a bed of chalcocite particles in aqueous solution of ethyl xanthate (1: 4.7×10^{-5} mol/L, from Heyes and Trahar, 1979; 2: 1.44×10^{-5} mol/L, from Richardson et al., 1984; 3 is a cyclic voltammogram: 5 mV/s, 1.9×10^{-5} mol/L KEX ,buffered at pH = 9.2 in 0.05 mol/L borate , from O'Dell et al., 1986)

However, the calculated potential for CuEX formation corresponding to a xanthate concentration of 4.7×10^{-5} mol/L are, respectively, 0.021V for Eq. (4-6), 0.082V for Eq. (4-7b) and −0.391V for Eq. (4-8b) (for pH = 11 and a sulphate ion concentration of 10^{-4} mol/L), which do not agree with the observed lower limiting flotation potential of Heyes and Trahar (1979) in Fig. 4.2 (curve 2). The calculated potential for reactions (4-1) also do not correspond with the results of curve 1 and 3 in Fig. 4.2.

Potential sweep measurements on a chalcocite particle bed electrodes in solutions containing ethyl xanthate (O'Dell et al., 1986) were found to display the same features as on a single particle electrode of natural chalcopyrite showing three anodic which were first reported by Kowal and Pomianowski (1973). The voltammogram and flotation results are compared in Fig. 4.3. It can be seen that an anodic current occurs at the same potential as the flotation edge. It is generally accepted that the first anodic peak in this system arises from the chemisorption of xanthate on the mineral surface. The second anodic peak is due to the formation of bulk copper ethyl xanthate. Therefore if the anodic current and the flotation edge correspond to the formation of CuX, there is a substantial difference between the observed and calculated potential. However, if the reaction is assumed to be Eq. (4-9), the calculated potential for CuX formation are −0.11V at pH = 8 and −0.24V at pH = 11, which are in reasonable agreement with the lower limiting flotation potential indicating that a reaction of the form of reaction (4-9a) may be the reaction of CuX formation.

Figure 4.3 also shows that there is an upper flotation edge and that its potential varies with the changes of pH. Heyes and Trahar (1984) attributed the cessation of flotation to the decomposition by the reactions of the type:

$$2CuEX + 4H_2O = 2HCuO_2^- + (EX)_2 + 6H^+ + 4e^- \qquad (4\text{-}10a)$$

$$E^0 = 1.402(V)$$

$$2CuEX + 4H_2O = 2Cu(OH)_2 + (EX)_2 + 4H^+ + 4e^- \qquad (4\text{-}10b)$$

$$E^0 = 0.891(V)$$

The calculated decomposition potential are 0.783 V at pH = 5, +0.516 V at pH = 8, and +0.251 V at pH = 11 for $[HCuO_2^-] = 10^{-6}$ mol/L for reaction (4-10a). Neither fits the data in Fig. 4.2. The calculated decomposition potential however are 0.59 V at pH = 5, 0.411 V at pH = 8 and 0.23 V at pH = 11 for reaction (4-10b), which are in excellent agreement with Basilio's results in Fig. 4.2, although the results of Heyes and Trahar in Fig. 4.2 show that there is no upper limiting potential for chalcocite at pH = 8. Thus the fall in recovery will commence at the potential predicted by the reaction of the type (4-10b). The reaction (4-10) also gives an evidence that dixanthogen dose not confer floatability on chalcocite in the region of potential in Fig. 4.2.

2. Chalcopyrite

The influence of pulp potential on the floatability of chalcopyrite is shown in Fig. 4.4 for an initial concentration of 2×10^{-5} mol/L ethyl xanthate and butyl xanthate. The lower flotation potential is −0.1V for KBX and 0V for KEX. The hydrophobic entity is usually assumed to be dixanthogen (Allison et al., 1972; Woods, 1991; Wang et al., 1992) by the reaction (1-3). The calculated potential in terms of reaction (1-3), are, however, 0.217 V and 0.177 V, respectively, for ethyl and butyl xanthate oxidation to dixanthogen for a concentration of 2×10^{-5} mol/L, which corresponds to the region of maximum recovery but not to the lower limiting potential for flotation, indicating that some other surface hydrophobicity to the mineral. Richardson and Walker (1985) considered that ethyl xanthate flotation of chalcopyrite may be induced by the reaction:

$$CuFeS_2 + X^- = CuX + FeS_2 + e^- \qquad (4\text{-}11)$$

$$E^0 = -0.096(V)$$

For $[EX^-] = 2 \times 10^{-5}$ mol/L, CuX would be produced at 0.18V. Again, it does not correspond with the observed lower limiting flotation potential.

Voltammograms obtained with chalcopyrite electrode (see Fig. 4.5) show that the anodic current appears at the potential corresponding to the formation of dixanthogen by the reaction of the form (1-3). Therefore, it is reasonable to assume

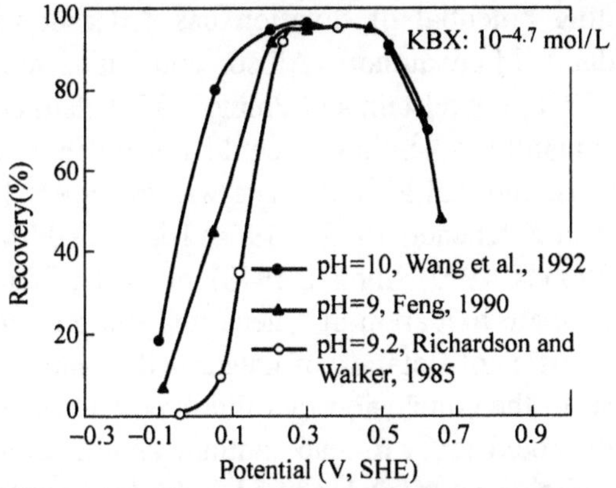

Figure 4.4 Flotation recovery of chalcopyrite as a function of pulp potential

that dixanthogen may be responsible for the hydrophobicity of chalcopyrite in xanthate-induced flotation. Recently, Cheng and Iwasaki (1992) reported that the flotation of chalcopyrite commenced only at oxidizing potential (above 0.4V) with sodium isopropyl xanthate as collector at near neutral pH under nitrogen aeration, which is obviously attributed to dixanthogen.

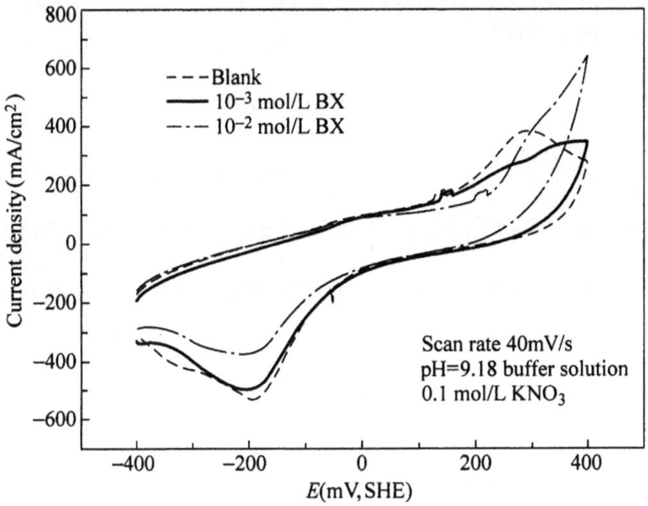

Figure 4.5 Voltammograms for chalcopyrite electrode in borate solution at pH = 9.2

4.1.2 Lead Sulphide Minerals

1. Galena

There is lack of consensus in the published electrochemical studies of flotation of galena and the nature of the important hydrophobic species responsible for flotation. The limiting potential of flotation has different values for different pretreatment procedures. Many authors (Allison and Finkelstein 1971; Finkelstein et al., 1975; Poling, 1976; Finkelstein and Poling, 1977) claimed that lead xanthate was formed predominantly, dixanthogen could not be formed in the xanthate-oxygen-galena systems, and that lead xanthate was extremely hydrophobic. Other researchers (Tolun and Kitchener, 1963—1964; Toperi and Tolun, 1969; Woods, 1971; Gardner and Woods, 1977; Ahmed, 1978), however, found that dixanthogen was a product of xanthate oxidation at galena and the presence of dixanthogen was essential for the efficient flotation of galena with xanthate collectors, which led these researchers to the conclusion that the interaction mechanism of galena with xanthate was the coexistence of lead xanthate and dixanthogen.

In the case hydrophobic entity is assumed to be lead xanthate, lead xanthate would be formed by the reaction of the form (1-4) including:

$$PbS + 2X^- + 4H_2O = PbX_2 + SO_4^{2-} + 8H^+ + 8e^- \quad (4\text{-}12a)$$

$$2PbS + 4X^- + 3H_2O = 2PbX_2 + S_2O_3^{2-} + 6H^+ + 8e^- \quad (4\text{-}12b)$$

$$PbS + 2X^- = PbX_2 + S + 2e^- \quad (4\text{-}12c)$$

According to the date in literature (Du Tietz, 1975; Garrels and Christ, 1965; Vanghan and Craig, 1978)

$$PbX_2 = Pb^{2+} + 2X^-$$

$$K_{sp,EX} = [Pb^{2+}][EX^-]^2 = 10^{-16.7}, \qquad K_{sp,BX} = [Pb^{2+}][BX^-]^2 = 10^{-18} \quad (4\text{-}13)$$

$$\Delta G^{0\prime} = -RT\ln K_{sp,EX} = -2.303RT\lg K_{sp,EX} = 95.288(\text{kJ/mol}) \quad (4\text{-}14a)$$

$$\Delta G^{0\prime\prime} = -RT\ln K_{sp,BX} = -2.303RT\lg K_{sp,BX} = 102.705(\text{kJ/mol}) \quad (4\text{-}14b)$$

Since

$$\Delta G^{0\prime} = 2\Delta G^0_{EX^-} + \Delta G^0_{Pb^{2+}} - \Delta G^0_{PbEX_2} \quad (4\text{-}14c)$$

Thus

$$\Delta G^0_{PbEX_2} - 2\Delta G^0_{EX^-} = \Delta G^0_{Pb^{2+}} - \Delta G^{0\prime}$$
$$= -24.309 - 95.288 = -119.597(\text{kJ/mol}) \quad (4\text{-}14d)$$

$$\Delta G^0_{PbBX_2} - 2\Delta G^0_{BX^-} = \Delta G^0_{Pb^{2+}} - \Delta G^{0\prime\prime}$$
$$= -24.309 - 102.705 = -127.014(\text{kJ/mol}) \quad (4\text{-}14e)$$

The change of standard free energy of reaction (4-12a) is

$$\Delta G^0 = \Delta G^0_{PbX_2} + \Delta G^0_{SO_4^{2-}} - 2\Delta G^0_{X^-} - \Delta G^0_{PbS} - 4\Delta G^0_{H_2O} \quad (4\text{-}15a)$$

For ethyl xanthate

$$\Delta G^0 = 179.852(\text{kJ/mol}), \quad E^0 = 0.233(\text{V}) \tag{4-15b}$$

For butyl xanthate

$$\Delta G^0 = 172.435(\text{kJ/mol}), \quad E^0 = 0.223(\text{V}) \tag{4-15c}$$

Similarly, the change of standard free energy of reaction (4-12b) is
For ethyl xanthate

$$\Delta G^0 = 125.525(\text{kJ/mol}), \quad E^0 = 0.163(\text{V}) \tag{4-16a}$$

For butyl xanthate

$$\Delta G^0 = 110.691(\text{kJ/mol}), \quad E^0 = 0.143(\text{V}) \tag{4-16b}$$

The change of standard free energy of reaction (4-12c) is
For ethyl xanthate

$$\Delta G^0 = -26.921(\text{kJ/mol}), \quad E^0 = -0.14(\text{V}) \tag{4-17a}$$

For butyl xanthate

$$\Delta G^0 = 34.334(\text{kJ/mol}), \quad E^0 = -0.178(\text{V}) \tag{4-17b}$$

The xanthate-induced flotation of galena with an initial xanthate concentration of 10^{-4} mol/L is presented in Fig. 4.6 and Fig. 4.7. It can be seen from Fig. 4.6 that flotation ranges from 0 V to + 0.41 V, the floatability of galena is bad when the potential is lower than –50 mV, while potential ranges from –50 mV to 0 mV, the recovery increase rapidly. The potential upper limit of flotation is about 410 mV (Sun et al., 2002). Figure 4.7 further shows that the potential range of galena flotation is different at different conditions such as pH and reagent concentration. At a given pH, the flotation potential range is wider at higher concentration of xanthate and at given concentration, the flotation potential range is wider in weak acidic pH media.

Figure 4.8 shows the voltammograms of galena electrode in the presence of xanthate at natural pH value. It shows that the lower limiting potential for the flotation of galena is corresponding to the initial oxidation potential, which would be formed by the reaction of the form (4-12). When the concentration of butyl xanthate is 10^{-4} mol/L, the calculated equilibrium potential for reactions (4-12a), (4-12b) and (4-12c) with the concentration of all dissolved species of 10^{-6} mol/L are respectively –63mV, –84mV, –58mV, at pH = 6.8, which are quite close to the lower edge of flotation, indicating that the anode oxidation of galena attribute to the reaction (4-12).

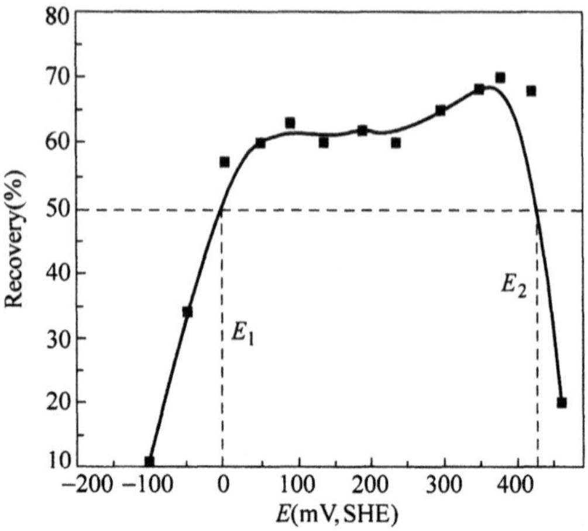

Figure 4.6 Relation between recovery of galena and pulp potential in the presence of condition (BX: 10^{-4} mol/L, KNO_3: 0.1 mol/L; flotation time: 2 min, at natural pH)

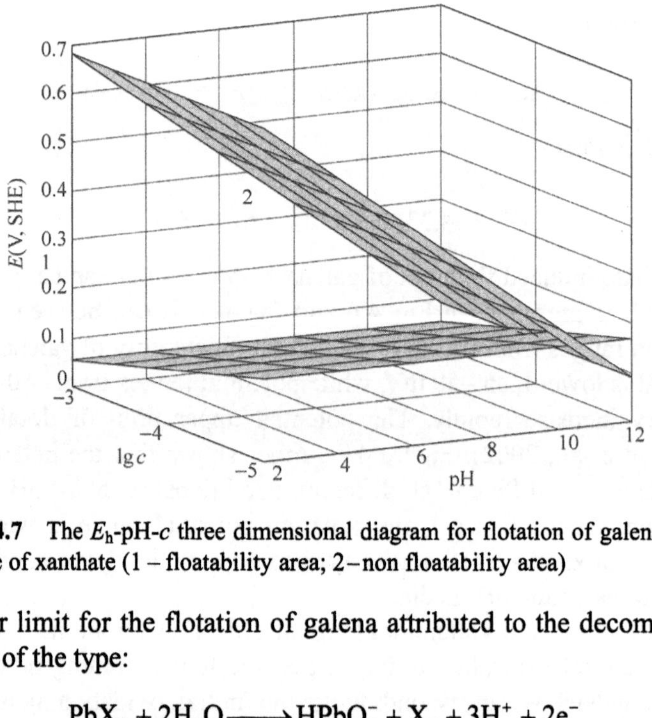

Figure 4.7 The E_h-pH-c three dimensional diagram for flotation of galena in the presence of xanthate (1 – floatability area; 2 – non floatability area)

The upper limit for the flotation of galena attributed to the decomposition by the reaction of the type:

$$PbX_2 + 2H_2O \longrightarrow HPbO_2^- + X_2 + 3H^+ + 2e^- \qquad (4\text{-}18)$$

$$E^0 = 1.225(V)$$

$$PbX_2 + 2H_2O \longrightarrow Pb(OH)_2 + X_2 + 2H^+ + 2e^- \qquad (4\text{-}19)$$

$$E^0 = 0.8(V)$$

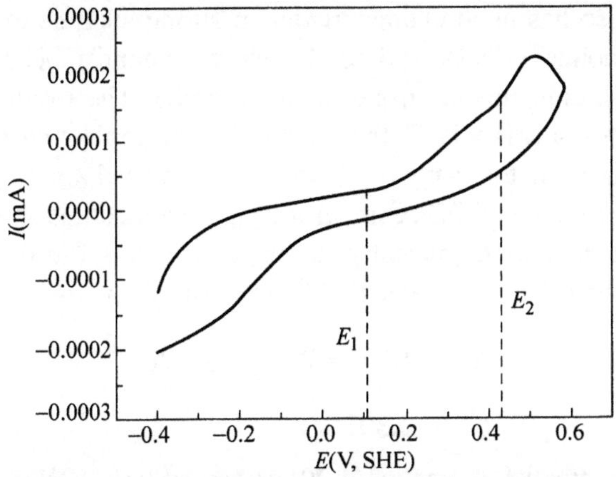

Figure 4.8 Voltammograms for galena electrode in the presence of xanthate at natural pH (BX: 10^{-4} mol/L, KNO_3: 0.1 mol/L, scan rate: 0.5 mV/s)

The calculated equilibrium potential for reactions (4-18) and (4-19) with the concentration of dissolved species 10^{-6} mol/L are respectively 417 mV and 409 mV. When the potential is higher than 400 mV, the recovery of galena fall gradually due to the reaction (4-18) and (4-19).

Trahar et al. (Trahar, 1984; Guy and Trahar, 1984) presented convincing evidence that galena floats rapidly under the conditions in which lead xanthate is present on the mineral surface, but not under the conditions in which dixanthogen should be present on the mineral surface when galena is ground in sodium sulphide solution. They reported that the xanthate-induced flotation of galena at pH = 8 and 11 with an initial xanthate concentration of 2.3×10^{-5} mol/L for grinding in sodium sulphide extended from 0 V to +0.3 V at pH = 8 and from −0.1 V to +0.1 V at pH = 11. When galena is ground in an oxidizing environment (ceramic mill), flotation occurred at a much wider and lower potential, the flotation was rapid over a wide potential range extending from −0.35 to +0.35 V at pH = 9.2 for a butyl xanthate concentration of 2×10^{-5} mol/L and −0.5 to 0.35 V at pH = 8 for a ethyl xanthate concentration of 2.3×10^{-5} mol/L. Evidently, the potential of formation of lead xanthate according to reaction (4-12c) is independent of pH, which is not consistent with the flotation results. Whereas, the calculated equilibrium potential for reactions (4-12a) and (4-12b) in 2.3×10^{-5} mol/L ethyl xanthate and 10^{-6} mol/L sulphate or thiosulphate are respectively −0.97mV, −98 mV at pH = 8 and −229 mV, −231 mV at pH = 11, indicating that reaction (4-12b) seems to fit their flotation data, i.e. the formation of lead xanthate and the presence of thiosulphate due to the oxidation reaction.

Electrochemical investigations on the galena/oxygen/diethyl dithiocarbamate (DDTC) system have also been conducted and indicated that lead diethyl dithiocarbamate (PbD_2) was formed which conferred the mineral hydrophobic (Yarar et al., 1969; Wang et al., 1992) .Wang found that the interaction of DDTC with galena is characterized by fast rate. The PbD_2 can be firmly adsorbed on galena. DDTC also has good characteristics in strong alkaline media for galena flotation, Gu Guohua (1998) studied the electrochemical behavior of galena using DDTC as a collector in strong alkaline media. The results show that, in strong alkaline media (pH > 12.5) the potential of electrochemical adsorption of DDTC on galena is in the range of 0 to 0.2 V. The voltammogram of galena electrode in the presence of DDTC in strong alkaline media is shown in Fig. 4.9. It is evident that the anodic current peak arises at about 0 to 0.2 V in different scan potential ranges. This indicates that there exists an anodic reaction:

$$PbS + 2D^- = PbD_2 + S + 2e^- \qquad (4\text{-}20)$$

$$E^0 = -0.301(V)$$

The reversible potential of reaction(4-20) corresponding to DDTC concentration of 4×10^{-5}mol/L is calculated to be −0.044 V which is quite close to the initial oxidation potential in the voltammogram. The anodic peak of the formation of PbD_2 almost coincides with the peak of galena self-oxidation (Gu and Wang, 2000). In this case, because the products of the oxidation of galena itself are $HPbO_2^-$ and S, there exists a vis-reaction of the formation of PbD_2:

$$HPbO_2^- + 2D^- + 3H^+ = PbD_2 + 2H_2O \qquad (4\text{-}21)$$

During the cathodic scan, the first cathodic peak is corresponding to the formation of PbS again, the other is the reduction of surface surplus sulfur.

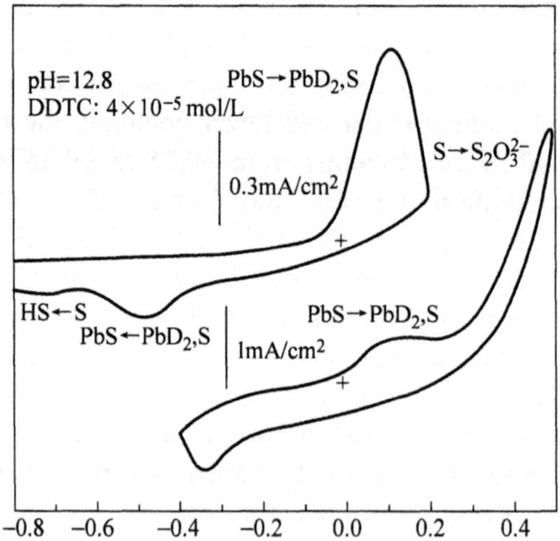

Figure 4.9 Cyclic voltammograms of galena electrode at potential scan of 20 mV/s (Background solution: pH = 12.8 buffer solution plus 0.5 mol/L KNO_3 at 25℃, 4×10^{-5} mol/L DDTC)

When the upper limit of the potential scan is above 0.4V, the anodic current peak of formation of PbD_2 occurs due to two reactions. One is that of reaction (4-20). The other reaction of the formation of PbD_2 may be:

$$2PbS + 3H_2O + 4D^- = 2PbD_2 + S_2O_3^{2-} + 6H^+ + 8e^- \qquad (4\text{-}22)$$

$$E^0 = 0.082(V)$$

Figure 4.10 represents voltammograms at different DDTC concentration at pH = 12.8. The upper limit of potential scan is 0.2 V and 0.5 V respectively. The anodic current peak of the formation of PbD_2 increases with the increase of DDTC concentration, meanwhile the anodic current peak of the oxidation of galena decreases.

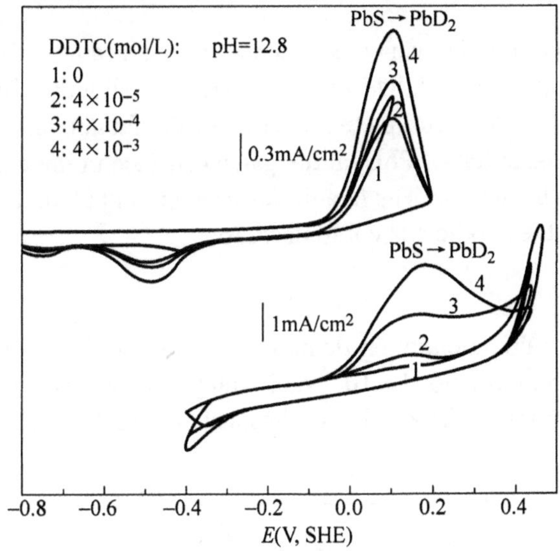

Figure 4.10 Cyclic voltammograms of galena electrode in different DDTC concentration at potential scan of 20 mV/s (Background solution: pH = 12.8 buffer solution plus 0.5 mol/L KNO_3 at 25℃)

It can be concluded that there exists a critical concentration of DDTC between 0 and 4×10^{-3} mol/L at given pH. Above this concentration, PbD_2 is formed by anodic reaction (4-20). Below the critical value, the oxidation of galena occurs as the major anodic process. It also follows from Fig. 4.10 that the degree of the cathodic current peak increased is lower than that of the anodic peak rising. This indicates that PbD_2 is not reduced completely in the cathodic process.

Figure 4.11 shows three cycles of the voltammograms of the galena electrode at pH = 12.8 with DDTC concentration 4×10^{-3} mol/L. Because of the high DDTC concentration, the oxidation of galena itself is depressed. The anodic current peak of PbD_2 arises evidently at 0 – 0.2V. However, the anodic currents decrease in

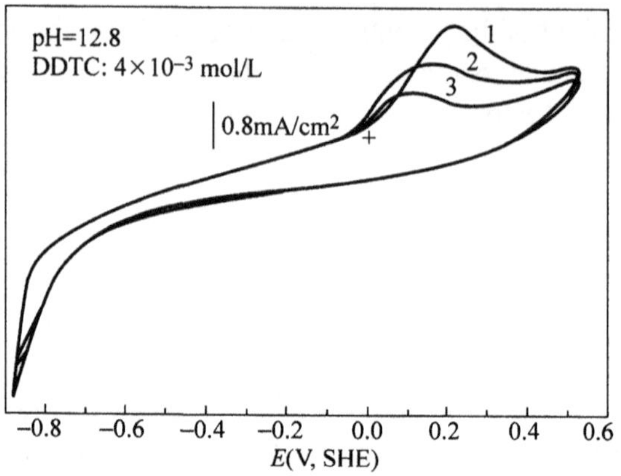

Figure 4.11 First three consecutive cycling voltammograms of galena electrode at potential scan of 20 mV/s (Background solution: pH = 12.8 buff solutions plus 0.5 mol/L KNO_3 at 25℃, 4×10^{-3} mol/L DDTC)

turn when the electrode is cyclically scanned within a range of −0.9V and 0.6V and the anodic peak potential tends to be negative slightly. This phenomenon may be attributed to the incomplete reduction or the faradic de-sorption of PbD_2. The incomplete reduction of PbD_2 at the galena surface in the cathodic scan may hold back the anodic process. The faradic de-sorption of PbD_2 in the strong reducing condition causes the anodic current to decrease.

2. Jamesonite

The influence of pulp potential on floatability of jamesonite is shown in Fig. 4.12 for an initial concentration of 1×10^{-4} mol/L ethyl xanthate (EX), dithiocarbamate (DDTC) and ammonium dialkyl dithio-phosphate (ADDP). If the upper and lower

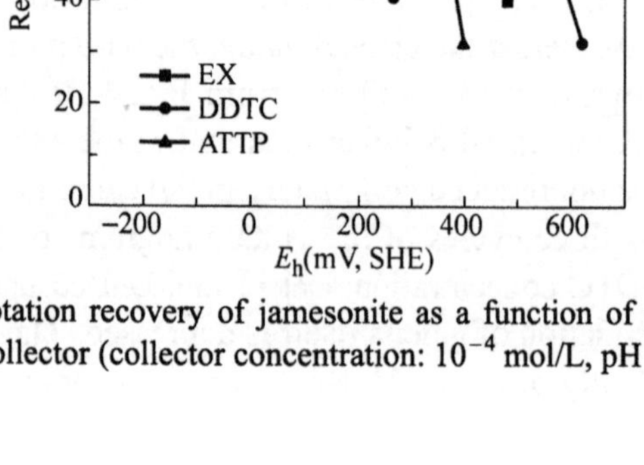

Figure 4.12 Flotation recovery of jamesonite as a function of pulp potential in the presence of collector (collector concentration: 10^{-4} mol/L, pH = 8.8)

potential limit for flotation was defined at 50% flotation recovery, it has been shown that jamesonite could be floated at a wide range of pulp potential at a given pH. The potential of flotation is in the range of 0.12V to 0.47V for EX. Using DDTC as a collector the potential range of flotation is from 0.27V to 0.57V. Jamesonite has a wider potential range of flotation with ADDP as a collector, the lower limit flotation potential is −0.18V, the upper limit is 0.37V.

Figure 4.13 presents the lower (E_h^L) and upper (E_h^U) limiting flotation potential of jamesonite as a function of pH with collector concentration of 1×10^{-4} mol/L. It can be seen that the lower (E_h^L) and upper (E_h^U) limiting flotation potential is changed with the pH value. The flotation of jamesonite may occur only at a range of pulp potential $E_h^L < E_h < E_h^U$. The flotation potential with DDTC as a collector is higher than that with EX as a collector.

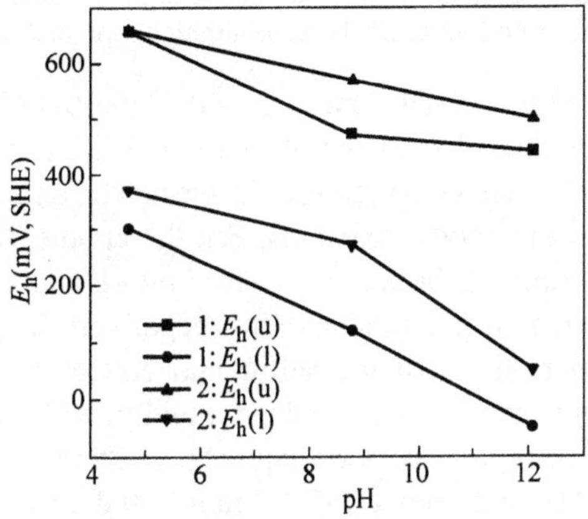

Figure 4.13 The lower (E_h^L) and upper (E_h^U) limiting flotation potential of jamesonite as a function of pH (1–EX, 2–DDTC, collector concentration: 1×10^{-4} mol/L)

Figure 4.14 is the Tafel curves of jamesonite under the conditions of different concentration of DDTC in natural pH solution. Obviously, the corrosive potential moves negatively and its corrosive current decreases with the DDTC concentration increasing. DDTC can obviously inhibit the anodic corrosion of jamesonite due to its chemisorption.

In cathodic area, the Tafel slope in the presence of DDTC is bigger than that in the absence of DDTC, and the cathodic curves under the conditions of different DDTC concentration are almost parallel and their Tafel slopes only change a little. These demonstrate that the chemisorption of DDTC on the surface of jamesonite electrode also inhibits the cathodic reaction, but the chemisorption amount of DDTC is a little and almost not affected by the DDTC concentration due to their negatively electric properties of DDTC anion and the electrode surface. This reveals that there is a little DDTC chemisorption on the mineral even if the potential is lower (i.e., negative potential).

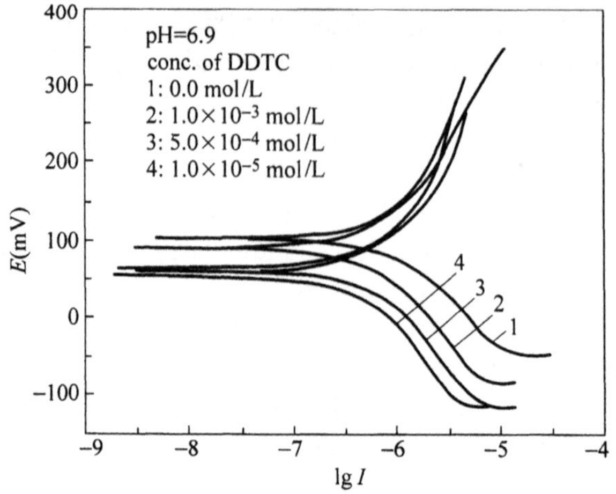

Figure 4.14 Tafel curves of jamesonite electrode in 0.1 mol/L KNO_3 solution under the conditions of different DDTC concentration (unit of I: A/cm^2)

In anodic area, the anodic Tafel slope at 1.0×10^{-4} mol/L of DDTC concentration is remarkably bigger than that in the absence of DDTC, so the adsorption of DDTC anion is quite intensive on jamesonite electrode, and further results in the passivation to inhibit the anodic reactions. But the anodic Tafel slope decreases when DDTC concentration is between 5.0×10^{-4} mol/L and 1.0×10^{-3} mol/L. This results from the corrosive potential moving negatively to promote the anodic reaction. With the deposition of the sulfur and collector metal salts from the anodic reaction, corrosive currents reduce rapidly with the enhancement of electrode potential continuously. Consequentally, there must be an optimum concentration of DDTC between 1.0×10^{-4} mol/L and 5.0×10^{-4} mol/L to inhibit more effectively anodic corrosion reaction.

Figure 4.15 is CV diagrams of jamesonite electrode in 0.1 mol/L KNO_3 buffer solution containing different concentration of DDTC. The collector-metal-salts from the electrochemical reactions of jamesonite with the DDTC solutions of different concentration can effectively adhere to the surface of the mineral electrode to inhibit anodic process so that their ap_1 peaks are nearly the same. But the inhibition actions for anode electrochemical reactions are different at ap_2、ap_3、ap_4 peaks when the DDTC concentration is different: (1) the current intensity of ap_3、ap_4 peaks increases with the enhancement of the DDTC concentration; (2) not only are the ap_2 peaks most obviously passive but also the peak current intensity of ap_3 and ap_4 (dot line) at 5.0×10^{-4} mol/L of the DDTC concentration is lower than that (dash line) at 1.0×10^{-3} mol/L of the DDTC concentration. Consequentially, there must be a critical concentration of DDTC. DDTC and its metal salts can inhibit the electrochemical reactions on the mineral in an extensive potential range when the DDTC concentration is lower than the critical value. But DDTC can promote the electrochemical reaction on the mineral when the DDTC concentration is higher than the critical value. It means

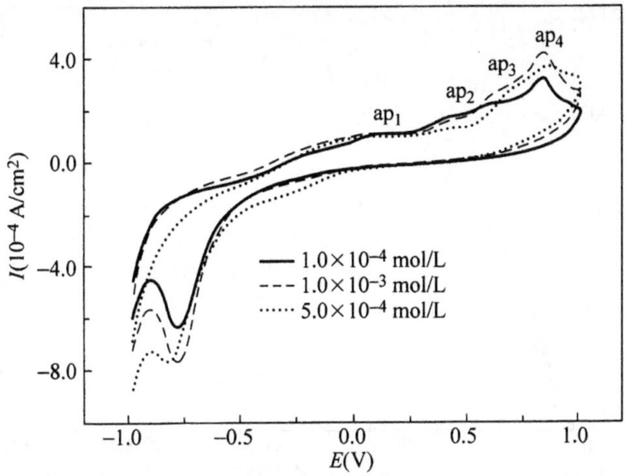

Figure 4.15 Voltammograms for jamesonite electrode in 0.1 mol/L KNO_3 buffer solution of pH = 6.86 containing different concentration of DDTC (scan rate: 50 mV/s)

that the collector-salts can not cover the surface of the mineral electrode in this case. DDTC can effectively inhibit the electrochemical reaction from 0 to 300 mV whatever the experimented concentration of DDTC is, but in the point of flotation practice, the lower the collector concentration, the lower the reagent cost. So, the optimum concentration of the collector must be some value in the range from 1.0×10^{-4} to 5.0×10^{-4} mol/L. This result is unanimous with the results discussed in Fig. 4.14.

Figure 4.16 is the electrochemistry impedance spectrum (EIS) of jamesonite electrode under the conditions of different DDTC concentration, and its EIS parameters are shown in Table 4.2. It follows that:

(1) The electrochemical resistance is smaller in the absence of DDTC, but increases four times in presence of DDTC. So the adsorption of DDTC on jamesonite results in reducing the reaction rate of the corrosive electrochemistry in open circuit potential.

(2) The electrochemical impedance gradually decreases but does not vary very much with the DDTC concentration increasing. It indicates that DDTC takes part in the electrochemical reaction and the reaction rate increases as the DDTC concentration enhances. As contrasted with it, the passivation of the collector-salts of reaction production on the mineral electrode further inhibits the anodic reaction so that the electrochemical resistance is wholly much bigger than that of self-corrosive reaction in the absence of DDTC.

(3) Capacitance impedance loop is smaller and its interfacial capacitance of mineral/solution is bigger in the absense of DDTC. But the capacitance impedance loop obviously enlarges and the interfacial capacitance becomes small in the presence of DDTC. With the DDTC concentration increasing, there is no obvious change of the capacitance impedance loop, but its interfacial capacitance increases.

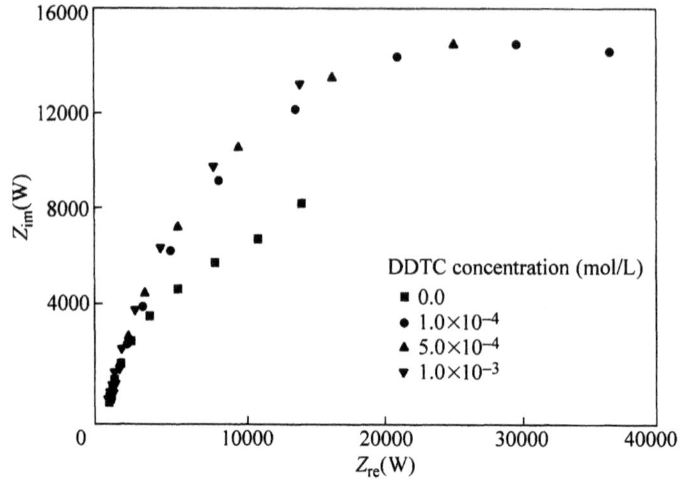

Figure 4.16 EIS of jamesonite electrode in 0.1 mol/L KNO_3 solution containing different concentration of DDTC

Usually, as the organic matter with small electrical medium constant can push aside H_2O molecule with big electrical medium constant, ε becomes small and C_d rises according to the following formula:

$$C_d = 4\pi\varepsilon/d \tag{4-23}$$

Table 4.2 EIS parameters of jamesonite electrode under the conditions of different conc. of DDTC

Direct current volt, V_d (V)	0			
Conc. of DDTC (mol/L)	0	1.0×10^{-4}	5.0×10^{-4}	1.0×10^{-3}
Electrochemical resistance, R_r (kΩ)	11.80	45.60	42.79	39.95
Interfacial capacitance, C_d (μF/cm^2)	274.4	107.6	154.3	186.5

Here, ε is the electrical medium constant of the adsorbed matter, and d is the diameter of the adsorbed molecule.

Consequentally, the interfacial capacitance in the presence of DDTC is smaller than that in absence of DDTC, so mineral hydrophobicity will increase when an organic matter is adsorbing on a mineral electrode. It reasonably explains the experimental appearance in Fig. 4.12.

The interfacial capacitance increases with the DDTC concentration added. The relationship among potential difference ψ_1 of diffusion layer, the electric charge density q on the surface of an electrode and the concentration c of a solution according to Gouy, Chapman and Stern model theory is as follows.

$$q = \sqrt{8cRT\varepsilon_0\varepsilon_r}\,\sinh\left(\frac{|Z|\psi_1 F}{2RT}\right) \tag{4-24}$$

As DDTC adsorbs on jamesonite electrode chemically, the double electric charge layer is treated as a plate capacitor, the capacitance C_1 of the tight layer as a constant, and the change of the capacitance of the double electric charge layer is designated to the capacitance C_2 of the diffusion layer. Thereby, the tight layer and the diffusion layer are looked upon as two series capacitances according to the method from Cooper and Harrison, then:

$$\varphi_a = \psi_1 + \frac{1}{C_1}\sqrt{8cRT\varepsilon_0\varepsilon_r}\,\sinh\left(\frac{|Z|\psi_1 F}{2RT}\right) \tag{4-25}$$

Here, φ_a is the potential of the whole double electric charge layer.

When the electric charge density q on the surface of an electrode and the concentration c of a solution are very small, namely, the absolute value of $\psi_1 F$ is small far from RT, the formula can be spread out in terms of a progression, and its high grade terms can be ignored, then:

$$\varphi_a = \psi_1 + \frac{1}{C_1}\sqrt{\frac{2c\varepsilon_0\varepsilon_r}{RT}}F\psi_1 \tag{4-26}$$

c is very small in a dilute solution; the second term can be ignored because it is much smaller than the first term in the formula above, then, $\varphi_a \approx \psi_1$. This indicates that the capacitance of the diffusion layer is just the capacitance of the whole double electric charge layer. Because

$$C_1 = \frac{q}{\varphi_a - \psi_1} \tag{4-27}$$

The formulas (4-26) and (4-27) are transferred into:

$$C_d \approx C_2 = \frac{q}{\psi_1} = \frac{C_1(\varphi_a - \psi_1)}{\psi_1} = \sqrt{\frac{2c\varepsilon_0\varepsilon_r}{RT}}F \tag{4-28}$$

The formula of a plate capacitor is as follows:

$$C_d = \frac{\varepsilon_0\varepsilon_r}{l} \tag{4-29}$$

Compare the formulas (4-28) and (4-29), then

$$l = \frac{1}{F}\sqrt{\frac{RT\varepsilon_0\varepsilon_r}{2c}} \tag{4-30}$$

Therefore, the effective thickness l is in inverse proportion to $c^{1/2}$ the square root of adsorbed ion concentration. It reasonably explains the appearance that the capacitance increases with the DDTC concentration added and also indicates that the adsorption behavior of DDTC on the surface of jamesonite is similar to ion characteristic adsorption. The adsorption of DDTC is very tight on jamesonite.

DDTC is an organic anion with comparatively big ε and variable shape. After adsorption and bonding with mineral electrode, it can be effectively close to the electrode surface so that l becomes small and ε becomes big. So its capacitance increases with the DDTC concentration added, amount of adsorption increasing.

Therefore, the corrosive potential of jamesonite moves negatively with the DDTC concentration added. It promotes the anodic reaction of jamesonite. On the contrary, DDTC metal salt of the reaction production covered tightly the electrode surface inhibits its anodic reaction to result in the decrease of the corrosive current. There must be an optimum concentration of collector DDTC between 1.0×10^{-4}—5.0×10^{-4} mol/L.

4.1.3 Zinc Sulphide Minerals

1. Sphalerite

Only limited studies on the electrochemical behavior of sphalerite have been reported, perhaps due to its high electrical resistivity. The Relation between recovery of sphalerite and pulp potential is presented in Fig. 4.17 with an initial butyl xanthate concentration of 10^{-4} mol/L. It can be seen from Fig. 4.17 that flotation begins at 0 V, the upper limit potential is 0.31 V.

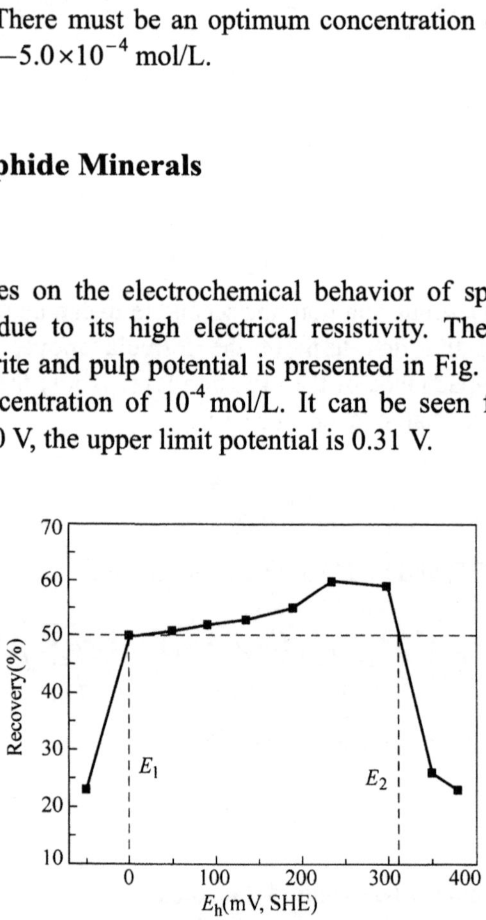

Figure 4.17 Relation between recovery of sphalerite and pulp potential in the presence of BX (BX: 10^{-4} mol/L, KNO_3: 0.1 mol/L, flotation time: 2 min, at natural pH)

Figure 4.18 shows the voltammograms of sphalerite electrode in the presence of xanthate at natural pH. The initial oxidation potential is -10 mV and the peak potential of further oxidation is 300 mV. For collector flotation of sphalerite, the initial flotation potential of sphalerite may be corresponding to the oxidation potential. The reaction is as following:

$$ZnS + 2X^- \longrightarrow ZnX_2 + S + 2e^- \quad (4\text{-}31)$$

$$E^0 = -0.05(V)$$

$$2ZnS + 4X^- + 3H_2O \longrightarrow 2ZnX_2 + S_2O_3^{2-} + 6H^+ + 8e^- \quad (4\text{-}32)$$

$$E^0 = 0.256(V)$$

The potential is 200 mV based on reaction (4-31) and 10 mV on reaction (4-32) at natural pH. It can be inferred that the initial reaction of sphalerite corresponds to reaction (4-32).

For the collector flotation of sphalerite, the upper limit potential of flotation may be corresponding to the following decomposition reactions:

$$ZnX_2 + 2H_2O \longrightarrow H_2ZnO_2^- + 3H^+ + X_2 + 2e^- \quad (4\text{-}33)$$

$$E^0 = 0.956(V)$$

$$ZnX_2 + 2H_2O \longrightarrow Zn(OH)_2 + X_2 + 2H^+ + 2e^- \quad (4\text{-}34)$$

$$E^0 = 0.743(V)$$

Assuming the concentration of all dissolved species to be 10^{-6}mol/L, the reaction potential are 0.37 V and 0.32 V respectively, corresponding to reactions (4-33) and (4-34) and the oxidation peak potential in Fig. 4.18. Therefore the upper limit potential of flotation of sphalerite may depend on reaction (4-33) or (4-34), i.e. the decomposition of zinc xanthate and the formation of zinc hydroxide or oxy-zinc species.

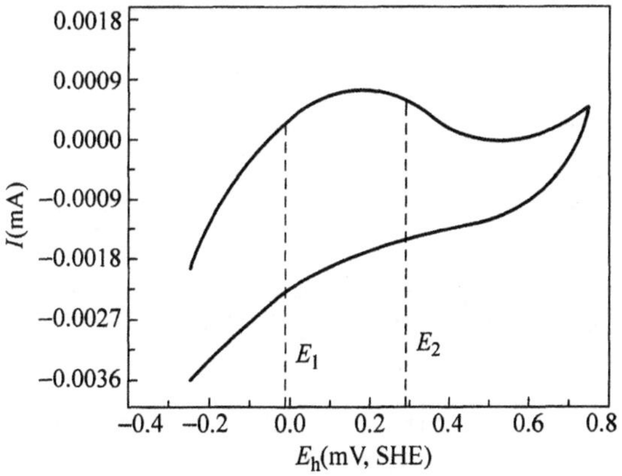

Figure 4.18 Cyclic voltammetric curves for sphalerite electrode at natural pH in the presence of condition (BX: 10^{-4} mol/L, KNO_3: 0.1 mol/L, scan rate: 0.5 mV/s)

Figure 4.19 shows the relationship of E_h-pH-c of sphalerite in the presence of xanthate. The floatable region of sphalerite is dependent on the pH, potential and the concentration of reagent addition. It can be seen that there exists an appropriate potential range for zinc sulphide flotation for a given pH and reagent concentration. Within this range, minerals represent good floatability. Above this range, minerals floatability descends, and the limiting potential of flotation always corresponds to specified reactions.

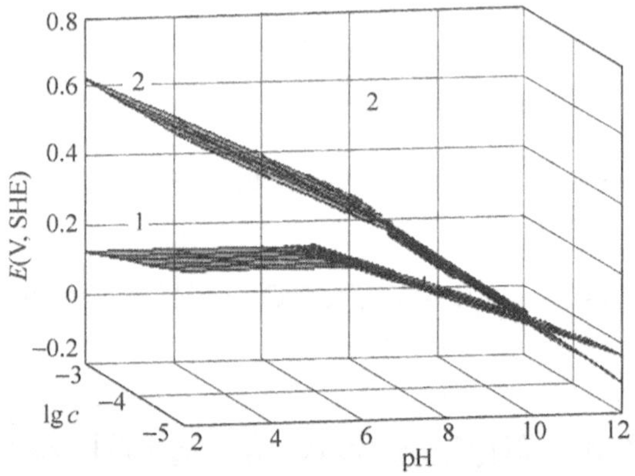

Figure 4.19 The relationship of E_h-pH-c of sphalerite in the presence of xanthate
1–floatability area; 2–non floatability area

2. Marmatite

The influence of pulp potential on the flotation of marmatite at different pH is given in Fig. 4.20 using ethyl xanthate as a collector. In acidic pH media, marmatite exhibits a wide floatable potential range and the upper potential limit of flotation

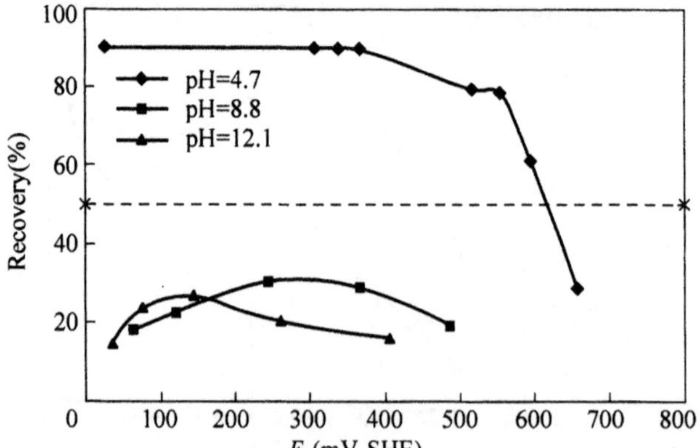

Figure 4.20 Flotation recovery of marmatite as a function of pulp potential with ethyl xanthate as a collector at different pH (KEX: 10^{-4} mol/L)

of marmatite is 620 mV. In alkaline pH media, the floatability of marmatite is very poor. The initial potential of ethyl xanthate flotation of marmatite as low as 0 V suggests the formation of zinc xanthate according to similar reaction (4-32) to sphalerite because the formation potential of ethyl dixanthogen is 0.17 V for concentration 10^{-4} mol/L based on the following reaction.

$$2EX^{-} \longrightarrow (EX)_2 + 2e^{-} \tag{4-35a}$$

$$\begin{cases} E^0 = -0.06(\text{V}) \\ E_{\text{h}} = -0.06 - 0.059\lg[EX^{-}] \end{cases} \tag{4-35b}$$

However, the decomposition potential of zinc xanthate into dixanthogen is above 0.3 V according to reactions (4-33) or (4-34) and the upper potential limit of flotation of marmatite extends to 620 mV, which indicates the coexistence of dixanthogen on marmatite in this condition. The difference of flotation behavior and hydrophobic entity between sphalerite and marmatite may be due to the existence of iron in marmatite.

Figure 4.21 reveals the influence of pulp potential on the flotation of marmatite at different pH using dithiocarbamate as a collector. Figure 4.21 shows that marmatite has good floatability with a maximum recovery of about 90% in acidic pH media with a potential range 300 – 750 mV when DDTC is used as a collector. In alkaline pH media, the floatability of marmatite becomes very poor in various potential regions. If the collector metal salt was formed into marmatite like sphalerite, the reaction between marmatite and dithiocarbamate may be written simply as follows:

$$ZnS + 2D^{-} \longrightarrow ZnD_2 + S + 2e^{-} \tag{4-36a}$$

$$E^0 = -0.21(\text{V})$$

$$2ZnS + 4D^{-} + 3H_2O \longrightarrow 2ZnD_2 + S_2O_3^{2-} + 6H^{+} + 8e^{-} \tag{4-36b}$$

$$E^0 = -0.128(\text{V})$$

The calculated potential are, respectively 26 mV and −260 mV for 10^{-4} mol/L DDTC at pH = 4.7, which are not consistent with the flotation beginning potential in Fig. 4.21. On the other hand, the reaction of dithiocarbamate oxidation to thiouram disulphide can be written as

$$2DDTC^{-} = (DDTC)_2 + 2e^{-} \tag{4-37a}$$

$$\begin{cases} E^0 = -0.015(\text{V}) \\ E_{\text{h}} = -0.015 - 0.059\lg[DDTC] \end{cases} \tag{4-37b}$$

The calculated potential is 221 mV for 10^{-4} mol/L DDTC, which reasonably corresponds to the initial potential of marmatite flotation in Fig. 4.21. Therefore, the hydrophobic entity of dithiocarbamate on marmatite may be assumed to be disulphide.

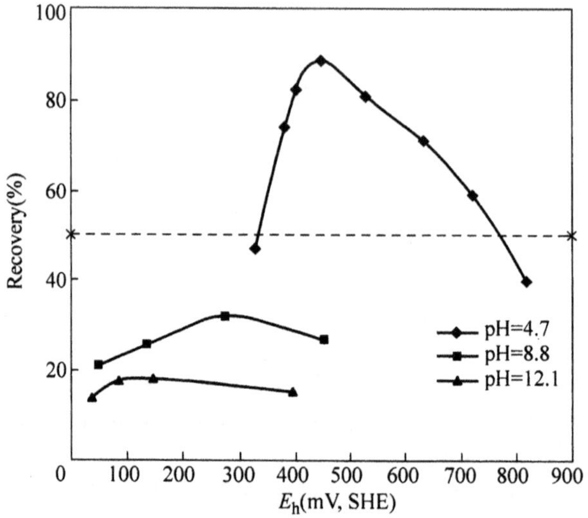

Figure 4.21 Flotation recovery of marmatite as a function of pulp potential with dithiocarbamate as a collector at different pH (DDTC: 10^{-4} mol/L)

The voltammograms for marmatite electrode in different pH buffer solutions are presented in Fig. 4.22. It can be seen from Fig. 4.22(a) that the first anodic peak occurred at about 200 mV, which may be due to the oxidation of dithiocarbamate to disulphide. The anodic oxidation peaks at higher potential may be attributed to the further oxidation of marmatite to form oxy-sulphur and zinc/iron hydroxide resulting in flotation descending.

Figure 4.22(b) shows that no anodic oxidation peaks occurred at lower potential region in alkaline media for marmatite electrode in xanthate and dithiocarbamate solution indicates no formation of collector species on marmatite. When the potential increases to higher potential region, the occurrence of anodic oxidation peak may be due to the self-oxidation of marmatite. It accounts for no flotation of marmatite in alkaline media either using xanthate or dithiocarbamate as collectors.

4.1.4 Iron Sulphide Minerals

1. Pyrite

The influence of potential on the floatability of pyrite with butyl xanthate as a collector has been determined and the result is given in Fig. 4.23. It follows that flotation begins at 0.1 V for an initial KBX concentration of 10^{-4} mol/L. The flotation potential ranges from 0.10 V to +0.31 V.

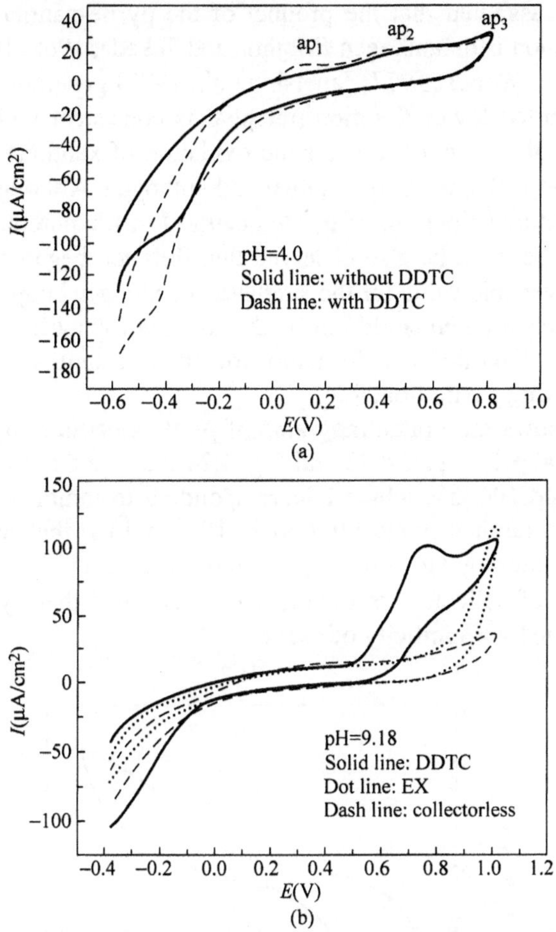

Figure 4.22 Voltammograms for marmatite electrode in different pH buffer solution (a) in 0.1 mol/L Na_2SO_4 media at scanning rate of 50mV/s; (b) in 0.1 mol/L KNO_3 media at scanning rate of 20 mV/s

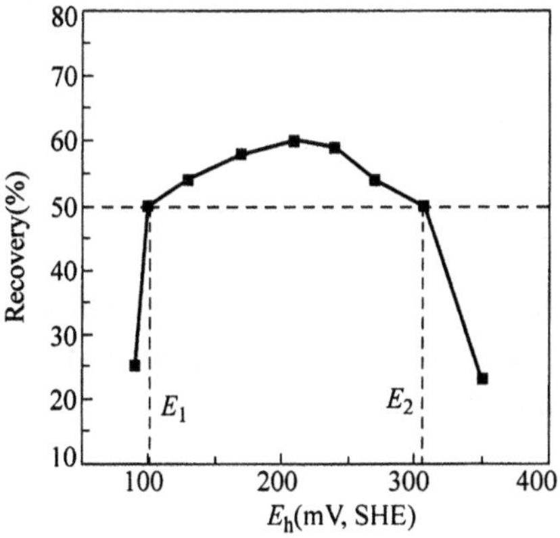

Figure 4.23 Recovery of pyrite as a function of pulp potential in the presence of condition (BX: 10^{-4} mol/L, KNO_3: 0.1 mol/L, flotation time: 2 min)

It is generally assumed that the product of the pyrite-xanthate interaction in air-saturated solution is dixanthogen (Majima and Takeda, 1968; Ball and Rickard, 1976; Gardner and Woods, 1977; Janetski et al., 1977) generated by the reaction (4-35). The observed lower flotation potential is consistent with the calculated reversible potential of about 0.1 V for the oxidation of xanthate to dixanthogen, taking E^0 to be −0.128 V for butyl xanthate/dixanthogen. Although Trahar (1984) found that the optimal flotation of pyrite occurred in the potential region where dixanthogen was formed, he also observed that flotation began at −0.15 V much lower than the reversible value for the xanthate/dixanthogen couple, which led him to suggest that dixanthogen could not be the sole hydrophobic entity responsible for the flotation of pyrite. The formation of an iron xanthate species may be possible in rendering pyrite floatable.

Figure 4.24 shows the voltammograms of pyrite electrode in the presence of xanthate at natural pH. Figure 4.23 and Fig. 4.24 indicate that the initial flotation potential is around 106 mV, which is corresponding to initial potential of anode oxidation. When xanthate concentration is 10^{-4} mol/L, the potential is about 100 mV. The flotation upper limit edge of pyrite is 300 mV, which is near to the upper limit edge of collectorless flotation. It is assumed that hydroxide may be responsible for the hydrophilicity of pyrite.

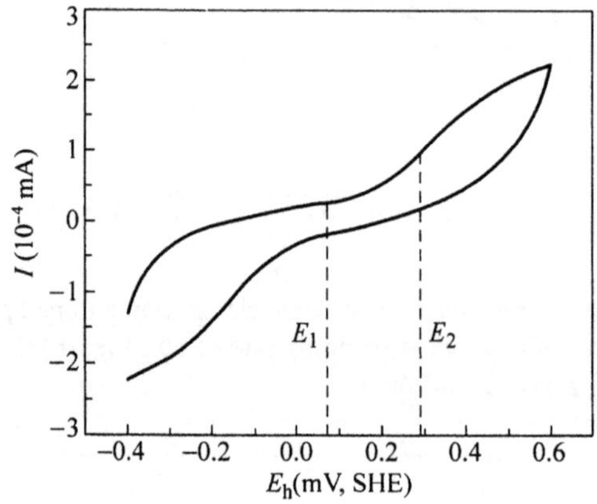

Figure 4.24 Cyclic voltammetric curves for pyrite electrode at natural pH in the presence of condition (BX: 10^{-4} mol/L, KNO_3: 0.1 mol/L, scan rate: 0.5 mV/s)

Figure 4.25 further shows the relationship of E_h-pH-c of pyrite in the presence of xanthate. It can be seen that the range of flotation potential becomes narrow with the increase of pH and the decrease of concentration. Under the condition of alkaline media, floatability of pyrite is very poor in various concentration conditions and potential range.

2. Pyrrhotite

The relationship between pyrrhotite flotation recovery and pulp potential is presented in Fig. 4.26. It can be shown that the flotation of pyrrhotite has different

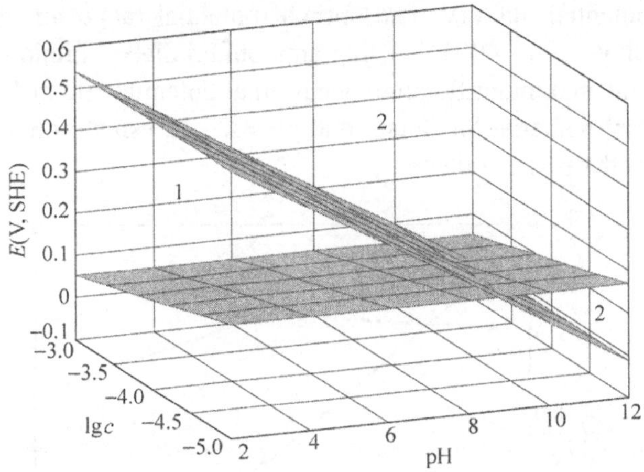

Figure 4.25 The relationship of E_h-pH-c of pyrite in the presence of xanthate (1: floatability area; 2: non floatability area)

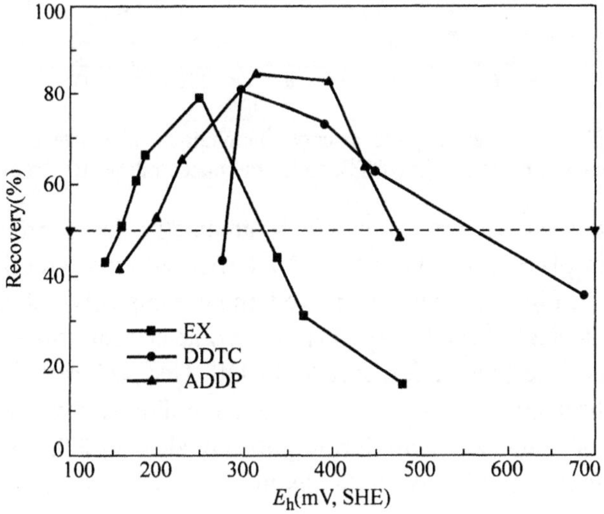

Figure 4.26 Relationship between recovery of pyrrhotite and pulp potential in the presence of collector (collector concentration: 10^{-4} mol/L, pH = 8.8)

suitable pulp potential for different collectors. If the flotation potential upperand lower limits were defined according to a flotation recovery of 50% respectively ($R = 50\%$), the relationship of pyrrhotite flotation upper (E_h^U) and lower (E_h^L) limit potential is presented in Fig. 4.27. It can be seen that the flotation of pyrrhotite may occurs only in some rang of the pulp potential at various pH. At pH = 8.8, the optimal potential range for the flotation of pyrrhotite is about 150 – 320 mV using ethyl xanthate as a collector and the maximum flotation occurred at potential 250 mV. The optimal potential range for the flotation of pyrrhotite is about 280 – 550 mV using dithiocarbamate as a collector and the maximum flotation occurred at potential 300 mV. The optimal potential range for the flotation of pyrrhotite is about 200 – 400 mV using ammonium dialkyl dithio-phosphate as a collector and the maximum flotation occurred at potential 300 mV. Pyrrhotite has a wider range of potential for flotation at pH = 4 – 12 using dithiocarbamate than using ethyl xanthate as a collector.

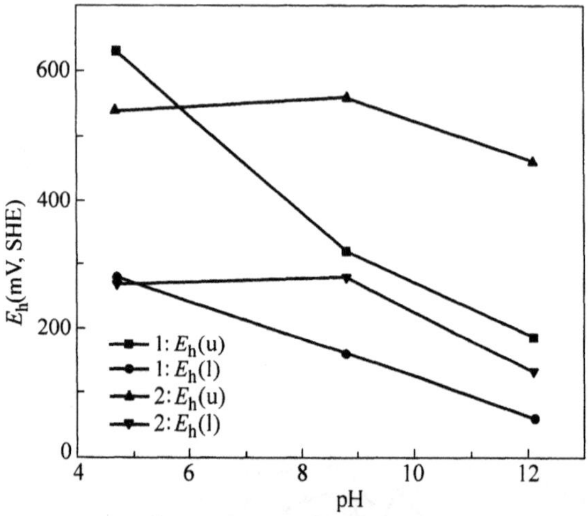

Figure 4.27 The lower (E_h^L) and upper (E_h^U) limiting flotation potential of pyrrhotite as a function of pH (1: EX, 2: DDTC, collector concentration: 10^{-4} mol/L)

The anodic scan section of cyclic voltammetry for pyrrhotite electrode, respectively, in pH = 7, 8.8, 11, 12.1, 12.7 buffer solution with dithiocarbamate are presented in Fig. 4.28. The cyclic voltammograms curve at pH = 8.8 is also presented in Fig. 4.28. It can be seen that the anodic current peak emerges at about 0.1V. As pH increases, the peak moves to the left. This peak may correspond to the formation of disulphide. When the concentration of dithiocarbamate is 10^{-4} mol/L, the oxidation of dithiocarbamate forming disulphide is $E_h = 0.22$V, which agrees with the results in Fig. 4.26. When the hydrophobic entity disulphide was formed, the flotation of pyrrhotite could be possible.

It can be also seen from Fig. 4.28 that the place of anodic peaks moves towards the left with the increase of pH. When pH = 12.7, the anodic peak disappears, and the anodic current increases rapidly. It indicates that only the oxidation of pyrrhotite takes place on the surface of pyrrhotite. In addition, no disulphide was formed at pH = 12.7. So the flotation of pyrrhotite could not be carried out.

3. Arsenopyrite

Arsenopyrite exhibits similar xanthate induced floatability to pyrite. The result obtained by Feng (1989) is presented in Fig. 4.29. It follows that flotation occurs at potential where dixanthogen is formed, confirming the mechanism of formation of dixanthogen from Allison et al. (1972).

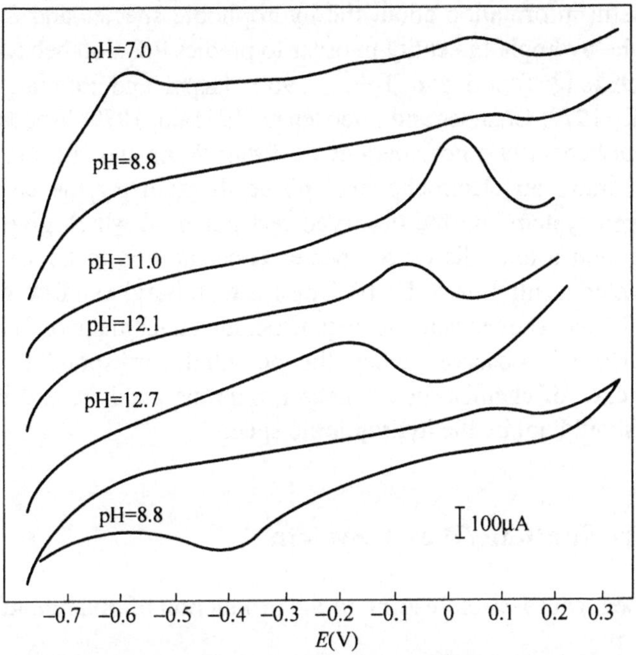

Figure 4.28 The anodic scan section of cyclic voltammetry for pyrrhotite electrode in different pH buffer solution at potential scan of 10 mV/s (Background solution: buffer solution plus 0.1 mol/L KNO_3, DDTC: 10^{-4}mol/L)

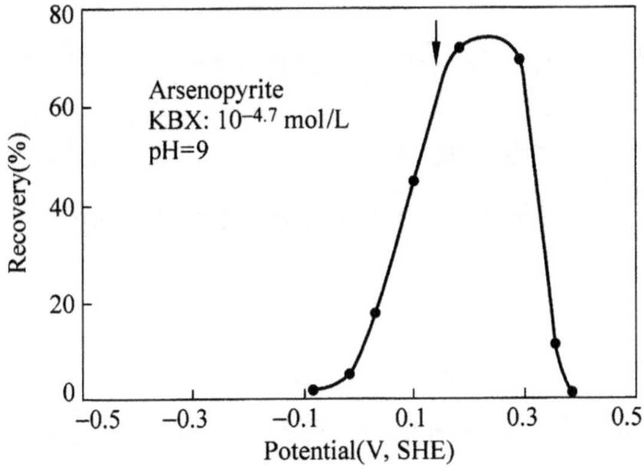

Figure 4.29 Flotation recovery of arsenopyrite as a function of pulp potential (Arrow indicates the reversible potential of butyl xanthate/dixanthogen couple)

4.2 E_h-pH Diagrams for the Collector/Water/Mineral System

Electrochemical phase diagrams have been used to investigate the collector water mineral system in which the experimental potential for flotation is compared with thermodynamic equilibriums for reactions in mineral/oxygen/collector system to obtain the useful information about the hydrophobic species and E_h-pH area of formation of the hydrophobic entity in order to predict flotation behaviors (Pritzker and Yoon, 1984a,b; Toperi and Tolun, 1969; Hepel and Pomianowski, 1977; Abramov et al., 1977; Chander and Fuerstenau, 1975a,b, 1979; Woods et al., 1990; Wang and Forssberg, 1991; Forssberg et al., 1984; Wang and Hu, 1989). Abramov et al. (1977) draws an electrochemical phase diagram for the chalcocite/butyl xanthate/oxygen system and the observed rest potential which give firm bubble attachment, assumes total dissolved species concentration to be 10^{-6} mol/L and total carbonate concentration to be 10^{-4} mol/L with butyl xanthate concentration of 10^{-4} mol/L and shows that the experimental attachment of bubbles to the chalcocite surface is observed when the potential correspond to the field of stable coexistence of chemisorbed collector, cuprous xanthate and indicates that dixanthogen should not be the hydrophobic species.

4.2.1 Butyl Xanthate/Water System

The electrochemical reaction leading to the production of dixanthogen is

$$2BX^- = (BX)_2 + 2e^- \tag{4-38a}$$

$$\begin{cases} E^0 = -0.1(V) \\ E_h = -0.1 - 0.059\ \lg[BX^-] \end{cases} \tag{4-38b}$$

$$2HBX = (BX)_2 + 2H^+ + 2e^- \tag{4-39a}$$

$$\begin{cases} E^0 = 0.201(V) \\ E_h = 0.201 - 0.059\ \lg[HBX^-] - 0.059pH \end{cases} \tag{4-39b}$$

$$HBX = H^+ + BX^- \tag{4-40a}$$

$$K_a = 7.9 \times 10^{-6} \tag{4-40b}$$

The upper E_h limit for water stability is defined by the reaction:

$$H_2O = \frac{1}{2}O_2 + 2H^+ + 2e^- \tag{4-41a}$$

$$\begin{cases} E^0 = 1.23(V) \\ E_h = 1.22 - 0.059pH \end{cases} \tag{4-41b}$$

The lower E_h limit of water stability is defined by the reaction:

$$H_2 = 2H^+ + 2e^- \tag{4-42a}$$

$$E_h = -0.059\text{pH} \tag{4-42b}$$

Thus, the various equilibrium lines, together with the experimental potential at which flotation recovery is greater than 50%, are shown in Fig. 4.30 and Fig. 4.31. It can be seen from Fig. 4.30 that the E_h-pH area for the flotation of galena should be in the same area as the formation of lead butyl xanthate, whereas the lower limiting potential of chalcopyrite are corresponding to that of the formation of cuprous xanthate but the floatability is extended to wider E_h range where dixanthogen is dominant. Figure 4.31 shows that the E_h-pH region of the flotation of pyrite and arsenopyrite occurs only in the area where dixanthogen is found.

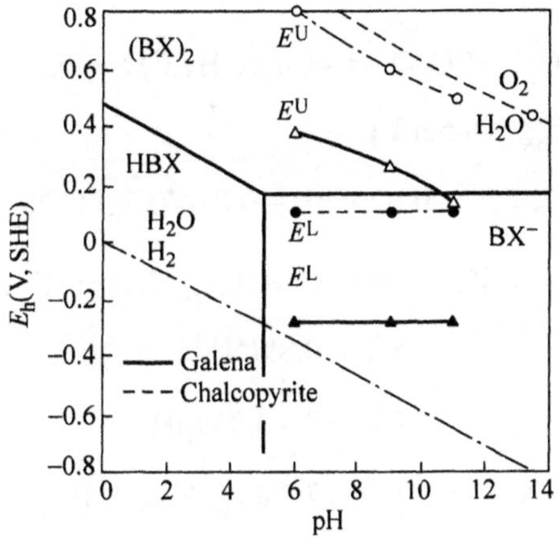

Figure 4.30 Electrochemical phase diagram for the butyl xanthate/oxygen system and the observed lower (E^L) and upper (E^U) limiting flotation potential of galena and chalcopyrite at which flotation recovery is greater than 50% (BX^-: 2×10^{-5} mol/L)

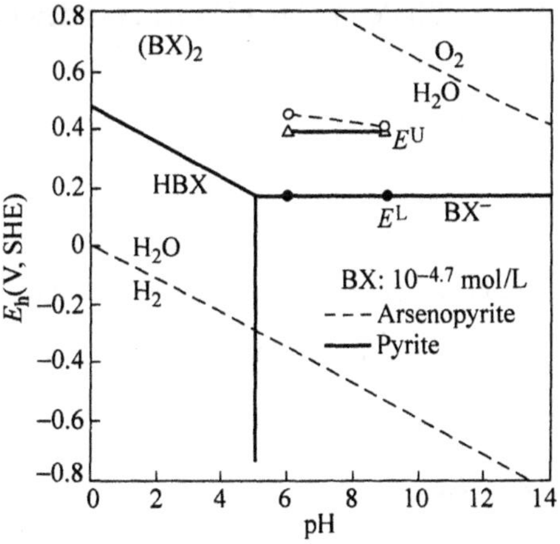

Figure 4.31 Electrochemical phase diagram for the butyl xanthate/oxygen system and the observed lower (E^L) and upper (E^U) limiting flotation potential of pyrite and arsenopyrite at which flotation recovery is greater than 50% (BX^-: 2×10^{-5} mol/L)

4.2.2 Chalcocite-Oxygen-Xanthate System

The electrochemical reaction of formation of cuprous xanthate is defined by the following equation

$$2Cu_2S + 4X^- + 3H_2O = 4CuX + S_2O_3^{2-} + 6H^+ + 8e^- \quad (4\text{-}43a)$$

$$\begin{cases} E_{EX}^0 = 0.147(V) \\ E_h = E^0 - 0.04425pH + 0.007375\lg[S_2O_3^{2-}] - 0.0295\lg[X^-] \end{cases} \quad (4\text{-}43b)$$

The decomposition of cuprous xanthate by oxidation is defined by the following reactions:

$$2CuX + CO_3^{2-} + 2H_2O = Cu_2(OH)_2CO_3 + X_2 + 2H^+ + 4e^- \quad (4\text{-}44a)$$

$$\begin{cases} E_{EX}^0 = 0.53(V) \\ E_h = E^0 - 0.0295pH - 0.01475\lg[CO_3^{2-}] \end{cases} \quad (4\text{-}44b)$$

$$2CuX + 4H_2O = 2Cu(OH)_2 + X_2 + 4H^+ + 4e^- \quad (4\text{-}45a)$$

$$\begin{cases} E_{EX}^0 = 0.89(V) \\ E_h = E^0 - 0.059pH \end{cases} \quad (4\text{-}45b)$$

$$2CuX + 4H_2O = 2CuO_2^{2-} + X_2 + 8H^+ + 4e^- \quad (4\text{-}46a)$$

$$\begin{cases} E_{EX}^0 = 1.8(V) \\ E_h = E^0 + 0.0295\lg[CuO_2^{2-}] - 0.118pH \end{cases} \quad (4\text{-}46b)$$

The stability of cuprous xanthate in high pH media is determined by the reaction of the type:

$$CuX + 4OH^- = CuO_2^{2-} + X^- + 2H_2O \quad (4\text{-}47a)$$

$$\begin{cases} K_{EX}^0 = 1.45 \times 10^{-6} \\ \lg[OH^-] = (1/4)(\lg[X^-][CuO_2^{2-}] - \lg K) \end{cases} \quad (4\text{-}47b)$$

The dissolution of copper hydroxide at high pH is

$$Cu(OH)_2 = CuO_2^{2-} + 2H^+ \quad (4\text{-}48a)$$

$$K = 1.57 \times 10^{-31}$$

$$Cu_2(OH)_2CO_3 + 2OH^- = 2Cu(OH)_2 + CO_3^{2-} \quad (4\text{-}48b)$$

$$K = 1.57 \times 10^3$$

The oxidation of xanthate ion into dixanthogen is defined by the Eq. (4-35), in which $E^0_{EX/(EX)_2} = -0.06$.

In drawing the equilibrium lines, the concentration of dissolved species is assumed to be 10^{-6} mol/L. The electrochemical phase diagrams for the chalcocite/xanthate/oxygen system are shown in Fig. 4.32 with an ethyl xanthate concentration of 4.7×10^{-5} mol/L. Figure 4.32 shows that the lower and upper limiting potential pH area of flotation of chalcocite at which the recovery is greater than 50% (Basiollio et al., 1985; Heyes and Trahar, 1977) is located in the region where cuprous ethyl xanthate is stable species, indicating that the hydrophobic entity is cuprous ethyl xanthate. The flotation ceases at higher potential by the oxidation into $Cu_2(OH)_2$, CuO_2^{2-} and dixanthogen, and at high pH conditions by the decomposition into CuO_2^{2-}, $S_2O_3^{2-}$.

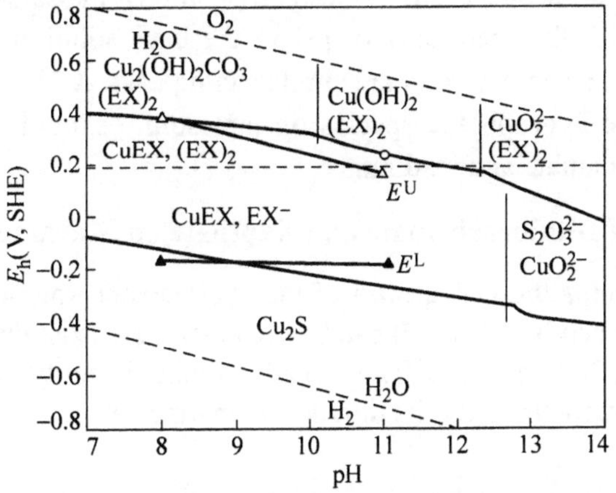

Figure 4.32 Electrochemical phase diagram for the chalcocite/ethyl xanthate/oxygen system and the observed lower (E^L) and upper (E^U) limiting flotation potential at which flotation recovery is greater than 50%, dashed line is the formation of dixanthogen

4.3 Surface Analysis

Some instrumental methods have been used for the investigation of sulphide mineral-thio-collector system such as infra-red (IR) spectroscopy (Mielezarski and Yoon, 1989; Leppinen, 1990; Persson et al., 1991; Laajalehto et al., 1993) and X-ray photoelectron spectroscopy (XPS) (Pillai et al., 1983; Page and Hazell, 1989; Grano et al., 1990; Laajalehto et al., 1991). These surface sensitive spectroscopic techniques can be applied for the direct determination of the surface composition at the conditions related to flotation.

4.3.1 UV Analysis of Collector Adsorption on Sulphide Minerals

1. Adsorption of Dithiocarbamate on Jamesonite

In order to characterize the adsorption species on mineral surface, DDTC is oxidized into the dimmer by adding definite H_2O_2 into the DDTC solution, which then is extracted by cyclohexane to determine its UV spectrum. As seen from the UV spectrum in Fig. 4.33, there are three UV absorbance peaks at 230 nm, 261 nm, 280 nm respectively. The maximum absorbance peak is at 230 nm, the next peak is at 260 nm, and the weak peak is at 280 nm. The peak at 230 nm can serve as a characteristic absorbance peak, and the peak at 260nm results from absorbance overlapping of diethyl dithiocarbamate and its dimmer.

The UV spectra of the cyclohexane solution extracted from jamesonite acted by DDTC solution, shown in Fig. 4.34, indicates that the adsorption of diethyl dithiocarbamate on the surface of jamesonite is almost the same at pH = 4 and 7, but decreases with the increasing of pH in the base solution. Because the UV spectra in Fig. 4.33 and Fig. 4.34 are similar with peaks at 230 nm and 260 nm, it indicates that the hydrophobic species on jamesonite should be the mixture of diethyl dithiocarbamate and its dimmer.

2. Adsorption of Dithiocarbamate and Xanthate on Marmatite

Figure 4.35 presents the UV spectra of the cyclohexane solution extracted from marmatite after acted by DDTC. It can be seen that there exist three UV absorption peaks at 230 nm, 260 nm and 280 nm. The hydrophobic entity may be the mixture of diethyl dithiocarbamate and its dimmer on marmatite.

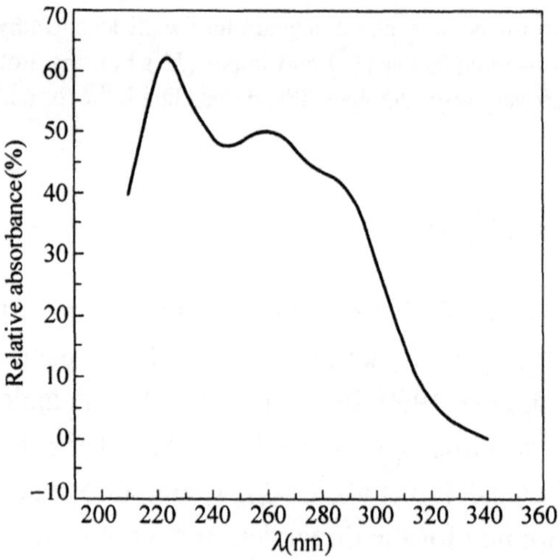

Figure 4.33 UV spectrum of the cyclohexane solution extracted from the oxidized DDTC solution by H_2O_2

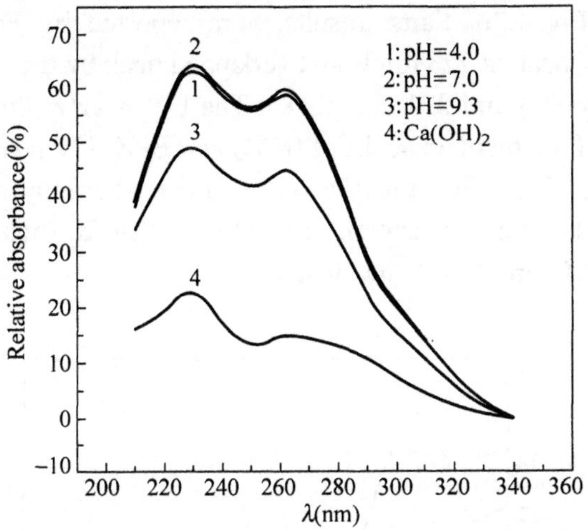

Figure 4.34 UV spectra of the cyclohexane solution extracted from jamesonite acted by DDTC solution

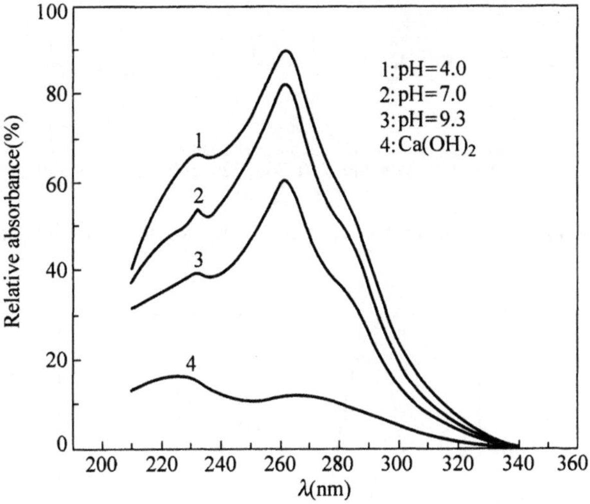

Figure 4.35 UV spectra of the cyclohexane solution extracted from marmatite acted by DDTC solution

Figure 4.36 shows the UV absorbance spectra of diethyl dixanthogen $(EX)_2$, diethyl monothiocarbonate (MTC^-) and diethyl thiocarbonate (EPX). The UV adsorption peaks of $(EX)_2$ lie in about 238 nm (strong) and 286 nm (weak). The ratio of adsorption intensity of two peaks is 2. UV peak of MTC lies in about 225 nm (strong), EPX lies in about 348 nm (strong) and 221 nm (weak).

After ethyl xanthate interacts with marmatite, the UV spectra of the hydrophobic species on marmatite surface extracted by cyclohexane are shown in Fig. 4.37. There are three UV peaks, lying in 228 nm, 263 nm and 286 nm respectively. It is very similar to Fig. 4.36. Same results were reported by Fornasiero (1995). Therefore, the UV peak at 228 nm is an overlapped peak by these peaks of 238 nm $(EX)_2$, 225 nm (MTC) and 221 nm (EPX). The UV peak at 263nm results from the overlapping of the mediate peaks of $(EX)_2$ and EPX. UV peak at 286nm is just the peak of EX_2 in Fig. 4.36. It may be derived that after ethyl xanthate interacts with marmatite, the main adsorption product on its surface may be dixanthogen with a little of EPX^- and MTC^- metal salt.

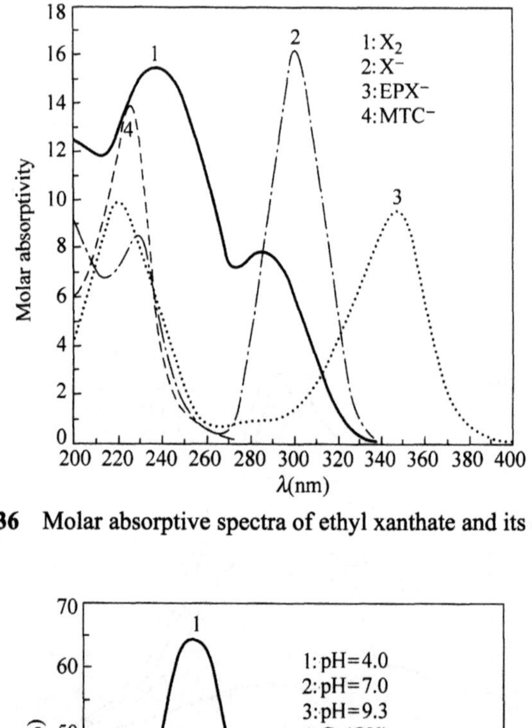

Figure 4.36 Molar absorptive spectra of ethyl xanthate and its derivatives

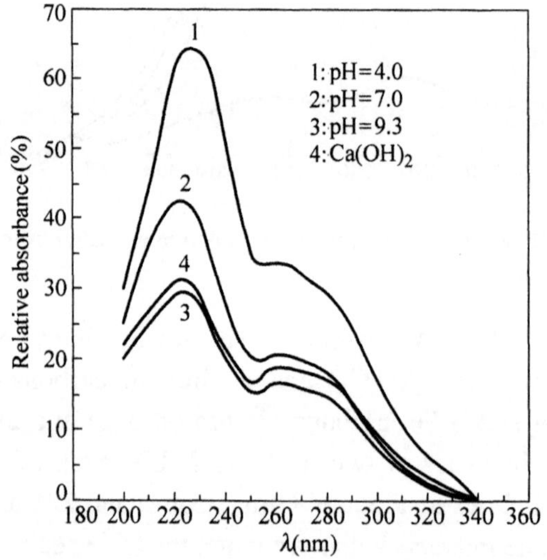

Figure 4.37 UV spectra of the cyclohexane solution extracted from marmatite acted by EX solution

From the flotation results in Fig. 4.21 and the voltammogram in Fig. 4.22, it is derived that the hydrophobic entity of collector on marmatite is mainly of disulphide of xanthate and dithiocarbamate, which is further confirmed by the above UV analysis. However, the UV analysis also suggested the coexistence of collector salts. We propose that the initial oxidation products of xanthate and dithiocarbamate on marmatite should be mainly of disulphide. At higher potential, the adsorbed disulphide may be decomposed and react with surface zinc species to form some parts of collector salts as in the following reactions:

$$Zn(OH)_2 + D_2 + 2H^+ + 2e^- = ZnD_2 + 2H_2O \tag{4-49a}$$

$$\begin{cases} E^0 = 0.8337(V) \\ E_h = 0.8337 - 0.059pH = 0.2921(V) \end{cases} \tag{4-49b}$$

$$Zn(OH)_2 + (EX)_2 + 2H^+ + 2e^- = Zn(EX)_2 + 2H_2O \tag{4-50a}$$

$$\begin{cases} E^0 = 0.5563(V) \\ E_h = 0.5563 - 0.059pH = 0.01468(V) \end{cases} \tag{4-50b}$$

4.3.2 FTIR Analysis of Adsorption of Thio-Collectors on Sulphide Minerals

Persson et al. (1991) used diffuse reflection infrared Fourier transform (DRIFT) spectroscopy to study the interactions between galena, pyrite sphalerite and ethyl xanthate. They provided the evidence that the DRIFT spectrum of oxidized galena treated with an aqueous solution of potassium ethyl xanthate is practically identical with that of solid lead (II) ethyl xanthate, which can be formed as the only detectable surface species on oxidized galena. Dialkyl dixanthogen is formed as the only surface species in the reaction between oxidized pyrite and aqueous solution of potassium alkyl xanthate.

Leppinen (1990) used FTIR-ATR techniques to study the adsorption of ethyl xanthate on pyrite, pyrrhotite and chalcopyrite. He found that the FTIR signals of the xanthate species on pyrite occurred at approximately the same position as those of bulk dixanthogen together with a surface product, the signal of which coincided well with that of ferric ethyl xanthate. The adsorption product of KEX on pyrrhotite was much the same as that on pyrite but no signal of iron xanthate species was observed on the surface of pyrrhotite. On chalcopyrite the signals of diethyl dixanthogen and a metal xanthate compound closely resembling copper (I) xanthate were observed. These results led Leppinen (1990) to conclude that the main adsorption product of ethyl xanthate on pyrite and pyrrhotite is diethyl dixanthogen. An iron xanthate surface compound is also observed on pyrite at monolayer coverage. A mixture of copper xanthate compound and diethyl dixanthogen occurs on chalcopyrite. Leppinen et al. (1989) also studied the adsorption of ethyl xanthate on chalcocite and galena and examined the influence of pulp potential. He reported that on chalcopyrite the xanthate adsorption at pH = 9.2 initially resulted in the formation of dixanthogen at potential above the reversible potential for the xanthate/dixanthogen couples. Copper ethyl xanthogen is at high potential.

1. Adsorption of Ethyl Xanthate on Pyrrhotite

The FTIR reflection spectra of ethyl xanthate, and iron ethyl xanthate were reported to show the following characteristic absorption bands of ethyl xanthate: the stretching vibration bands of the C—O—C at 1100 – 1172 cm^{-1} and C═S at 1049 cm^{-1} and 1008 cm^{-1}. When dixanthogen was formed, the stretching vibration bands changed. The characteristic absorption bands are 1260 cm^{-1}, 1240 cm^{-1}, 1020 cm^{-1} and 1105 cm^{-1}. When iron ethyl xanthate was formed, the stretching vibration band of C═S shifted to 1029 cm^{-1} and 1005 cm^{-1} (Mielezarski, 1997; Leppinen, 1990).

FTIR reflection spectra of ethyl xanthate adsorption on pyrrhotite are shown in Fig. 4.38. It can be seen that the characteristic absorption bands of diethyl dixanthogen which at 1023 cm^{-1}, 1242 cm^{-1}, and 1263 cm^{-1} appear on the surface of pyrrhotite, indicate that the dominating hydrophobic species on pyrrhotite surface is dixanthogen. Further, the effect of pulp potential on the adsorption of ethyl xanthate on pyrrhotite was determined and the results are presented in Fig. 4.39. It follows that at pH = 8.8 the ethyl xanthate adsorption on pyrrhotite is mainly of dixanthogen independent of potential in the range of 140 – 250 mV due to the occurrence of almost the same dixanthogen characteristic band. However, the intensity of the IR signals changes at various potential. It demonstrates the intensity of the IR signals of the characteristic dixanthogen peaks and hence

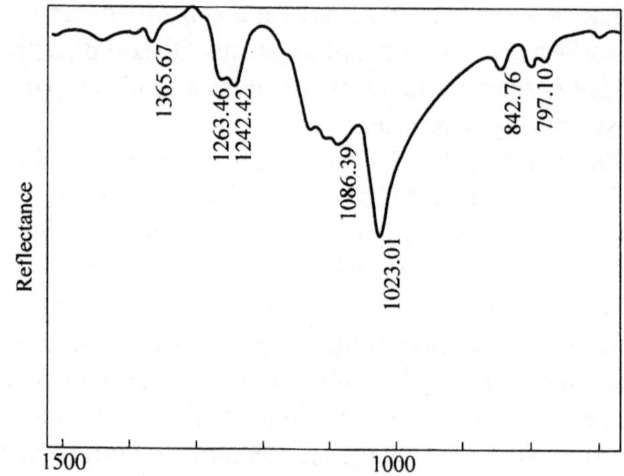

Figure 4.38 FTIR reflection spectra of ethyl xanthate adsorption on pyrrhotite (pH = 7.0, KEX: 5×10^{-3} mol/L; E_h = 297 mV)

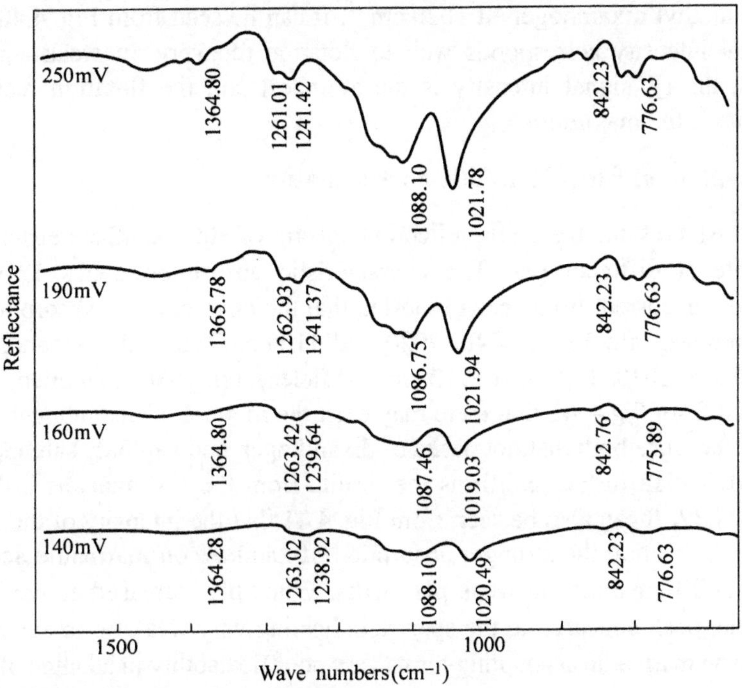

Figure 4.39 FTIR reflection spectra of ethyl xanthate adsorption on pyrrhotite at different pulp potential (pH = 8.8, KEX: 5×10^{-3} mol/L)

dixanthogen adsorption on pyrrhotite decreases with the decreasing potential from 250 to 140 mV. At pH = 8.8, flotation recovery and dixanthogen adsorption on pyrrhotite decrease with the decrease of pulp potential from 250 to 140 mV as shown in Fig. 4.26. The effect of pH on the FTIR signal intensity of xanthate adsorption on pyrrhotite and pyrrhotite flotation recovery is plotted in Fig. 4.40. The signal intensity has been measured based on the characteristic absorption

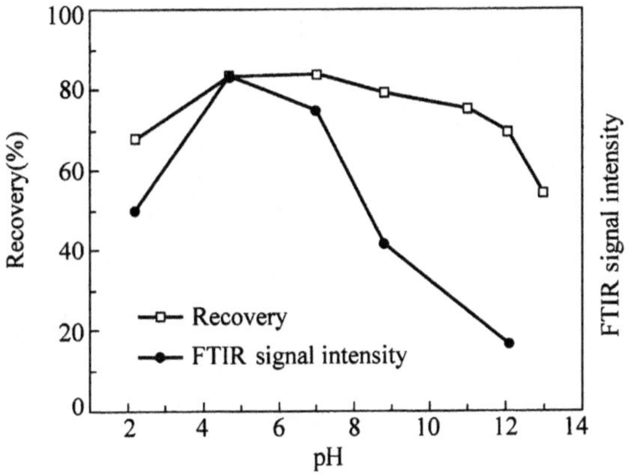

Figure 4.40 The effect of pH on FTIR signal intensity of pyrrhotite adsorption xanthate and on pyrrhotite flotation recovery

peak of diethyl dixanthogen at 1020 cm^{-1}. It can be seen from Fig. 4.40 that the IR signal intensity corresponds well to flotation recovery. In weak acidic pulp solution, the IR signal intensity is the strongest and the flotation recovery of pyrrhotite is the maximum.

2. Adsorption of Ethyl Xanthate on Marmatite

Figure 4.41 presents the FTIR reflection spectra of ethyl xanthate adsorption on marmatite at different pH. The characteristic absorption peaks 1210, 1108, 1025 cm^{-1} occurred. It has been reported that the characteristic absorption bands of dixanthogen are 1260, 1240, 1020 and 1105 cm^{-1} and those bands of zinc xanthate are: 1030, 1125 and 1212 cm^{-1} (Mielezarski, 1986; Leppinen, 1990). It is derived from Fig.4. 41 that there may exist the mixture of dixanthogen and zinc xanthate because both distinct peaks of dixanthogen and xanthate salt appeares in Fig. 4.41, which further confirms the results from the UV analysis in Fig. 4.36 and Fig. 4.37. It can also be seen from Fig. 4.41 that the intensity of the IR peaks is strong indicating the stronger adsorption of xanthate on marmatite accounting for its good floatability in weak pH media. When pH increased above 7, only a very weak peak appeared in the spectra indicating very weak or no adsorption of xanthate on marmatite accounting for its very poor floatability in alkaline pH media.

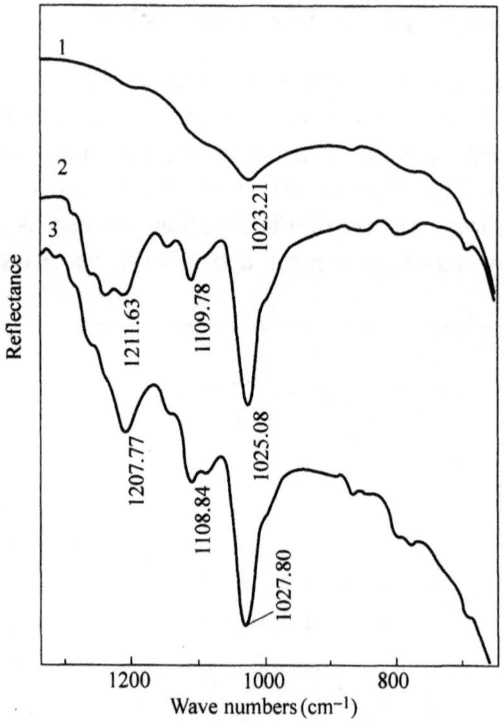

Figure 4.41 FTIR reflection spectra of ethyl xanthate adsorption on marmatite (1: pH = 7.0; 2: pH = 4.7; 3: pH = 2.2; KEX: 5×10^{-3} mol/L)

3. Adsorption of Ethyl Xanthate on Jamesonite

It was reported that when lead ethyl xanthate was formed, the stretching vibration band of C═S shifted to 1018 and 996 cm^{-1}. The stretching vibration band of C—O—C shifted to 1112 and 1207 cm^{-1} (Leppinen, 1990).

The FTIR reflection spectra of ethyl xanthate adsorption on jamesonite are shown in Fig. 4.42. It can be seen that the characteristic absorption bands of lead ethyl xanthate at 1020, 1112 and 1206 cm^{-1} appeared on the surface of jamesonite, indicating the primary hydrophobic species on jamesonite surface to be lead ethyl xanthate. It is possible that antimony ethyl xanthate was formed on jamesonite surface simultaneity like lead ethyl xanthate.

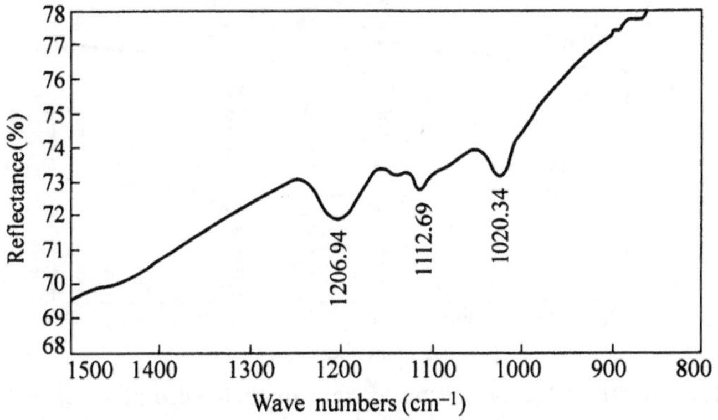

Figure 4.42 FTIR reflection spectra of jamesonite adsorption ethyl xanthate (pH = 7.0; KEX: 5×10^{-3} mol/L; E_h = 307 mV)

The effect of pulp potential on the adsorption of ethyl xanthate on jamesonite was further examined and the results are presented in Fig. 4.43. At pH = 4.7 the ethyl xanthate adsorption on jamesonite was mainly of metal ethyl xanthate independent of potential in the range of 280 – 385 mV due to the occurrence of the same metal ethyl xanthate characteristic band. However, the intensity of the IR signals was changed as the pulp potential varied. The intensity of the IR signals of the characteristic reaches peaks of metal ethyl xanthate and hence the metal ethyl xanthate adsorption on jamesonite decreases with the descending of the pulp potential from 385 – 280 mV. At pH = 4.7, the flotation recovery and metal ethyl xanthate adsorption on jamesonite decrease with the decrease of pulp potential from 385 to 280 mV as shown in Fig. 4.12. The effect of pH on FTIR signal intensity of xanthate adsorption on jamesonite, which was measured near 1206 cm^{-1}, and on flotation recovery is plotted in Fig. 4.44. It can be seen that the IR signal intensity corresponds reasonably well to the flotation recovery. In acidic pH media, the IR signal intensity was the strongest and jamesonite showed best flotation. The IR signal intensity gradually decreases and the flotation of jamesonite exhibits good flotation in weak alkaline pH media and decreases in strong alkaline pH media.

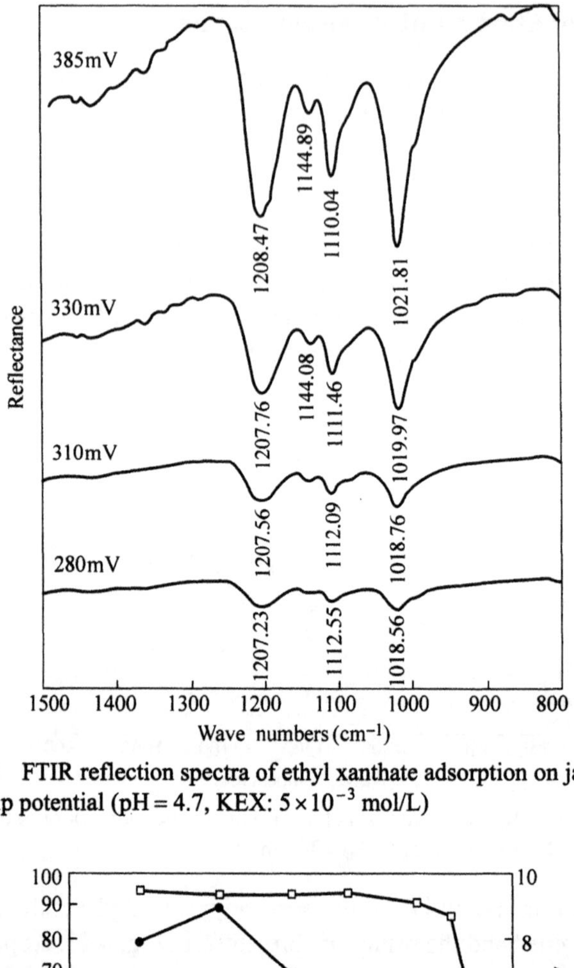

Figure 4.43 FTIR reflection spectra of ethyl xanthate adsorption on jamesonite at different pulp potential (pH = 4.7, KEX: 5×10^{-3} mol/L)

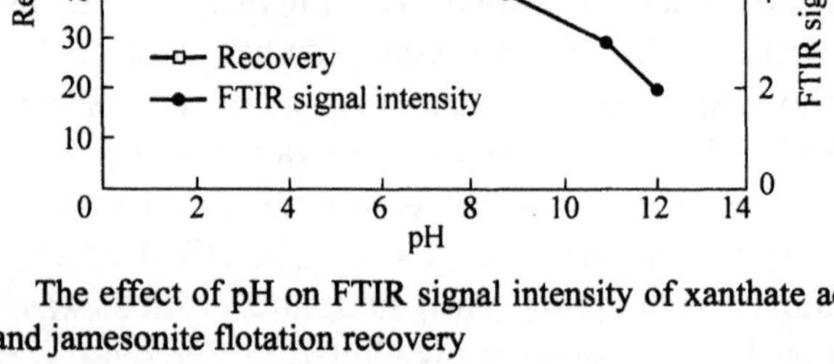

Figure 4.44 The effect of pH on FTIR signal intensity of xanthate adsorption on jamesonite and jamesonite flotation recovery

4. Adsorption of Dithiocarbamate on Pyrrhotite

The FTIR reflection spectra of DDTC (solid) and FeD_3 are shown in Fig. 4.45. Only at the 700 – 2000 cm^{-1} region most of the important functional groups vibrations of diethyl dithiocarbamate are observed. From Fig. 4.44, it can be seen that many reflection spectra peaks appeared. When C═S was connected to atom N, there were a lot of stronger reflection spectra appearing in a range of 1570 – 1395 cm^{-1}, 1420 – 1260 cm^{-1} and 1140 – 940 cm^{-1}. The characteristic absorption bands of diethyl dithiocarbamate (DDTC) are 1475, 1455, 1412, 1375, 1260, 1202 and 984 cm^{-1}. The characteristic absorption bands of iron diethyl dithiocarbamate (FeD_3) are: 1491, 1454, 1354, 1270, 1208, 1140 and 995 cm^{-1}. It was reported that the characteristic absorption bands of thiouram disulphide are 1464, 1424, 1350, 1270, 1200 and 1140 cm^{-1} (Pouchert, 1970).

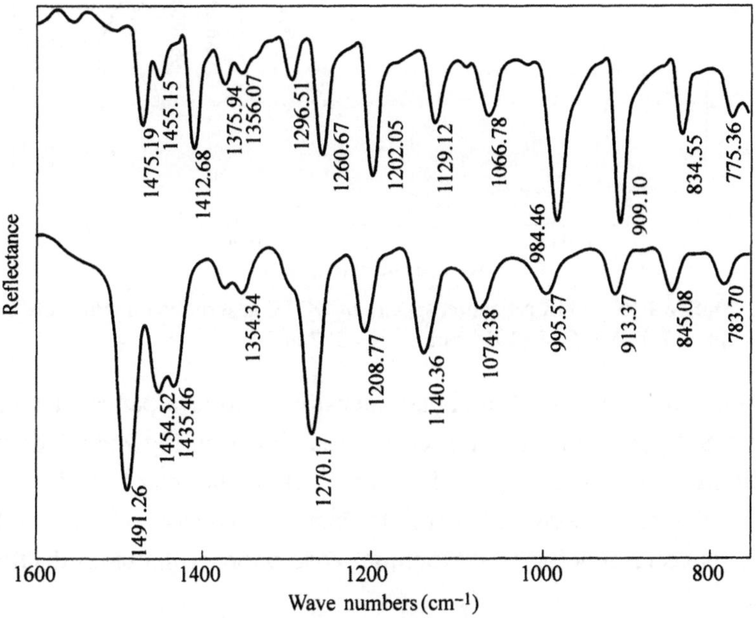

Figure 4.45 FTIR spectra of DDTC and FeD_3

FTIR reflection spectra of diethyl dithiocarbamate adsorption on pyrrhotite at pulp pH = 7.0 are demonstrated in Fig. 4.46. From Fig. 4.46, it can be seen that the characteristic absorption bands of dimmer of diethyl dithiocarbamate at 1468, 1425, 1350, 1269, 1201, 1139, 1064, 1005 and 968 cm^{-1} appear on the surface of pyrrhotite, indicating the dominating hydrophobic species on pyrrhotite surface to be disulphide of dithiocarbamate. Further, the effect of pulp potential on the adsorption of dithiocarbamate on pyrrhotite was examined and the results are presented in Fig. 4.47. It follows that at pH = 8.8 the dithiocarbamate adsorption on pyrrhotite is mainly of disulphide independent of potential in the range of 297 – 687 mV due to the occurrence of almost the same disulphide characteristic band. However, the intensity of the IR signals changes at various potential values. It demonstrates the intensity of the IR signals of the characteristic peaks of thiouram disulphide and hence its adsorption on pyrrhotite decreases with the increase of the potential from 297 – 687 mV. At pH = 8.8, flotation recovery and thiouram disulphide adsorption on pyrrhotite are decreased with the increase of pulp potential from 297 to 687 mV.

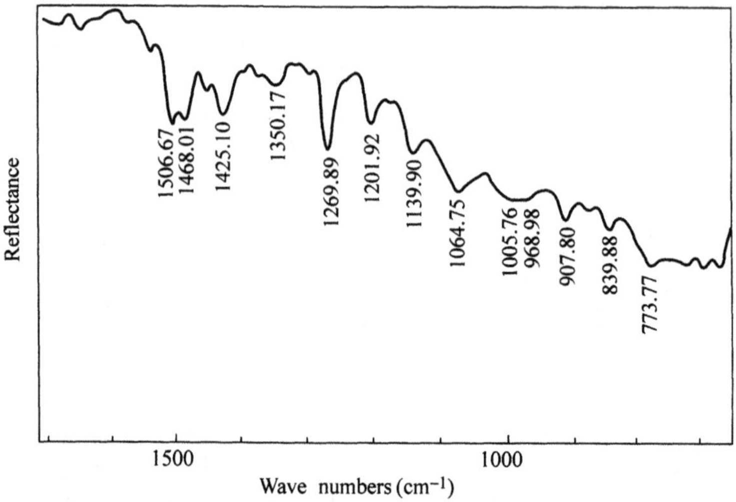

Figure 4.46 FTIR reflection spectra of DDTC adsorption on pyrrhotite (pH = 7.0; DDTC: 5×10^{-3} mol/L; $E_h = 377$ mV)

The effect of pH on the FTIR signal intensity of dithiocarbamate adsorption on pyrrhotite and pyrrhotite flotation recovery is plotted in Fig. 4.48. The intensity of the IR signal is based on peak 1270 cm^{-1}. It can be seen from Fig. 4.47 that the IR signal intensity corresponds well to flotation recovery. The optimal flotation pH range is corresponding to the optimal adsorption of thiouram disulphide on pyrrhotite.

5. Adsorption of Diethyl Dithiocarbamate on Jamesonite

The FTIR reflection spectra of lead diethyl dithiocarbamate was reported to show the characteristic absorption bands 900, 981, 1088, 1138, 1200, 1248, 1320, 1351, 1398 and 1460 cm^{-1} (Pouchert, 1970). The FTIR reflection spectra of sodium diethyl dithiocarbamate adsorption on jamesonite at pH = 7.0 are shown in Fig. 4.49. It can be seen that the characteristic absorption surface is lead diethyl dithiocarbamate. Further, the effect of pulp potential on the adsorption of diethyl dithiocarbamate on jamesonite was examined and the results are shown in Fig. 4.50. At pH = 4.7 the diethyl dithiocarbamate adsorption on jamesonite is mainly of lead diethyl dithiocarbamate independent of potential in the range of 485 – 680 mV due to the occurrence of almost the same lead diethyl dithiocarbamate characteristic band. However, the intensity of the IR signals changes at various potential values. It demonstrates the intensity of the IR signals of the characteristic peak lead diethyl

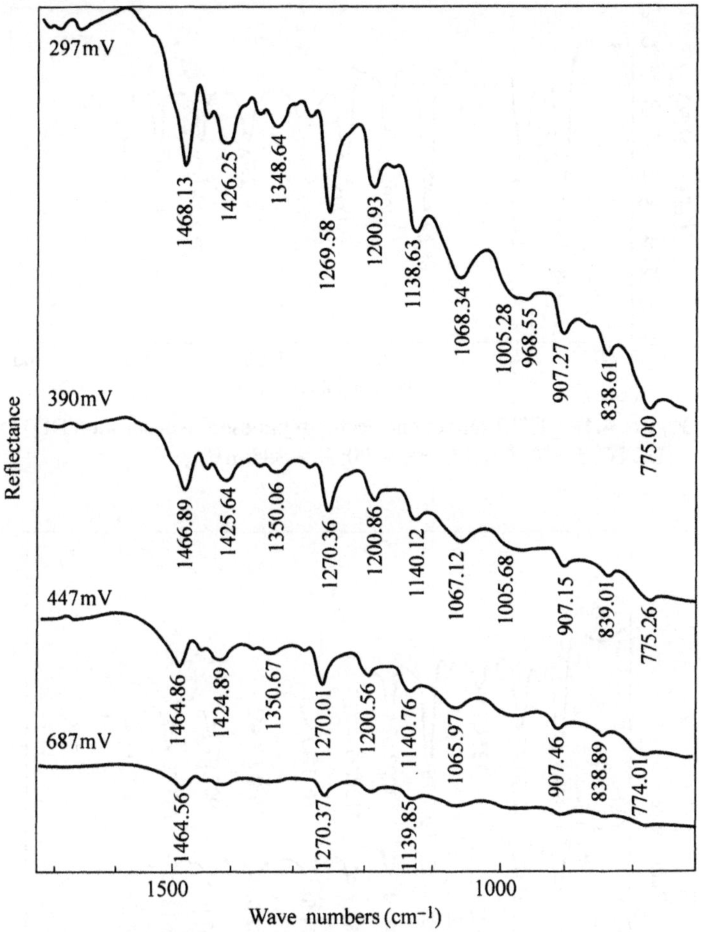

Figure 4.47 FTIR reflection spectra of DDTC adsorption on pyrrhotite at different pulp potential (pH = 8.8; DDTC: 5×10^{-3} mol/L)

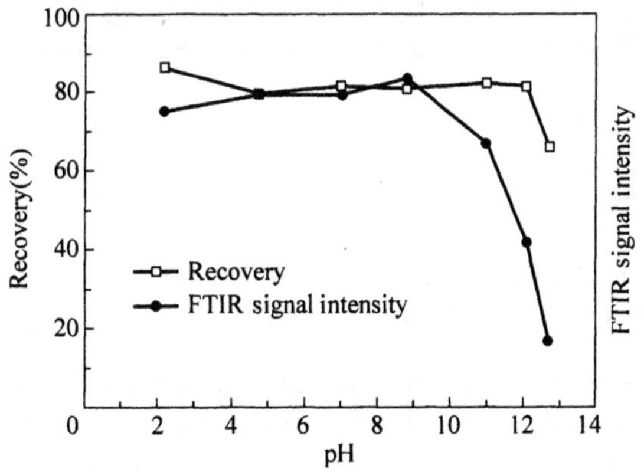

Figure 4.48 The effect of pH on the FTIR signal intensity of DDTC adsorption on pyrrhotite and pyrrhotite flotation recovery

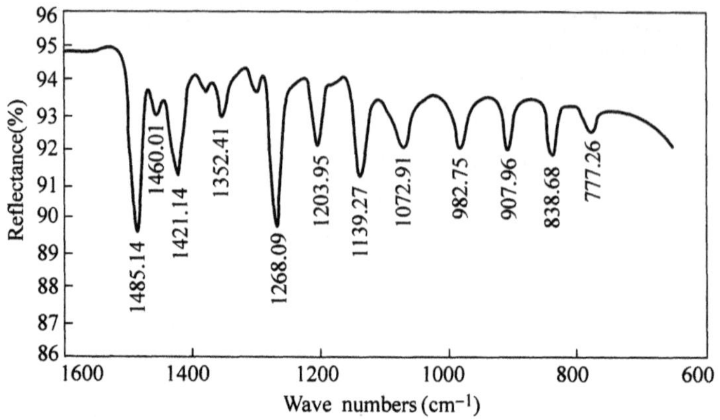

Figure 4.49 FTIR reflection spectra of jamesonite adsorption DDTC (DDTC: 5×10^{-3} mol/L; pH = 7.0; E_h = 385 mV)

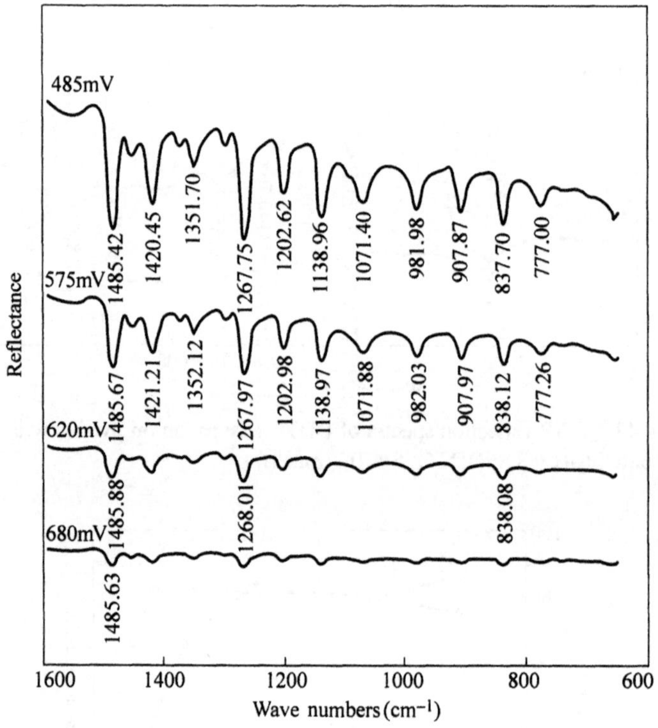

Figure 4.50 FTIR reflection spectra of DDTC adsorption on jamesonite at different pulp potential (pH = 4.7; DDTC: 5×10^{-3} mol/L)

dithiocarbamate and hence lead diethyl dithiocarbamate adsorption on jamesonite decrease with the decrease of the pulp potential from 485 – 680 mV. These results confirm the flotation and voltammograms results that the primary hydrophobic entity on jamesonite is collector salts.

4.3.3 XPS Analysis of Collector Adsorption on Sulphide Minerals

The X-ray photoelectron spectroscopy (XPS) identification of the products of xanthate sorption at the surface of galena by Laajalehto et al. (1993) showes that lead xanthate in the molecular form is the adsorbed entity. The distribution of xanthate at the surface is irregular and even at low surface coverage the formation of three dimensional aggregates of lead xanthate occurs.

Cases et al. (1990, 1991) studied the influence of the grinding media on the adsorption of xanthate on galena and pyrite using XPS and FTIR. They found that only lead xanthate, either monocoordinated or in the three dimensional form was formed on galena surface after dry grinding and that however, wet grinding led to heterogeneous adsorbed layer formed with monocoordinated lead xanthate and dixanthogen at low concentration; stoichiometric or non-stoichiometric lead xanthate, dixanthogen, and the dimmer of the monothiocarbonate for higher concentration. They also find that only dixanthogen is observed on pyrite surface after using stainless steel rods. Whereas, ferric iron xanthate and physically absorbed xanthate ions are formed when iron rods are utilized.

The adsorption of xanthate on jamesonite is further determined by using XPS. Figs. 4.51 and 4.52 are respectively the XPS of $Pb(4f^5)$ on jamesonite in the presence and absence of ethyl xanthate at pH = 8.8. It has been found that the electron binding energy of $Pb(4f^5)$ on jamesonite surface is 138.07 eV, which shifted to $Pb(4f^5)$ 137.47 eV with the difference $\Delta E_{Pb(4f^5)} = 138.07 - 137.47 = 0.6$ eV in the presence of ethyl xanthate. The chemical shifting of electron binding energy of $Pb(4f^5)$ suggested the adsorption of ethyl xanthate on jamesonite and the formation of lead xanthate, which further confirmed the results revealed by the UV and FTIR analyses.

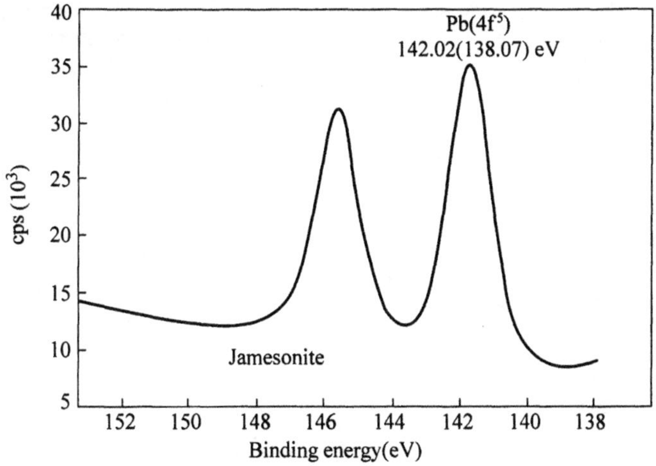

Figure 4.51 The XPS of $Pb(4f^5)$ of jamesonite at pH = 8.8

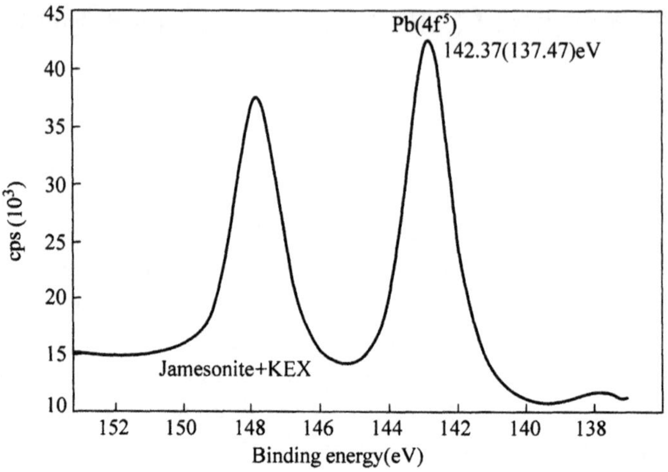

Figure 4.52 The XPS of $Pb(4f^5)$ on jamesonite surface in the presence of ethyl xanthate at pH = 8.8

Figures 4.53 and 4.54 are respectively the XPS of $Sb(3d^5)$ on jamesonite in the presence and absence of ethyl xanthate at pH = 8.8. It is obvious that the electron binding energy of $Sb(3d^5)$ on jamesonite surface is 529.83 eV, which shifted to $Sb(3d^5)$ 529.15 eV with the difference $\Delta E_{Sb(3d^5)} = 529.83 - 529.15 = 0.68$ eV in the presence of ethyl xanthate. The chemical shifting of electron binding energy of $Pb(3d^5)$ suggested the adsorption of ethyl xanthate on jamesonite and the formation of antimony xanthate. It has been reported that the electron binding energy of $Sb(3d^5)$ are, respectively, 529.50 eV in Sb_2S_3 and 529, 20 eV in Sb_2S_5 (Wang Jianqi, 1992). And thus, it may be derived that the antimony xanthate salt with high valence antimony is formed on jamesonite.

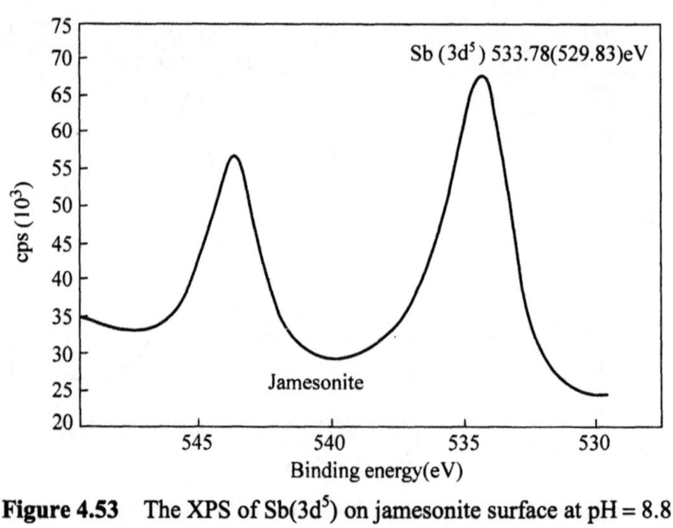

Figure 4.53 The XPS of $Sb(3d^5)$ on jamesonite surface at pH = 8.8

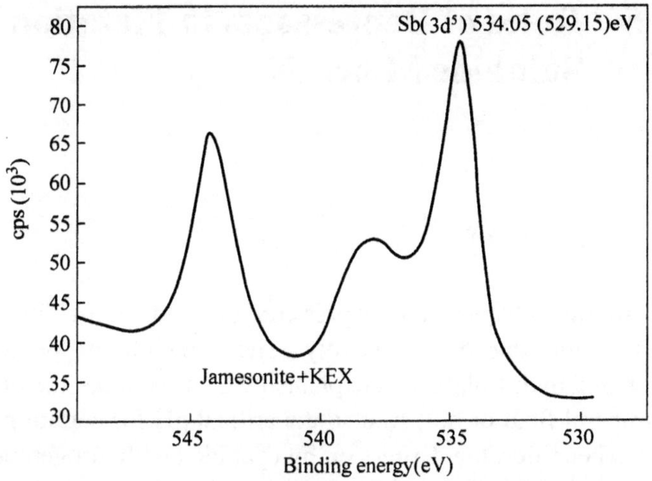

Figure 4.54 The XPS of Pb(3f^5) on jamesonite surface in the presence of ethyl xanthate at pH = 8.8

Because of the strong hydrolysis property of antimony salt, it is very difficult to prepare antimony xanthate salt to obtain its IR spectrums. Therefore, the formation of antimony xanthate is difficult to be identified by the UV and FTIR analysis, which has been determined by using XPS. Finally, it can be concluded that the interaction mechanism between ethyl xanthate and jamesonite are attributed to the formation of lead and antimony xanthate on the surface in the light of flotation results, voltammogram measurement, UV and FTIR as well as XPS analyses.

Chapter 5 Roles of Depressants in Flotation of Sulphide Minerals

Abstract In this chapter, the depression mechanism of five kinds of depressants is introduced respectively. The principle of depression by hydroxyl ion and hydrosulphide is explained which regulates the pH to make the given mineral float or not. And so the critical pH for certain minerals is determined. Thereafter, the depression by cyanide and hydrogen peroxide is narrated respectively which are that for cyanide the formation of metal cyanide complex results in depression of minerals while for hydrogen peroxide the decomposition of xanthate salts gives rise to the inhibitation of flotation. Lastly, the depression by the thio-organic such as polyhydroxyl and poly carboxylic xanthate is accounted for in detail including the flotation behavior, effect of pulp potential, adsorption mechanism and structure-property relation.

Keywords depressant; hydroxyl ion; hydrosulphide ion; cyanide; hydrogen peroxide; thio-organic

Modifiers in the flotation of sulphide minerals mainly include depressants and activators. A depressant is defined as a reagent which inhibits the adsorption of a collector on a given mineral or adsorbed on the mineral to make the surface hydrophilic, and includes inorganic depressants such as lime, sodium cyanide, sulphur dioxide, zinc sulphate, sodium sulphide etc., and organic depressants such as sulfhydryl acetic acid, polyacrylamide polymers containing various functional groups etc. In this chapter, roles of depressants in the flotation sulphide minerals will be discussed and some new organic depressants will be introduced.

5.1 Electrochemical Depression by Hydroxyl Ion

From the point of view of electrochemistry of flotation, a depressant is, however, defined as a reagent by the addition of which the oxidation of the mineral surface occurs at lower potential than collector oxidation or formation of metal collector salt which may be also decomposed under the conditions given in the discussions which follow. Under these conditions, the mixed potential model becomes one of mineral oxidation and oxygen reduction, the oxidation of the thio collector or the formation of the metal collector is suppressed, and the mineral will remain hydrophilic. Thus, hydroxide, sulphide, cyanide and other redox agents are usually used as depressants of sulphide minerals.

5.1.1 Depression of Galena and Pyrite

The underlying principle of depression by hydroxyl ion is that for each concentration of collector there is a pH below which any given mineral will float and above which it will not float. This is the so-called critical pH which can be determined according to the relative extent of oxidation of the mineral surface and collector or the formation of the metal collector salt.

For the case that the hydrophobic entity is disulphide, the mineral will be depressed when the reaction of the type (2-3) or (2-4) occurs before the reaction (1-3). Thus for the pyrite /diethyl dithiophosphate (DTP) system, pyrite will be depressed if the oxidation reaction

$$FeS_2 + 11H_2O = Fe(OH)_3 + 2SO_4^{2-} + 19H^+ + 15e^- \tag{5-1a}$$

$$\begin{cases} E^0 = 0.402(\mathrm{V}) \\ E_{h1} = 0.402 - 0.07473\mathrm{pH} + 7.87\times10^{-3}\lg[SO_4^{2-}] \end{cases} \tag{5-1b}$$

takes place prior to the oxidation reaction

$$2DTP^- = (DTP)_2 + 2e^- \tag{5-2a}$$

$$E_{h2} = 0.252 - 0.059\lg[DTP^-] \tag{5-2b}$$

The critical condition is

$$E_{h1} \leqslant E_{h2} \tag{5-3a}$$

Assuming that the sulphur could be oxidized to sulphate at overpotential 0.5 V and the concentration of sulphate is 10^{-6} mol/L, then Eqs. (5-1b), (5-2b) and (5-3a) become

$$0.8548 - 0.07473\mathrm{pH} \leqslant 0.252 - 0.059\lg[DTP^-] \tag{5-3b}$$

Thus, the critical pH above which pyrite will not float is

$$\mathrm{pH}_{Py}^{d} \geqslant \frac{0.6028 + 0.059\lg[DTP^-]}{0.07473} \tag{5-4}$$

In the case of the galena/diethyl/dithiophosphate system, if the reaction Eq. (2-22a) occurs before the following reaction

$$2PbS + 4DTP^- + 3H_2O = 2Pb(DTP)_2 + S_2O_3^{2-} + 6H^+ + 8e^- \tag{5-5a}$$

$$\begin{cases} E^0 = 0.245(\text{V}) \\ E_h = 0.245 - 0.04425\text{pH} - 0.0295\lg[\text{DTP}^-] + 0.007375\lg[\text{S}_2\text{O}_3^{2-}] \end{cases} \tag{5-5b}$$

where E^0 is calculated in the light of the standard free energy change of reaction (5-5a) and the solubility product of $Pb(DTP)_2$ is 7.5×10^{-12} from reference (Karkovsky, 1957).

The critical pH condition determined by the reaction Eqs. (2-22) and (5-5) is

$$\text{pH}_{\text{Ga}}^{\text{d}} \geqslant \frac{0.366 + 0.0295\lg[\text{DTP}^-]}{0.02945} = 12.43 + \lg[\text{DTP}^-] \tag{5-6}$$

If lead dithiophosphate is decomposed by the following reaction

$$\text{Pb(DTP)}_2 + 2\text{OH}^- = \text{HPbO}_2^- + 2\text{DTP}^- + \text{H}^+ \tag{5-7}$$

$$K = 7.73\times10^{-12}$$

Then, galena will be depressed. The critical pH determined by decomposition reaction is defined by

$$\text{pH}_{\text{Ga}}^{\text{d}} \geqslant 11.05 + \frac{2}{3}\lg[\text{DTP}^-] \tag{5-8}$$

Similar results are obtained for pyrite/ethyl xanthate from the Eqs. (5-1) and (1-3).

$$\text{pH}_{\text{Py}}^{\text{d}} \geqslant \frac{0.9418 + 0.059\lg[\text{EX}^-]}{0.07473} = 12.24 + 0.79\lg[\text{EX}^-] \tag{5-9}$$

For the galena/ethyl xanthate system critical pH determined by Eqs. (2-22) and (4-12b) is

$$\text{pH}_{\text{Ga}}^{\text{d}} \geqslant 15.2 + \lg[\text{EX}^-] \tag{5-10}$$

The critical pH determined by the decomposition of lead xanthate is

$$\text{Pb(EX)}_2 + 2\text{OH}^- = \text{HPbO}_2^- + 2\text{EX}^- + \text{H}^+ \tag{5-11a}$$

$$\begin{cases} K = 2.1\times10^{-17} \\ \text{pH} = 12.89 + \dfrac{2}{3}\lg[\text{EX}^-] \end{cases} \tag{5-11b}$$

Therefore, the critical pH of hydroxyl depression of pyrite and galena could be calculated using Eqs. (5-4) to (5-8) or Eqs. (5-9) to (5-11). The results are given in Fig. 5.1 for the diethyl dithiophosphate system and in Fig. 5.2 for the ethyl xanthate system. Also shown in the same figures are the so-called contact curves (solid lines) from Sutherland and Wark (1955). The dashed lines are the calculated theoretical contact curves. It can be seen from Fig. 5.1 and Fig. 5.2 that the calculated collector concentration critical pH curves are in reasonable agreement with those results obtained form contact curves or flotation data. It shows that the hydroxyl depression of pyrite is attributed to its greater ease of oxidation to iron hydroxide and oxy-sulphur species which renders pyrite surface hydrophilic than dithiophosphate oxidation to dithiophosphatogen.

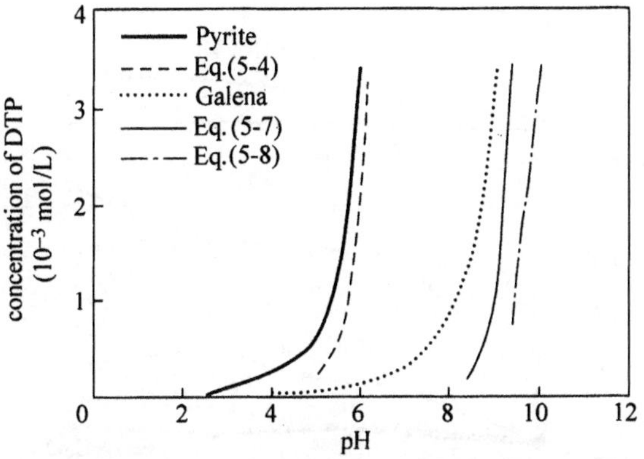

Figure 5.1 Relationship between concentration of sodium diethyl dithiophosphate and critical pH; dashed line is the result calculated based on Eqs. (5-4), (5-7) or (5-8); solid line is the data from Sutherland and Wark (1955)

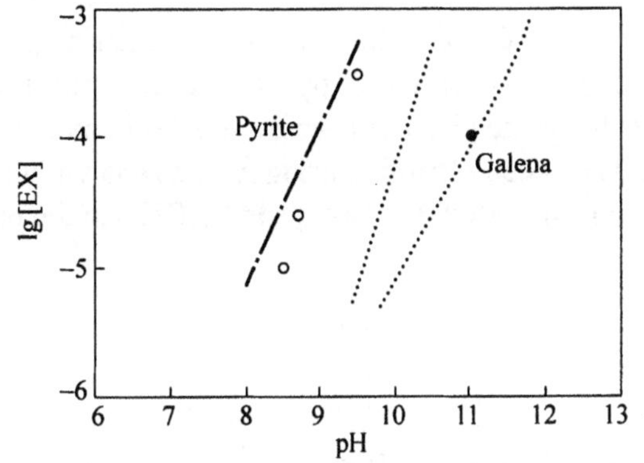

Figure 5.2 Relationship between concentration (mol/L) of potassium ethyl xanthate and critical pH; dashed line is the calculated results and symbols are the results from Fuerstenau et al. (1968)

Janetski et al. (1977) used voltammetry to study the oxidation of pyrite electrode in solution at different pH in the absence and presence of ethyl xanthate to demonstrate that the oxidation of pyrite itself increases as the pH is increased. At high pH condition, the oxidation of pyrite occurs at a potential cathodic to that for xanthate oxidation and hence, only the mineral will be oxidized at the mixed potential and flotation will be depressed.

Lime is a special kind of hydroxyl depressant and usually used for depressing pyrite in plant, which has stronger depression on pyrite than other alkaline depressants at the same pH condition (Fuerstenau, et al., 1968; Hu et al., 1995). Figure 5.3 shows that at pH = 12 modified by CaO and NaOH, galena is fairly floatable and pyrite is depressed. CaO is a stronger depressant than NaOH.

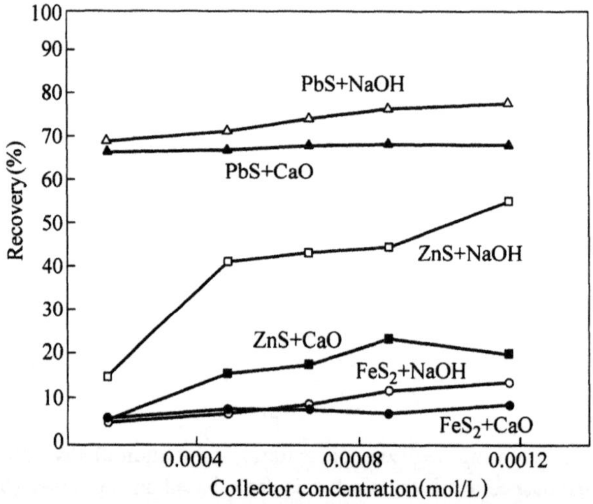

Figure 5.3 Flotation recovery of galena and pyrite as a function of butyl xanthate concentration by using lime and sodium hydroxide at pH = 12

Figure 5.4 shows the anodic oxidation of pyrite at different pH modified by CaO and NaOH. Obviously, the oxidation of pyrite is more rapid in CaO solution than in NaOH solution at the same pH. The calcium species in oxidized pyrite surface in the presence of CaO has been identified using XPS as shown in Fig. 5.5. The pyrite surface is oxidized to form $Ca(OH)_2$, $CaSO_4$ and $Fe(OH)_3$ in the presence of Ca lime.

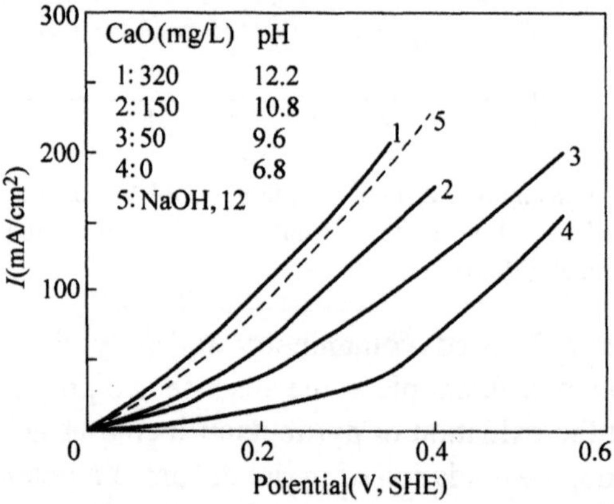

Figure 5.4 Voltammograms for a pyrite electrode in solutions at different pH conditions modified by CaO and NaOH (Linear potential sweeps at 20 mV/s)

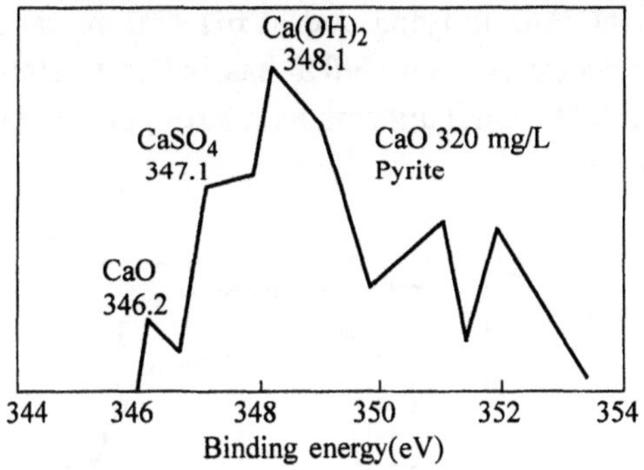

Figure 5.5 The extended XPS spectrum of Ca of pyrite surface treated in CaO solution

Voltammograms for a galena electrode in strong alkaline solution are presented in Fig. 5.6. It demonstrates that only the anodic oxidation of galena surface occurs in strong alkaline solution due to the reaction of Eq. (2-22) or (2-23) at the lower concentration of butyl xanthate or even high concentration of ethyl xanthate and thereby galena will be depressed. If the concentration of butyl xanthate is high enough to 2×10^{-3} mol/L, an evident anodic current which may be attributed to the formation of lead butyl xanthate occurs by the reaction of the type (4-12b).

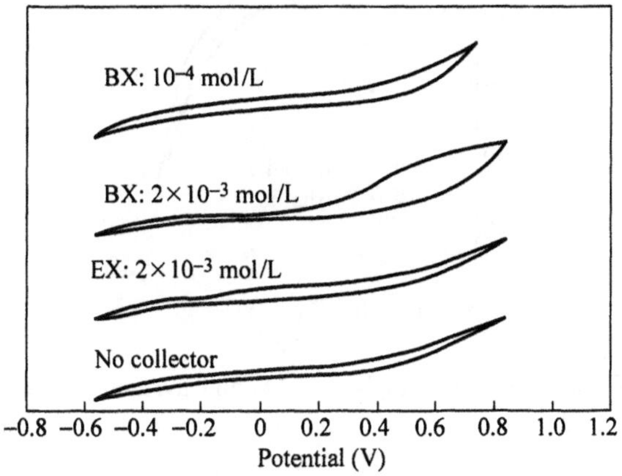

Figure 5.6 Voltammograms for a galena electrode in strong alkaline solution (CaO: 320 mg/L, pH = 12.2) at different concentration of xanthate (scan rate 20 mV/s)

5.1.2 Depression of Jamesonite and Pyrrhotite

With ethyl xanthate (10^{-4} mol/L) as a collector, the flotation recovery of jamesonite and pyrrhotite as a function of pH is presented in Fig. 5.7 with NaOH as pH modifier and in Fig. 5.8 with CaO as pH modifier. It can be seen from these two figures that both jamesonite and pyrrhotite exhibit good flotation response at pH < 12 in case of NaOH modifying and at pH < 11 in case of CaO modifying pH. Jamesonite with recovery above 90% has better floatability than pyrrhotite with recovery around 80%. CaO appears as a stronger depressant than NaOH on jamesonite and pyrrhotite.

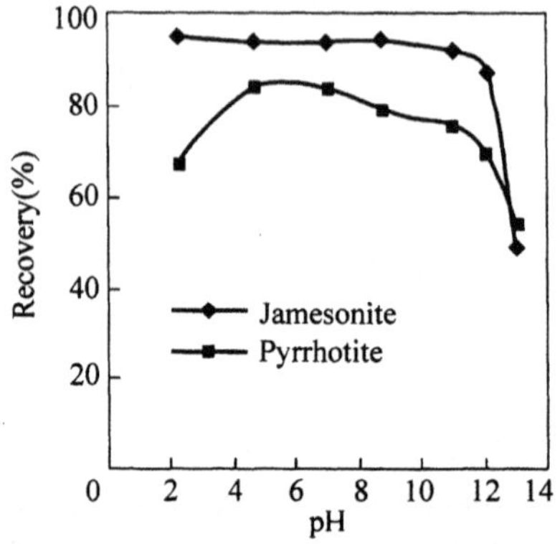

Figure 5.7 Flotation recovery of jamesonite and pyrrhotite as a function of pH modified with NaOH

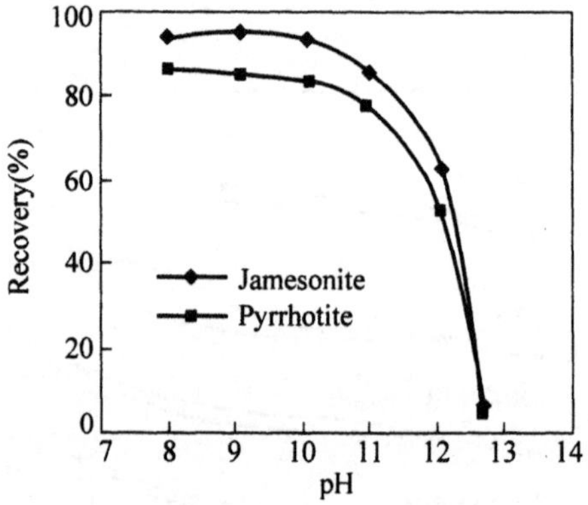

Figure 5.8 Flotation recovery of jamesonite and pyrrhotite as a function of pH modified with CaO

5.1.3 Interfacial Structure of Mineral/Solution in Different pH Modifier Solution

1. Interfacial Structure of Marmatite/Solution

In 0.1 mol/L KNO_3 solution, pH is adjusted to 10.0 by NaOH, $Ca(OH)_2$ and Na_2CO_3. The Tafel curves of marmatite electrode in the above solutions are determined as shown in Fig. 5.9. It follows from Fig. 5.9 that the electrochemical parameters of corrosive potential and current are almost not affected by the pH modifiers. The corrosive potential in Na_2CO_3 and $Ca(OH)_2$ solutions only change 3 mV and 13 mV negatively compared to that in NaOH solution, "ψ_1 effect" of Stern double electric charge layer occurs. So the pH modifier affects the structure of the double electric charge layer.

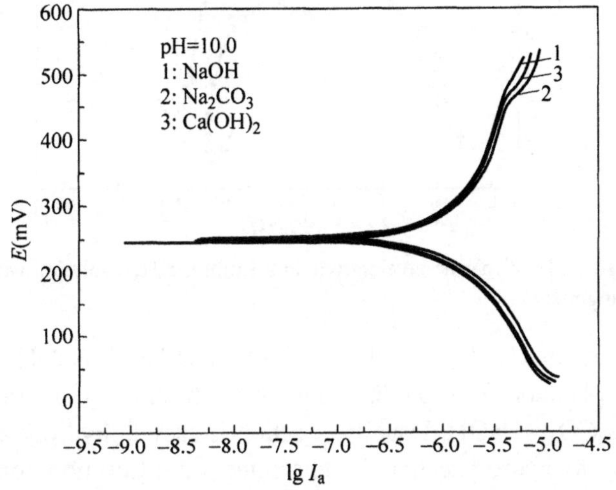

Figure 5.9 Polarization curves of marmatite electrode in 0.1 mol/L KNO_3 solution with different pH adjusting media (unit of I_a: A/cm^2)

If the effect of potential ψ_1 in diffusion layer is considered, then the electrochemical polarization equation is:

$$\begin{aligned}\eta_a &= \varphi - \frac{RT}{n\beta F}\ln nFKc_R + \frac{RT}{n\beta F}\ln I_a + \frac{z-n\beta}{n\beta}\psi_1 \\ &= \text{constant} + \frac{z-n\beta}{n\beta}\psi_1 + \frac{RT}{n\beta F}\ln I_a \end{aligned} \tag{5-12}$$

The relationship between the over-potential and lg I_a will deviate from the Tafel linear area due to the medium affecting the diffusion layer. The ψ_1 effect will gradually disappear and the polarization curves separate each other obviously when the potential is far from zero electric charge potential. This is the reason that CO_3^{2-} and $Ca(OH)^+$ ions have some surfactant action compared with OH^- ion to form characteristic adsorption more easily and to bring about the change of the capacitance of the double electric charge layer.

Figure 5.10 is EIS of marmatite electrode in 0.1mol/L KNO_3 solution with different pH modifiers at open circuit potential. This EIS is very complicated. Simple equivalent circuit can be treated as the series of electrochemical reaction resistance R_r with the capacitance impedance $C_a = (nF\Gamma^0)/(\kappa R_r)$ resulting from adsorbing action, and then parallel with the capacitance C_d of double electric

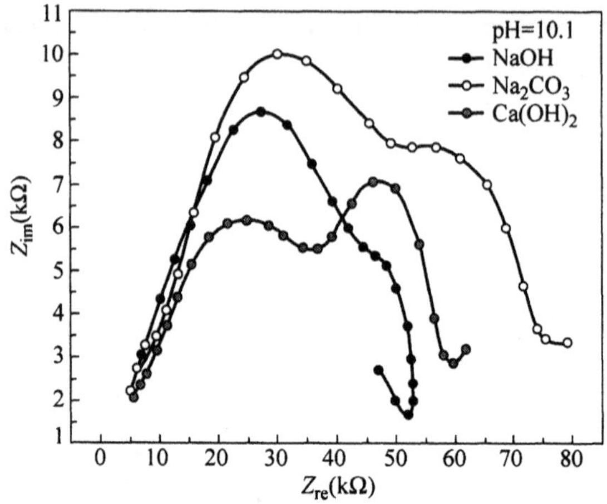

Figure 5.10 EIS of marmatite electrode in 0.1 mol/L KNO_3 solution with different pH adjusting media

charge layer. Obviously, $C_d(Ca(OH)_2) > C_d(Na_2CO_3) > C_d(NaOH)$. It reveals the effect of the pH modifiers on the interfacial structure of marmatite/solution: $Ca(OH)_2 > Na_2CO_3 > NaOH$. Usually, a cation is easy to hydrolyze and an anion is not so easy to hydrolyze that its characteristic adsorption forms. Adsorbed anion will enter inside H_2O polar layer and push aside H_2O to form tight structure on the mineral. The effective thickness of the tight layer will decrease and its capacitance will increase. $Ca(OH)_2$ in solution is very easy to ionize into $Ca(OH)^+$, which is not easy to hydrolyze and is able to form strong characteristic adsorption on the surface of the mineral.

2. Interfacial Structure of Jamesonite/Solution

In 0.1 mol/L KNO_3 solution, pH is adjusted to 9.98 by NaOH, $Ca(OH)_2$ and Na_2CO_3 respectively. Electrochemical impedance spectroscopy (EIS) of jamesonite electrode in 0.1 mol/L KNO_3 solution with different pH adjusting was measured and the results are given in Fig. 5.11. Because $R_{NaOH} > R_{Ca(OH)_2}$, $C_d = 1/(\omega R)$, then $C_d(Ca(OH)_2) < C_d(NaOH)$. It suggests that $Ca(OH)^+$ has stronger tendency to adsorb onto jamesonite surface than OH^-. $Ca(OH)_2$ is a more efficient depressant than NaOH in the flotation of sulphide minerals.

The Tafel curves of jamesonite electrode in the above solutions are determined and shown in Fig. 5.12. It may be seen from Fig. 5.12 that:

(1) The corrosive potential of jamesonite in NaOH, $Ca(OH)_2$ and Na_2CO_3 solution are 45 mV, 57 mV and 63 mV respectively. Three kinds of pH adjustors do not influence much on their corrosive potential, only a little positive moving occurs. It indicates that these adjustors have a similar effect on jamesonite electrode process. The promoting actions of these adjustors for the cathodic process become slightly strong in the order of NaOH, $Ca(OH)_2$ and Na_2CO_3.

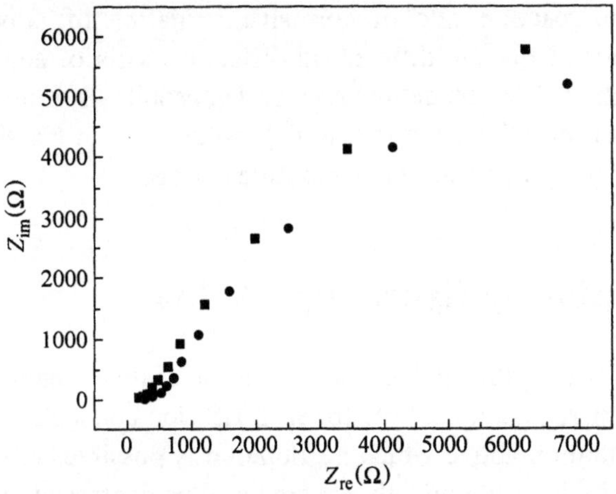

Figure 5.11 EIS of jamesonite electrode in 0.1 mol/L KNO_3 solution with different pH adjusting media (pH = 9.98; square dot: NaOH; ellipse dot: $Ca(OH)_2$)

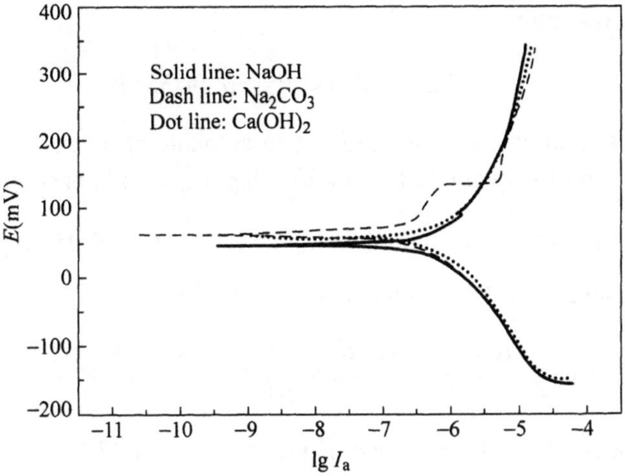

Figure 5.12 Tafel curves of jamesonite electrode in 0.1 mol/L KNO_3 solution with different pH adjusting media (pH = 9.98, unit of I_a: A/cm^2)

(2) The density of the corrosive current of jamesonite in NaOH solution is basically the same as that in $Ca(OH)_2$ solution , but it is minimal in Na_2CO_3 solution, about a fraction of the fourth of the former. There are obvious appearances of passivation and its breaking-down in strong polarization area in $NaCO_3$ solution· Because CO_3^{2-} ion is easier to form insoluble alkaline carbonate than OH^- ion, the carbonate salts are passive on the mineral surface to inhibit oxidation reaction.

(3) In the Tafel linear area, the anodic slope in the $Ca(OH)_2$ solution is slightly bigger than that in NaOH solution. It results from characteristic adsorption as the surface activity of $Ca(OH)^+$ ion is bigger than that of OH^- ion.

Therefore, the reaction rate of corrosive oxidation of jamesonite and its adsorption characteristics are different in different media of adjusting pH at the same pH condition. The oxidation rate of jamesonite is minimal in Na_2CO_3 solution. It is basically the same in $Ca(OH)_2$ solution as in NaOH solution, but $Ca(OH)^+$ is more easy to adsorb on jamesonite surface.

5.2 Depression by Hydrosulphide Ion

The depression by hydrosulphide ion is in a similar manner as hydroxyl depression, i.e. there is a critical pH for each HS^- ion concentration at a constant xanthate concentration above which no flotation is possible. In the case that the hydrophobic entity is disulphide, the mineral will be depressed when the reaction (3-5a) or (3-6a) occurs before the reaction (1-3). Thus for the pyrite /ethyl xanthate system, pyrite will be depressed if the oxidation reaction (3-5a) takes place prior to the oxidation reaction (4-35).

The critical condition is

$$E_h\,[\text{Eq. (3-5b)}] < E_h\,[\text{Eq. (4-35b)}] \tag{5-13}$$

Assuming that sulphur could be oxidized to sulphate at overpotential 0.5 V and the concentration of sulphate is 10^{-6} mol/L, then Eq. (5-13) become

$$0.5953 - 0.066\text{pH} - 0.007375\ \lg[HS^-] < -0.06 - 0.059\ \lg[EX^-] \tag{5-14}$$

Thus, the critical pH above which pyrite will not float is

$$\text{pH}^{\text{d}}_{\text{Py}} \geqslant \frac{0.6553 + 0.059\ \lg[EX^-] - 0.007375\ \lg[HS^-]}{0.066} \tag{5-15}$$

For the galena/ethyl xanthate system, if the reaction Eq. (3-5a) occurs before the reaction (4-12a), the critical pH condition determined by the reaction Eqs. (3-5b) and (4-12a) is

$$\text{pH}^{\text{d}}_{\text{Ga}} \geqslant \frac{0.071 + 0.01475\ \lg[EX^-] - 0.007375\ \lg[HS^-]}{0.015} \tag{5-16}$$

Therefore, the critical pH of hydrosulphide ion depression of pyrite and galena could be calculated using Eqs. (5-15) and (5-16). The results are plotted in Fig. 5.13. The same figure is the results of contact curves reported by Wark and Cox (1933). It can be seen from Fig. 5.13 that the critical pH condition defined by electrochemical equilibriums of Eqs. (5-15) and (5-16) is correlated reasonably well with the contact curves. The calculated concentration of sodium sulphide required for preventing the interaction between xanthate and galena or pyrite at various pH is quite close to the results from the contact curves reported by Wark and Cox (1933).

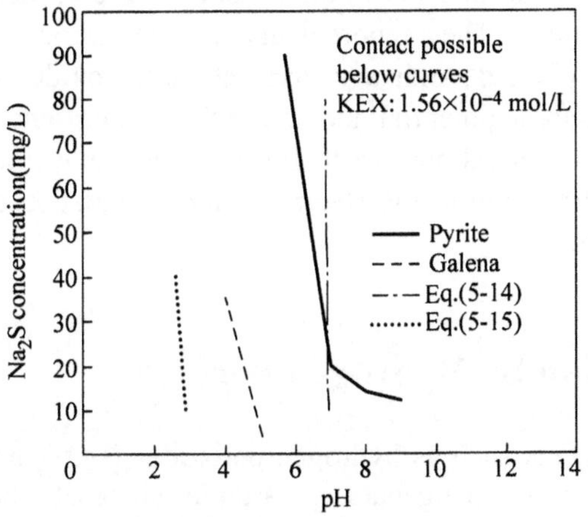

Figure 5.13 Relationship between pH and concentration of sodium sulphide necessary to prevent contact at surfaces of galena and pyrite

Janetski et al. (1977) used voltammetric method to study the electrochemical behavior of a pyrite electrode in ethyl xanthate solution containing various concentration of sodium sulphide. They observed an additional anodic wave due to the oxidation of the dissolved sulphide species present and that the wave appeared at potential cathodic to xanthate oxidation. Therefore, they concluded that the presence of sulphide in solution introduced an anodic process which occurred in preference to xanthate oxidation and hence dixanthogen would not be formed and the pyrite would not be rendered floatable.

5.3 Electrochemical Depression by Cyanide

The depression of pyrite by cyanide has been considered from the electrochemical viewpoint, the formation of ferric ferrocyanide in the surface of pyrite which was perhaps responsible for cyanide depression was proposed by Ball and Richard (1976) to occur by the following electrochemical reaction.

$$7Fe^{2+} + 18HCN = Fe_4[Fe(CN)_6]_3 + 18H^+ + 4e^- \qquad (5\text{-}17)$$

This equation represents the overall reaction which must occur in several steps. The mineral must produce some ferrous ion in solution which reacts with CN^- to form $Fe(CN)_6^{4-}$. Using measured oxidation potential and pyrite solubility values, Eligillani and Fuerstenau (1968)delineated the stability domains of the compound $Fe_4[Fe(CN)_6]_3$, $Fe(OH)_3$, Fe^{2+}, Fe^{3+}, HCN, for various concentration of KCN in the E_h-pH diagram that the regions of no flotation are predicted corresponding to the stability fields of $Fe_4[Fe(CN)_6]_3$ and $Fe(OH)_3$.

Janetski et al. (1977) also studied the behavior of a pyrite electrode in a solution of cyanide concentration in the absence and presence of xanthate using voltammetric technique. They reported that on increasing the concentration of cyanide at constant pH and xanthate concentration, the oxidation wave of xanthate is shifted to more anodic potential; indicating that the presence of cyanide, which may react with the mineral surface to form an insoluble iron cyanide complex will result in the inhibition of the electrochemical oxidation of xanthate and the depression of pyrite.

5.4 Depression by Hydrogen Peroxide

Many oxidizing agents such as hydrogen peroxide (H_2O_2), sodium hypochlorite (NaClO), potassium permanganate, potassium chromate have been used as depressants for pyrite, arsenopyrite and galena (Beattie and Poling, 1988; Shimoiizaka et al., 1976; Hoyack and Raghavan, 1987; Nagaraj et al., 1986; Wang, 1992).

Figures 5.14 and 5.15 demonstrate the influence of pulp potential and concentration of hydrogen peroxide on the flotation of galena and chalcopyrite. It can be seen from Fig. 5.14 that chalcopyrite has high flotation recovery in the potential range of 0.4 – 0.7 V modified by hydrogen peroxide and the floatability of galena sharply decreases with the potential enhancing. Figure 5.15 shows that galena is completely depressed and chalcopyrite remains floatable if the concentration of H_2O_2 is greater than 10^{-3} mol/L, suggesting that the conditions of flotation separation of chalcopyrite from galena in pulp potential is greater than 0.65 V at pH > 7.3 with 10^{-3} mol/L H_2O_2.

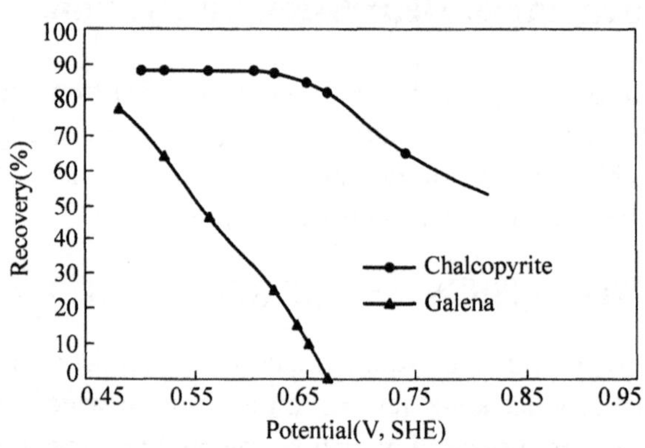

Figure 5.14 Flotation recovery of chalcopyrite and galena as a function of pulp potential modified by hydrogen peroxide (pH = 9.5; Wang, 1992)

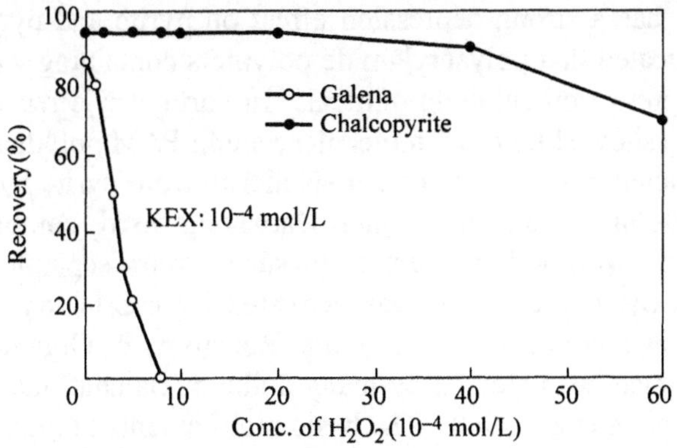

Figure 5.15 Flotation recovery of chalcopyrite and galena as a function of concentration of hydrogen peroxide (Wang, 1992)

The strong depressing action of H_2O_2 on galena may be attributed to its strong oxidizing action on lead xanthate in galena surface giving rise to the oxidation and decomposition of lead xanthate by the reaction of (5-18):

$$[Pb(EX)_2]_{ads} + H_2O_2 \longrightarrow Pb(OH)_2 + (EX)_2 + 2e^- \qquad (5\text{-}18)$$

$$E^0 = -0.94(V)$$

Therefore, at certain pH and H_2O_2 concentration, the complete decomposition of lead xanthate preadsorbed on galena renders galena surface hydrophilic and depression of galena, whereas the dixanthogen prefixed on chalcopyrite remains stable which confers chalcopyrite surface hydrophobic and floatable, which was proved by voltammogram method (Wang, 1992).

5.5 Depression of Marmatite and Pyrrhotite by Thio-Organic Depressants

Organic depressants were paid attention to by many scholars. The depressions of small molecular organic depressants in sulphide ore flotation were reported by some authors (Zhao et al., 1988; Lin et al., 1991; Cheng and Liu, 1994; Xu and Huang, 1995). The results show that, in sulfhydryl acetic acid, the sulfhydryl group acts as its mineral-philic group, while carboxyl group as its hydrophilic radical. Because of the relatively strong reducibility of sulfhydryl acetic acid, its inhibiting behavior fall into two parts. It can produce a relatively negative pulp potential to remove the xanthate adsorbed on the surface of the sulphide minerals or it is fixed on the surface of the solid by means of chemical adsorption. The stronger the reducibility, the better the inhibiting ability. Polymers are well-known depressants of both sulphide and non-sulphide minerals (Li et al., 2001; Xiong et al., 2004). Sun Wei et al. found that the new organic polymer depressant RC has a strong depression effect on pyrite and pyrrhotite. Nagaraj (1982) has indicated that polyacrylamide polymers containing various functional groups can depress iron sulphide minerals. In particular, a recent review by A. Boulton et al[11] showed that low molecular weight PAM polymers could be used to separate efficiently copper-activated sphalerite from pyrite by flotation in the presence of isobutyl xanthate. Xuan Daozhong used combined depressant calcium chloride with sodium humate to successfully separate the mixture of marmatite and pyrrhotite which was activated by cupric ions with potassium butyl-xanthate as a collector. Structure and Reactivity of Organic Depressant on Flotation have also been studied by many authors (Baldauf and Schubert, 1980; Wang, 1983; Cheng et al., 1998). The basic requirements of organic depressant is that there must be several (at least two) polar radical in molecular structure, which has a polar radicals to interact selectively with mineral surface, which is definitely stronger than the collector. It must have other polar radicals that render mineral surface hydrophilic.

The effect of DMPS on the flotation recovery of pyrrhotite and marmatite in the presence and absence of $CuSO_4$ with butyl xanthate is shown in Fig. 5.16. It follows that the flotation of pyrrhotite and marmatite is greatly affected by DMPS addition. In the absence of cupric ion the recovery of pyrrhotite hardly exceeded 40%. At pH = 2, the recovery of marmatite is more than 90%, but the recovery sharply decreases to below 20% with pH increasing. These results show that pyrrhotite and marmatite can not be separated in the absence of cupric ion with DMPS as depressant and xanthate as a collector. In the presence of cupric ions, marmatite flotation improves under wide pH condition. The flotation of pyrrhotite is activated only around pH = 2. The results demonstrate that flotation separation of copper-activated marmatite from pyrrhotite is possible in the presence of butyl xanthate and DMPS.

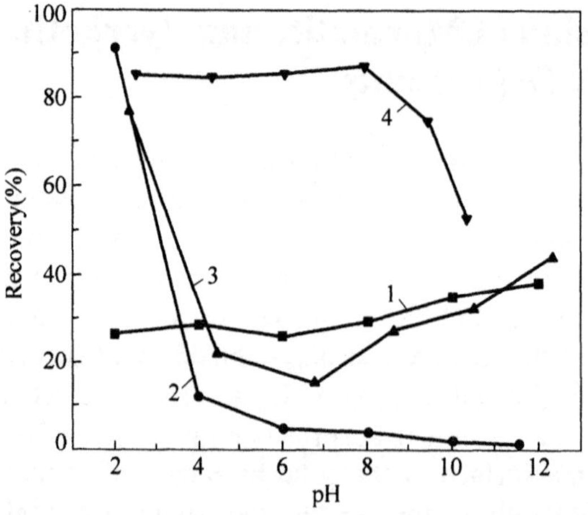

Figure 5.16 Flotation recovery of mineral as a function of pH using [KBX] = 10^{-4} mol/L as a collector and [DMPS] = 2×10^{-4} mol/L as a depressant (1–pyrrhotite; 2–marmatite; 3–pyrrhotite+[$CuSO_4$, 10^{-4} mol/L]; 4–marmatite+[$CuSO_4$, 10^{-4} mol/L])

At different pH, the flotation response of marmatite and arsenopyrite is presented in Fig. 5.17 and Fig. 5.18, using 125 mg/L glycerin-xanthate as a depressant and 1×10^{-4} mol/L butyl xanthate as a collector. At pH < 7, marmatite has a good floatability and glycerin-xanthate exhibits little effect on the flotation of marmatite. However, the flotation recovery of marmatite falls down sharply with pH increasing. At pH > 9, it is completely depressed in the presence and absence of glycerin-xanthate. Arsenopyrite shows a good flotation at pH < 4 in the absence of depressant, glycerin-xanthate has a stronger depression effect on arsenopyrite from pH = 4 – 12, the flotation recovery is below 20%. It indicates the possibility of flotation separation of marmatite from pyrrhotite with butyl xanthate as a collector, cupric ion as an activator and glycerin-xanthate as a depressant.

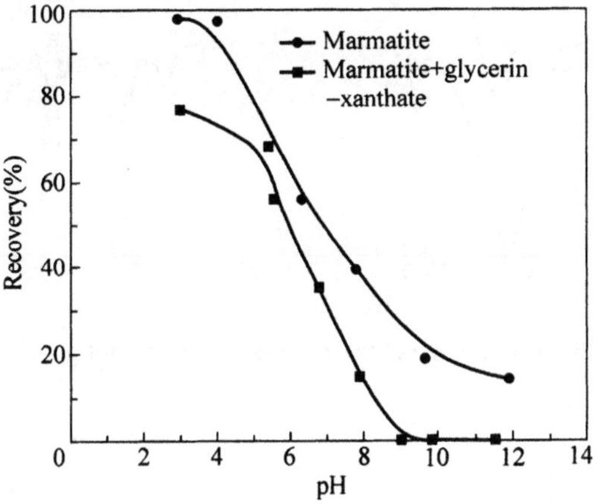

Figure 5.17 Flotation recovery of marmatite as a function of pH in the presence and absence of glycerin-xanthate

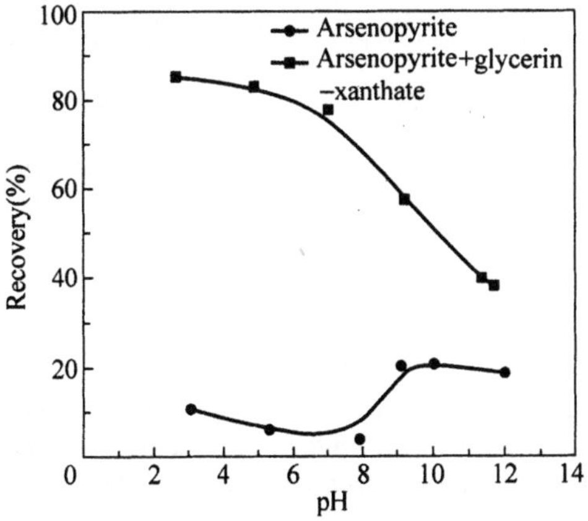

Figure 5.18 Flotation recovery of arsenopyrite as a function of pH in the presence and absence of glycerin-xanthate

The FTIR reflection spectra of organic depressant glycerin-xanthate $(HO)_n$—R O—CSSNa is given in Fig. 5.19, It shows the stretching vibration bands of the C═O at 1414 cm^{-1} and 1041 cm^{-1}, O—H at 3358 cm^{-1}, the dissymmetry stretching vibration bands of the C—H at 2942 cm^{-1}. The stretching vibration bands of the C═S at 1631 cm^{-1}, 1149 cm^{-1}, and 1043 cm^{-1} are the most important vibrations of xanthate group.

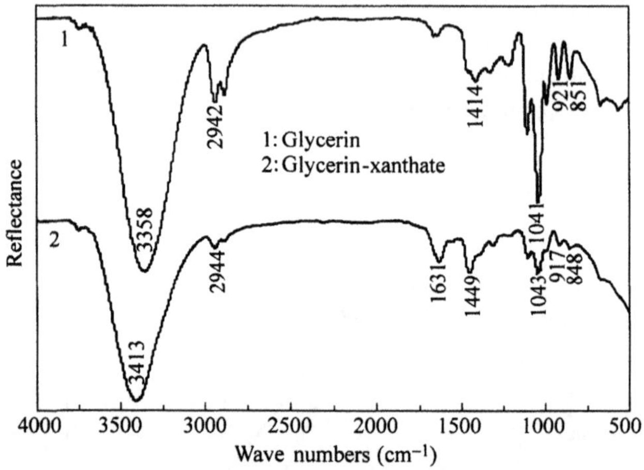

Figure 5.19 Infrared spectrum of organic depressant glycerin-xanthate

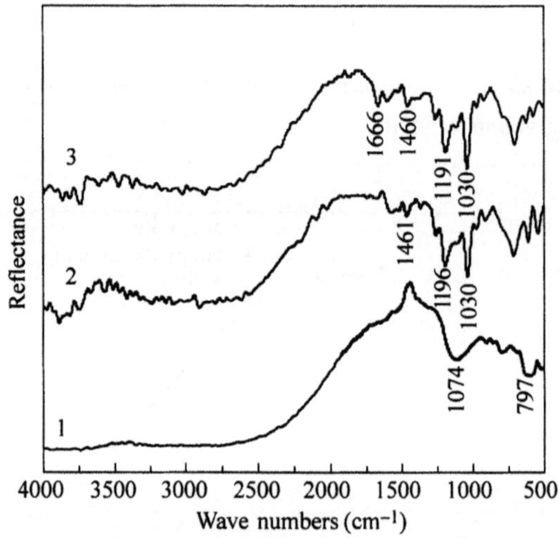

Figure 5.20 FTIR spectra of marmatite in the absence of reagents and in the presence of $CuSO_4$, butyl-xanthate, glycerin-xanthate (1: marmatite; 2: marmatite + $CuSO_4$ + butyl-xanthate; 3: marmatite + glycerin-xanthate + $CuSO_4$ + butyl-xanthate)

The FTIR reflection spectra of marmatite in the absence of reagents and in the presence of $CuSO_4$, butyl-xanthate, glycerin-xanthate are shown in Fig. 5.20. It follows that the absorption bands 1461, 1196, and 1030 cm^{-1} can be observed, after the marmatite was activated by Cu^{2+} and reacting with butyl-xanthate. It suggests that the metal butyl-xanthate and dixanthogen molecule absorb on the surface of marmatite. In the FTIR spectra of marmatite reacted with glycerin-xanthate in the presence of Cu^{2+} and butyl-xanthate, the characteristic absorption bands at 1460 cm^{-1} ,1191 cm^{-1} can be observed and the stretching vibration band of the —OH at 3200 – 3500 cm^{-1} does not exist. The results show that glycerin-xanthate can not prevent butyl-xanthate absorbing on the surface of marmatite, which remains hydrophobic and with good flotation.

5.6 Role of Polyhydroxyl and Poly Carboxylic Xanthate in the Flotation of Zinc-Iron Sulphide

5.6.1 Flotation Behavior of Zinc-Iron Sulphide with Polyhydroxyl and Polycarboxylic Xanthate as Depressants

With butyl xanthate (1.0×10^{-4} mol/L) as a collector and 2,3-dihydroxyl propyl dithiocarbonic sodium (GX2) as a depressant, the flotation recovery of marmatite, arsenopyrite and pyrrhotite is given in Fig. 5.21 as a function of pH.

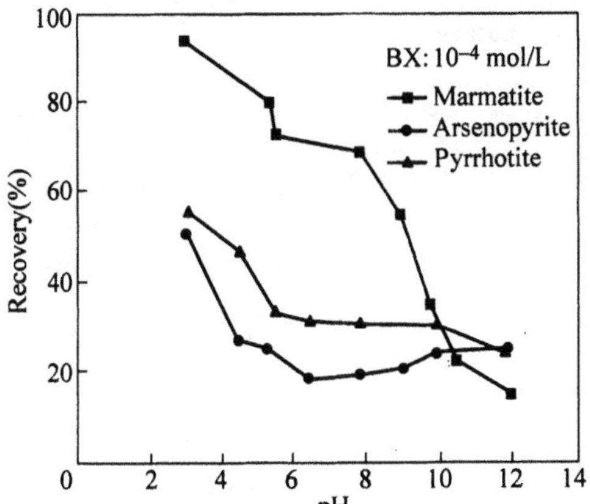

Figure 5.21 Flotation recovery of marmatite, arsenopyrite and pyrrhotite as a function of pH (BX: 1.0×10^{-4} mol/L, GX2: 120 mg/L)

It can be seen from Fig. 5.21 that the flotation recovery of marmatite, arsenopyrite and pyrrhotite decreases with the increase of pH. In the pH range of 4-8, 2,3-dihydroxyl propyl dithiocarbonic sodium (GX2) exhibits stronger depressing action on arsenopyrite and pyrrhotite than on marmatite. The recovery of marmatite is above 70% and the recovery of pyrrhotite and arsenopyrite is below 35%.

At pH = 6, the flotation recovery of marmatite, arsenopyrite and pyrrhotite as a function of depressant dosage GX2 is given in Fig. 5.22. With the increase of GX2 dosage, the flotation recovery of these three minerals decreases. However, marmatite remains with reasonably high flotation recovery of above 70%, and arsenopyrite and pyrrhotite exhibit poor flotation with recovery of below 35% when the concentration of GX2 is above 120 mg/L. It indicates the possibility for flotation separation of marmatite from arsenopyrite and pyrrhotite by using 2,3-dihydroxyl propyl dithiocarbonic sodium as a depressant and butyl xanthate as a collector.

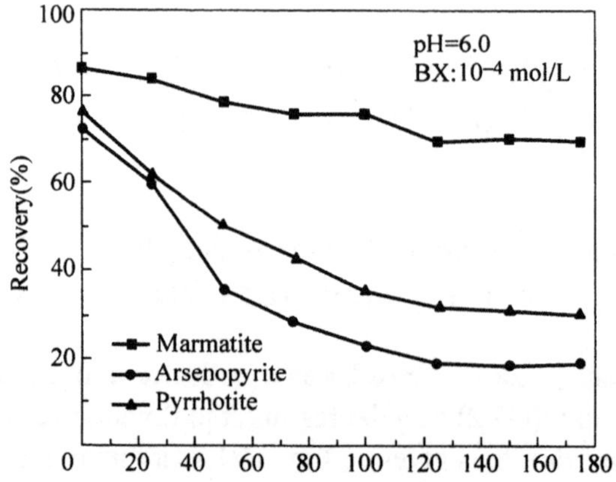

Figure 5.22 Flotation recovery of marmatite, arsenopyrite and pyrrhotite as a function of depressant dosage

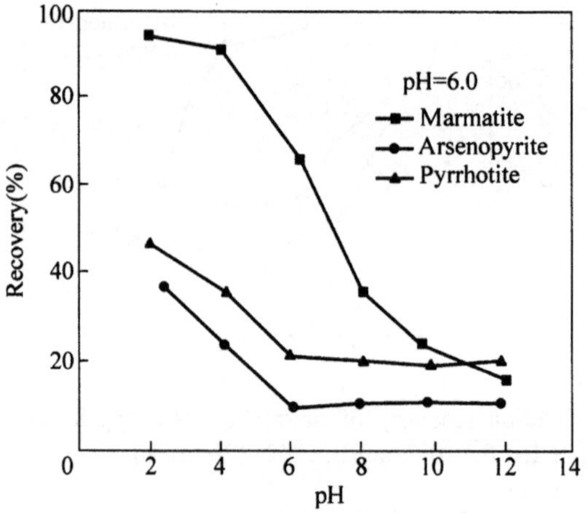

Figure 5.23 Flotation recovery of marmatite, arsenopyrite and pyrrhotite as a function of pH (BX: 1.0×10^{-4} mol/L, TX4: 120 mg/L)

Figure 5.23 and Fig. 5.24 demonstrate the flotation results of marmatite, arsenopyrite and pyrrhotite with butyl xanthate (1.0×10^{-4} mol/L) as a collector and (1-carbonic sodium-2-acetaic sodium) propanic sodium dithio carbonic sodium (TX4) as a depressant. It is obvious that similar to GX2, TX4 has stronger depression on arsenopyrite and pyrrhotite than on marmatite in the pH range of 2-6, at which the flotation recovery of marmatite can be above 70%, and that of arsenopyrite and pyrrhotite is below 30%. At pH = 6, with the increase of TX4 dosage, the flotation recovery of these three minerals decreases. However, marmatite still remains with reasonably high flotation recovery of above 60%, and the flotation of arsenopyrite and pyrrhotite is almost depressed with recovery of below 20% when the concentration of TX4 is above 120 mg/L. It also indicates that the flotation separation of marmatite from arsenopyrite and pyrrhotite may be possible with TX4 as a depressant and butyl xanthate as a collector.

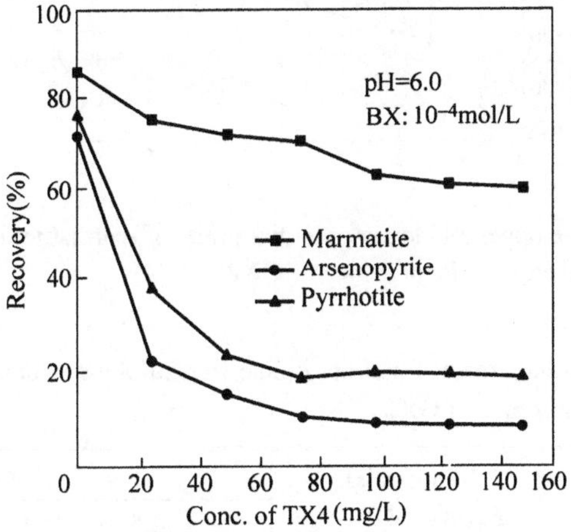

Figure 5.24 Flotation recovery of marmatite, arsenopyrite and pyrrhotite as a function of depressant dosage

5.6.2 Effect of Pulp Potential on the Flotation of Zinc-Iron Sulphide in the Presence of the Depressant

The influence of pulp potential on the flotation of marmatite, arsenopyrite and pyrrhotite with 10^{-4} mol/L butyl xanthate as a collector in the presence of 150 mg/L 2,3-dihydroxyl propyl dithiocarbonic sodium (GX2) has been tested. Taking the flotation recovery to be 50% as a criterion, above which the mineral is considered to be floatable and otherwise not floatable, the upper and lower potential limits of the flotation of marmatite, arsenopyrite and pyrrhotite at different pH are presented in Fig. 5.25 and Table 5.1. It is evident that marmatite is floatable in some range of potential at various pH, whereas arsenopyrite and pyrrhotite are not floatable in the corresponding conditions. It suggests that the flotation separation of marmatite from arsenopyrite and pyrrhotite may be accomplished with butyl xanthate as a collector and GX2 as a depressant by modifying pH and pulp potential. At acidic pH region, the potential range of marmatite flotation is wider, which is suitable for the flotation separation of marmatite from the other two minerals.

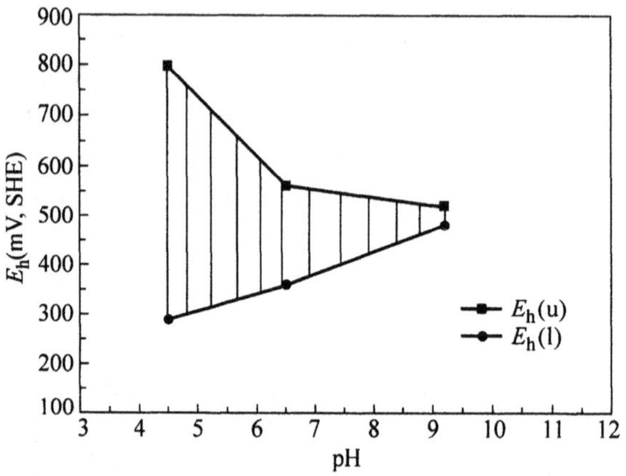

Figure 5.25 The upper and lower potential limits of marmatite flotation with butyl xanthate as a collector in the presence of GX2

Table 5.1 Potential area of flotation of zinc-iron sulphide with butyl xanthate as a collector in the presence of GX2

Minerals	Pulp potential E_h(mV, SHE)	pH		
		4.5	6.5	9.2
Marmatite	E_h(u)	> 800	560	520
	E_h(l)	290	360	480
Arsenopyrite	E_h(u)	No floatable		
	E_h(l)			
Pyrrhotite	E_h(u)	No floatable		
	E_h(l)			

Similarly, the influence of pulp potential on the flotation of marmatite, arsenopyrite and pyrrhotite with 10^{-4} mol/L butyl xanthate as a collector in the presence of 120 mg/L (1-carbonic sodium-2-acetaic sodium) propanic sodium dithio carbonic sodium (TX4) has been tested. The results are given in Fig. 5.26 and Table 5.2. It can be seen from Fig. 5.26 that at pH = 4.5 marmatite has wide floatable potential range from 0.3 V extended to above 0.7 V, at pH = 6.5 the floatable potential range is about 0.3 – 0.4 V, and at pH = 9.2 marmatite is not floatable. Table 5.2 demonstrates that in these conditions, arsenopyrite and pyrrhotite are not floatable. These results indicate that the flotation separation of marmatite from arsenopyrite and pyrrhotite may be accomplished using butyl xanthate as a collector and TX4 as a depressant in weak acidic pH media by modifying pulp potential.

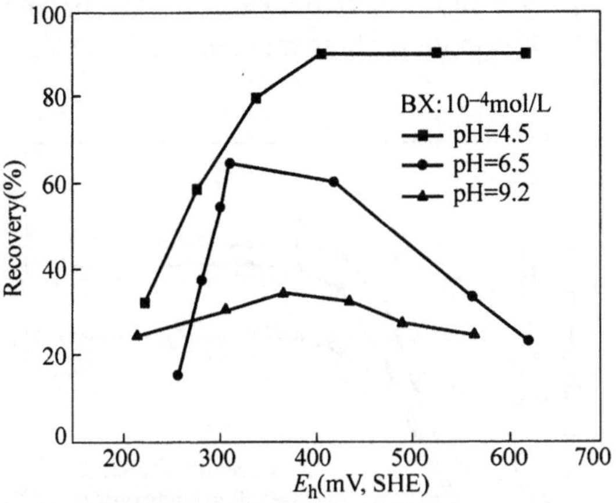

Figure 5.26 Flotation recovery of marmatite as a function of pulp potential with butyl xanthate as a collector in the presence of TX4

Table 5.2 Potential area of flotation of zinc-iron sulphide with butyl xanthate as a collector in the presence of TX4

Minerals	Pulp potential E_h(mV, SHE)	pH		
		4.5	6.5	9.2
Marmatite	E_h(u)	>800	490	No floatable
	E_h(l)	260	300	
Arsenopyrite	E_h(u)	No floatable		
	E_h(l)			
Pyrrhotite	E_h(u)	No floatable		
	E_h(l)			

5.6.3 Adsorption of Polyhydroxyl and Polycarboxylic Xanthate on Zinc-Iron Sulphide

Figure 5.27 presents the adsorption isotherms of 2,3-dihydroxyl propyl dithiocarbonic sodium (GX2) on marmatite, arsenopyrite and pyrrhotite. The adsorption of GX2 on these three zinc-iron sulphides is increased with its concentration until 100 mg/L, after which the adsorption reaches a plateau. The adsorption is increased in the order of arsenopyrite > pyrrhotite > marmatite, which is just the order of GX2 depression on these three minerals. The effect of pH on the adsorption GX2 on the three minerals is given in Fig. 5.28. It follows that the adsorption of GX2 increases with the increase of pH. The adsorption of GX2 increases rapidly at pH > 4 and reaches a plateau at pH > 6, which accounts for the depression action of GX2 on these three minerals in Fig. 5.21.

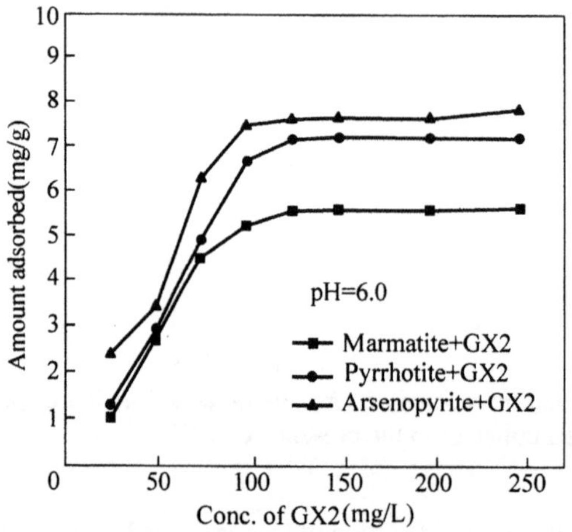

Figure 5.27 Adsorption of GX2 on marmatite, arsenopyrite and pyrrhotite as a function of initial concentration of GX2

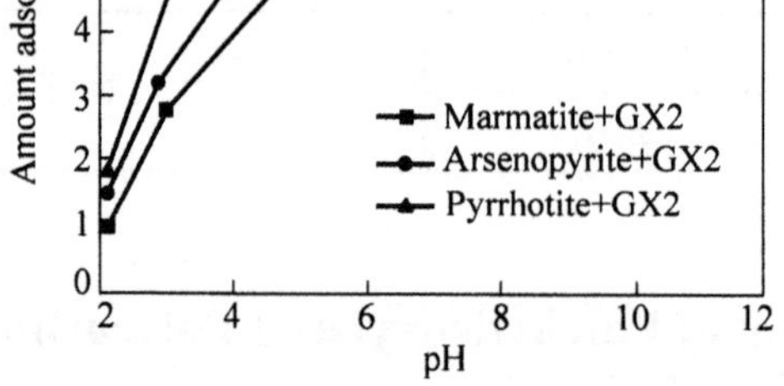

Figure 5.28 Adsorption of GX2 on marmatite, arsenopyrite and pyrrhotite as a function of pH (GX2: 120 mg/L)

Figure 5.29 and Fig. 5.30, respectively, show the adsorption of (1-carbonic sodium-2-acetaic sodium) propanic sodium dithio carbonic sodium (TX4) on marmatite, arsenopyrite and pyrrhotite as a function of concentration and pH. It can be seen from Fig. 5.29 that with the increase of TX4 concentration, the adsorption of TX4 on the three minerals is increased. At a concentration above 60 mg/L, the adsorption reaches a plateau. Adsorption magnitude is in the order of arsenopyrite > pyrrhotite > marmatite, which explains the depression order of TX4 on the three minerals in Fig. 5.24. Figure 5.30 shows that the adsorption of TX4 on the three minerals increases with the increase of pH. At pH > 4, the adsorption of TX4 on these three minerals reaches the maximum, which accounts for the depressing action of TX4 on the three minerals at pH > 6 in Fig. 5.23. The adsorption maximum is corresponding to the maximum depression.

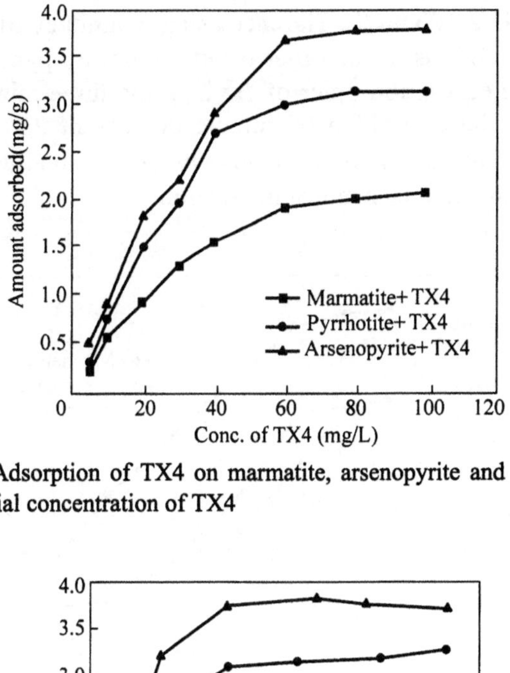

Figure 5.29 Adsorption of TX4 on marmatite, arsenopyrite and pyrrhotite as a function of initial concentration of TX4

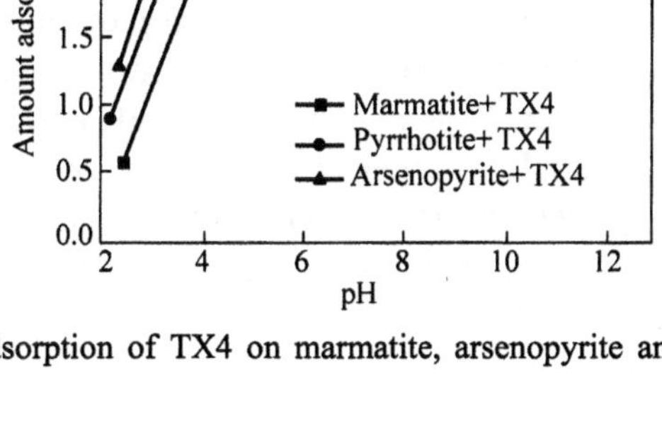

Figure 5.30 Adsorption of TX4 on marmatite, arsenopyrite and pyrrhotite as a function of pH

5.6.4 Effect of Polyhydroxyl and Polycarboxylic Xanthate on the Zeta Potential of Zinc-Iron Sulphide Minerals

Zeta potential of marmatite, arsenopyrite and pyrrhotite as a function of pH are presented in Fig. 5.31. It can be seen that these three minerals are negatively charged in wide pH range in water. No PZC are observed for arsenopyrite and pyrrhotite. In stronger acidic media, marmatite has a small positive zeta potential and the PZC occurs around pH = 5. The effects of polyhydroxyl and polycarboxylic xanthate on the zeta potential of the three minerals are given in Fig. 5.32 and Fig. 5.33 respectively. It can be seen from Fig. 5.32 that with the increase of GX2, the negative zeta potential of marmatite, pyrrhotite and arsenopyrite increase. The negative zeta potential reach the maximum and remained stable at the concentration of GX2 100 mg/L. The increasing magnitude of the zeta potential in the presence of GX2 is in the order of arsenopyrite > pyrrhotite > marmatite, which explains the adsorption order of GX2 on the three minerals. Figure 3.32 also suggests that the adsorption of anionic depressant 2,3-dihydroxyl propyl dithiocarbonic sodium (GX2) on negatively charged marmatite, arsenopyrite and pyrrhotite may be due to the chemical interaction.

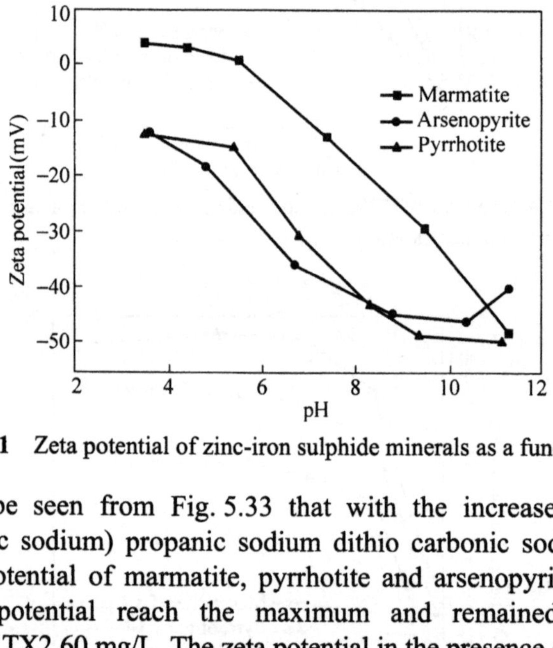

Figure 5.31 Zeta potential of zinc-iron sulphide minerals as a function of pH

It can also be seen from Fig. 5.33 that with the increase of (1-carbonic sodium-2-acetaic sodium) propanic sodium dithio carbonic sodium (TX4), the negative zeta potential of marmatite, pyrrhotite and arsenopyrite increase. The negative zeta potential reach the maximum and remained stable at the concentration of TX2 60 mg/L. The zeta potential in the presence of TX2 increases in the order of arsenopyrite > pyrrhotite > marmatite, which is corresponding to the adsorption order of TX2 on the three minerals. Figure 3.33 also suggests that the adsorption of anionic depressant TX2 on negatively charged marmatite, arsenopyrite and pyrrhotite may be due to the chemical interaction.

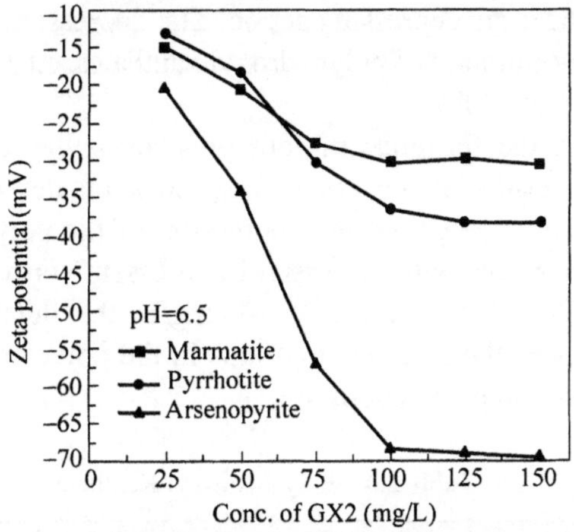

Figure 5.32 Zeta potential of zinc-iron sulphide minerals as a function of GX2 concentration

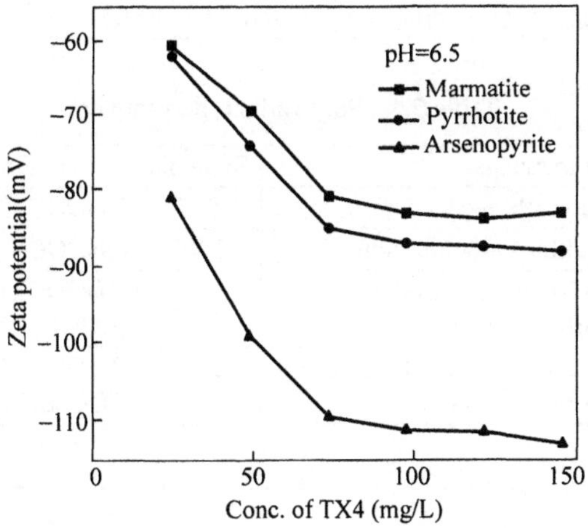

Figure 5.33 Zeta potential of zinc-iron sulphide minerals as a function of TX4 concentration

5.6.5 Structure-Property Relation of Polyhydroxyl and Polycarboxylic Xanthate

The influence of three kinds of polyhydroxyl xanthate and four kinds of polycarboxylic xanthate on arsenopyrite and pyrrhotite flotation are further examined in order to compare the effect of the depressant structure on their depressing action. The formulae of three polyhydroxyl xanthates are shown in Table 5.3 and their depression on arsenopyrite flotation is given in Fig. 5.34. It is obvious that the depressing ability of polyhydroxyl xanthates is in the order of GX3 > GX2 > GX1. The more the hydroxyl groups in the depressant polyhydroxyl xanthate, the stronger the depressing action. The depressing action increases with the increase of concentration of polyhydroxyl xanthates. At 120 mg/L, the flotation recovery can be below 30%.

Table 5.4 shows the formulae of four polycarboxylic xanthates. Figure 5.35 demonstrates the results of the polycarboxylic xanthates on the depression of arsenopyrite. It can be seen that the depression of the polycarboxylic xanthates on arsenopyrite increases with increase of depressant concentration and is in the order of TX4 > TX3 > TX2 > TX1. At 100 mg/L, the flotation recovery can be below 20%.The more the carboxylic groups in the polycarboxylic xanthates, the stronger the depression on arsenopyrite.

Table 5.3 Poly-hydroxyl Xanthates

Compounds	Symbols	Formulas	MW
2-hydroxyl ethyl dithio carbonic sodium	GX1	$(HO)C_2H_4OCSSNa$	160
2,3 dihydroxyl propyl dithio carbonic sodium	GX2	$(HO)_2C_3H_5OCSSN$	190
2,3,4,5,6, five hydroxyl heptyl dithio carbonic sodium	GX3	$(HO)_5C_6H_8OCSSN$	280

Table 5.4 Poly-carboxylic xanthates

Compounds	Symbols	Formulas	MW
Sodium acetate dithio carbonic sodium	TX1	$(NaOOC)CH_2OCSSNa$	196
Sodium propronate dithio carbonic sodium	TX2	$(NaOOC)C_2H_4OCSSNa$	210
(1-carbonic sodium-2-hydroxyl) Sodium propronate dithio carbonic sodium	TX3	(NaOOC)(HO)CHCH (COONa)OCSSNa	292
(1-carbonic sodium-2-sodium acetate) Sodium propronate dithio carbonic sodium	TX4	$(NaOOCCH_2)_2(NaOOC)$ COCSSNa	328

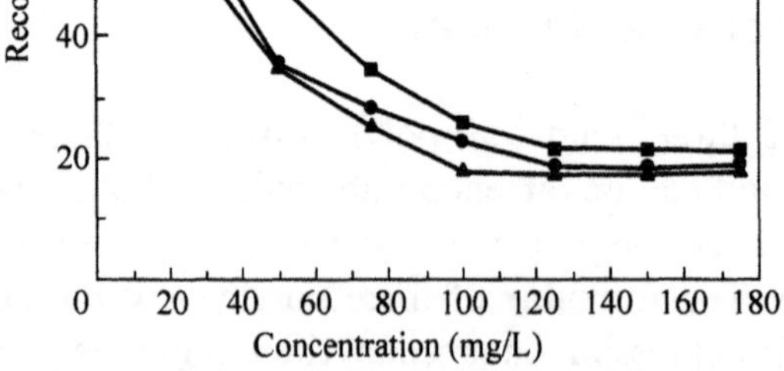

Figure 5.34 Flotation recovery of arsenopyrite as a function of polyhydroxyl xanthate concentration with butyl xanthate as a collector

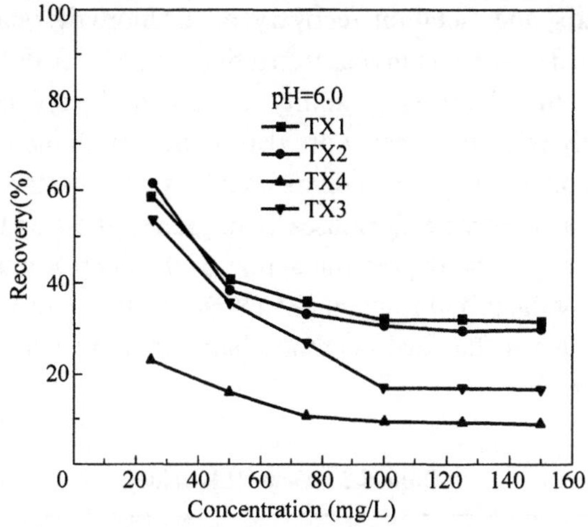

Figure 5.35 Flotation recovery of arsenopyrite as a function of polycarboxylic xanthate concentration with butyl xanthate as a collector

The depressing property of the organic depressant mainly depends on its structure. Some quantitative criteria have been used to determine the depressing ability of the organic depressants (Wang, 1982, 1997). The characteristic index of an organic depressant is defined as follows:

$$i_1 = \frac{\sum (x_g - x_H)^2}{\sum n\Phi} \tag{5-19}$$

or

$$i_2 = \sum (X_g - X_H)^2 - \sum n\Phi + K \tag{5-20}$$

where X_g is the group electronegativity of the polar head, for —OH, $X_g = 3.9$; for —COOH, $X_g = 4.1$; for —OCSS(H), $X_g = 2.7$; X_H electronegativity of hydrogen; n chain length of nonpolar group; $\Phi = 0.8$, $K = 20$.

The other quantitative criterion is hydrophilic-hydrophobic balance value defined as:

$$\begin{aligned} \mathrm{HLB}_1 = &\sum (\text{hydrophilic group value}) \\ &- \sum (\text{hydrophobic group value}) + 7 \end{aligned} \tag{5-21}$$

The i value and HLB value of polyhydroxyl and polycarboxylic xanthates are calculated and shown in Table 5.5. It is obvious that the polyhydroxyl xanthate with more hydroxyl groups and polycarboxylic xanthate with more carboxylic groups have greater i or HLB value. Taking the i or HLB values as x-axis and the flotation recovery of arsenopyrite or pyrrhotite at depressant concentration

120 mg/L as *y*-axis, the flotation recovery of arsenopyrite and pyrrhotite as a function of *i* or HLB values is given, respectively in Fig. 5.36 to Fig. 5.39. They demonstrate that the depressing ability of polyhydroxyl or polycarboxylic xanthates is closely related to their *i* or HLB values. With the increase of *i* value or HLB value of polyhydroxyl and polycarboxylic xanthates, the flotation recovery of pyrrhotite and arsenopyrite decreases. The greater the *i* or HLB value of the depressant, the stronger the depressing action of the depressants. In other words, when the mineral-philic group is given as xanthate radical, more hydroxylic groups or carboxylic groups in the hydrocarbon chain structure may produce stronger depressing actions.

Table 5.5 *i* and HLB value

Depressants	Formulas	*i*		HLB_1
		i_1	i_2	
GX1	$(HO)CH_2CH_2OCSSNa$	2.25	22.0	26.25
GX2	$(HO)_2C_3H_5OCSSNa$	2.85	24.44	27.68
GX3	$(HO)_5C_6H_8OCSSNa$	3.45	31.76	31.95
TX1	$(NaOOC)CH_2OCSSNa$	5.45	23.56	43.93
TX2	$(NaOOC)\ CH_2CH_2OCSSNa$	2.725	22.76	43.45
TX3	$(NaOOC)(HO)CHCH(COONa)OCSSNa$	7.25	30.0	64.45
TX4	$(NaOOCCH_2)_2(NaOOC)CCOCSSNa$	15.45	31.56	83.08

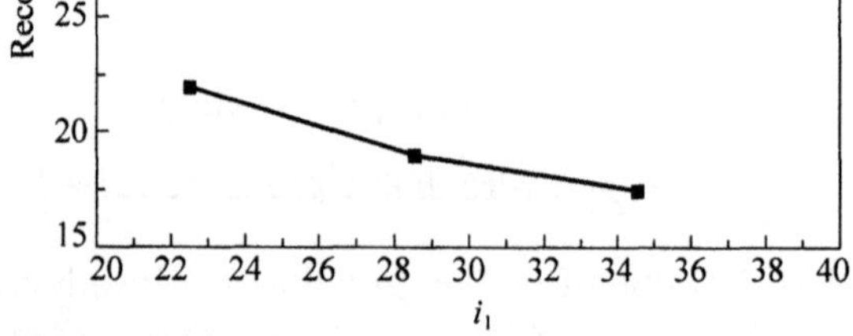

Figure 5.36 Flotation recovery of arsenopyrite and pyrrhotite as a function of i_1 value of polyhydroxyl xanthates

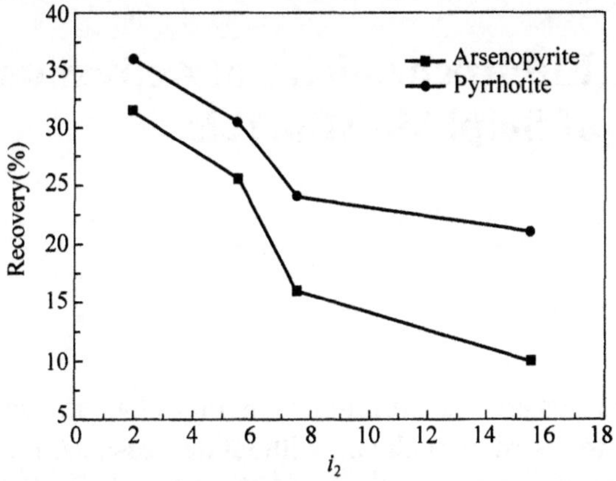

Figure 5.37 Flotation recovery of arsenopyrite and pyrrhotite as a function of i_2 value of polycarboxylic xanthates

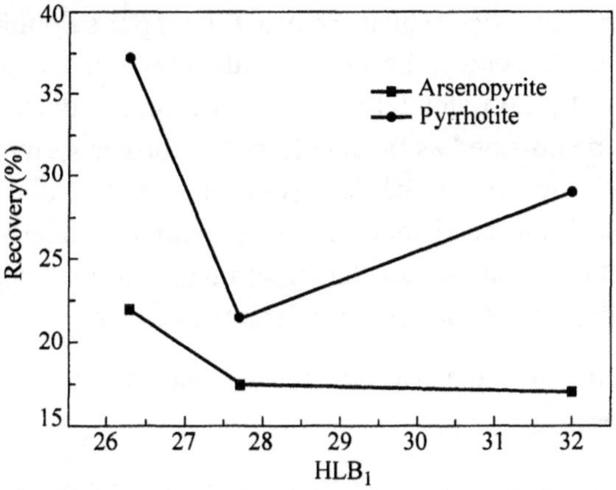

Figure 5.38 Flotation recovery of arsenopyrite and pyrrhotite as a function of HLB_1 value of polyhydroxyl xanthates

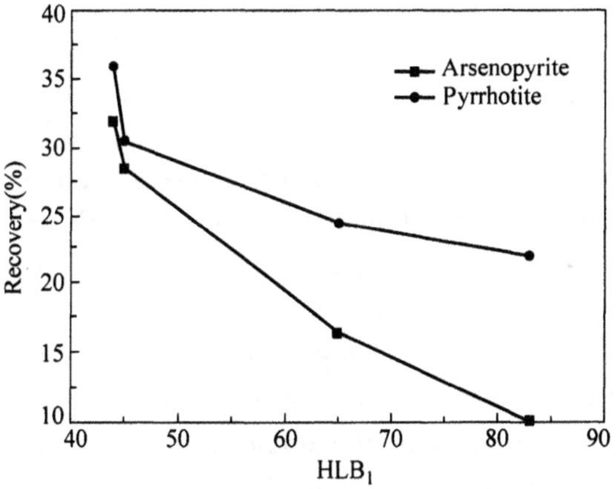

Figure 5.39 Flotation recovery of arsenopyrite and pyrrhotite as a function of HLB_1 value of polycarboxylic xanthates

Chapter 6 Electrochemistry of Activation Flotation of Sulphide Minerals

Abstract Two systems are discussed in this chapter, which are copper activating zinc-iron system with and without depressants. Firstly, the system in the absence of depressants is discussed. And it is obtained that at a specific pH the activation for each mineral occurs in a certain range. Through the electrochemical methods and surface analysis the entity contributing to the activation can be identified which are usually copper sulphides and vary for different minerals. Secondly, the system with depressants is researched. And also the effects of pulp potential on the activation are discussed. The same conclusion can be obtained as the one from the former system. Furthermore, zeta potential are involved in the discussion of activation and the mechanism can be explained from the changes of zeta potential. Similarly, the activation mechanism of this system is also studied through solution chemistry, bonding of activator with mineral surfaces and surface analysis.

Keywords activation; electrochemical mechanism; surface analysis; pulp potential; depressants

6.1 Electrochemical Mechanism of Copper Activating Sphalerite

Flotation recovery of sphalerite as a function of pH with butyl xanthate as a collector in the presence and absence of activator is presented in Fig. 6.1. Evidently, sphalerite exhibits good flotation in weak acidic media and the recovery decreases sharply with the increase of pH in the absence of activator. The flotation of sphalerite is activated in wide pH range in the presence of activators such as $CuSO_4$ or $Pb(NO_3)_2$. The recovery of sphalerite can be above 90% at $pH < 12$ using copper ion as an activator and at $pH < 10$ using lead ion as an activator.

The electrochemical mechanism of cupric sulphate activating sphalerite has been studied. The measured cyclic voltammogram curves in aqueous solution at different pH with sphalerite compound electrode are shown as solid lines in Fig. 6.1.

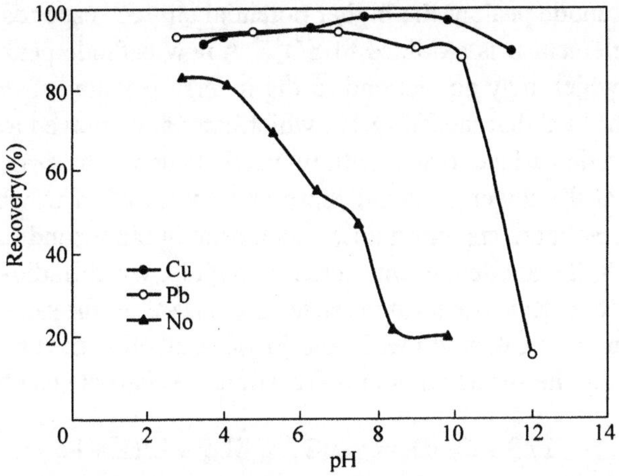

Figure 6.1 Flotation recovery of sphalerite as a function of pH with butyl xanthate (10^{-4} mol/L) as a collector (activator: 2×10^{-4} mol/L)

At pH = 6.0, the anode peak appearing in the first potential scan corresponds to the reaction:

$$ZnS = Zn^{2+} + S + 2e^{-} \quad (6\text{-}1)$$

$$E^0 = 0.265(V)$$

It may be derived that elemental sulphur is not the final product of surface oxidation of sphalerite, because the potential associated with the second anode peak corresponds to the thermodynamic potential of the reaction (3-11a) (taking potential barrier of SO_4^{2-} into consideration).

The first cathode peak appearing in inverse scan is corresponding to the reaction in which SO_4^{2-} is reduced to element S again. The second cathode peak with a lower potential contains two reactions.

$$ZnS + 2H^+ + 2e^- = Zn + H_2S \quad (6\text{-}2)$$

$$E^0 = -0.886(V)$$

$$Zn^{2+} + 2e^- = Zn \quad (6\text{-}3)$$

$$E^0 = -0.760(V)$$

In the second forward scan, at the negative potential, there appear two new anode peaks which correspond to the two inverse reactions above.

At pH = 9.5, the surface oxidation reaction of sphalerite is given by:

$$ZnS + 2H_2O = Zn(OH)_2 + S + 2H^+ + 2e^- \quad (6\text{-}4)$$

$$E^0 = 0.665(V)$$

The second anode peak at the higher potential still corresponds to the reaction in which the element S is oxidized to SO_4^{2-}. A new cathode peak appears in the inverse scan, which may correspond to the inverse reaction of reaction (6-4). It can probably be said that the $Zn(OH)_2$, which can not be moved away completely from the electrode surface, reacts with reduced element S to form ZnS. The third cathode peak at the lower potential represents the reaction in which the ZnS is reduced to Zn. So there appears a new anode peak in the second forward scan.

At pH = 12.8, the anode current increases rapidly, which indicates that a more intense oxidation reaction occurs at the sphalerite surface. At the same time two anode peaks appearing in weak acid media and in weak alkaline media merge together, which shows that the surface oxidation reaction of sphalerite can be given by:

$$ZnS + 6H_2O = ZnO_2^{2-} + SO_4^{2-} + 12H^+ + 8e^- \tag{6-5}$$

$$E^0 = 0.635(V)$$

Because ZnO_2^{2-} can not be completely dissolved on the anode surface, ZnO_2^{2-} can react with element S to form ZnS, so that the corresponding cathode peak appears in the inverse scan.

The cyclic voltammogram for sphalerite compound electrode changes greatly in aqueous solution with the $CuSO_4$ concentration of 1.1×10^{-4} mol/L, which is shown as dotted lines in Fig. 6.2. It can be seen that compared with the oxidation properties of sphalerite, the oxidation properties of those surface products change after sphalerite is activated by $CuSO_4$. In the term of initial oxidation potential, the surface products are superior to sphalerite in oxidation.

At pH = 6.0, the preset peak (0.25 V) appearing in the anode scan perhaps corresponds to the following two reactions.

$$Cu_2S + H_2S = 2CuS + 2H^+ + 2e^- \tag{6-6}$$

$$E^0 = 0.081(V)$$

$$CuS + H_2S = S - CuS + 2H^+ + 2e^- \tag{6-7}$$

$$E^0 = 0.148(V)$$

The anode peak (0.4 V) is due to the reactions in which Cu_2S and S – CuS are oxidized further.

$$Cu_2S = CuS + Cu^{2+} + 2e^- \tag{6-8}$$

$$E^0 = 0.408(V)$$

$$S - CuS + 4H_2O = CuS + SO_4^{2-} + 8H^+ + 6e^- \tag{6-9}$$

$$E^0 = 0.357(V)$$

When the SO_4^{2-} potential barrier is considered, because the thermodynamic reversible potential of reaction (6-8) is relatively high, it can be regarded that CuS is not oxidized further at the upper limit of the scan potential shown in Fig. 6.1. In the inverse scan, there appear three cathode peaks from right to left, which sequentially correspond to the back reaction of reactions (6-8), (6-9) and the reduction reaction of element sulphur.

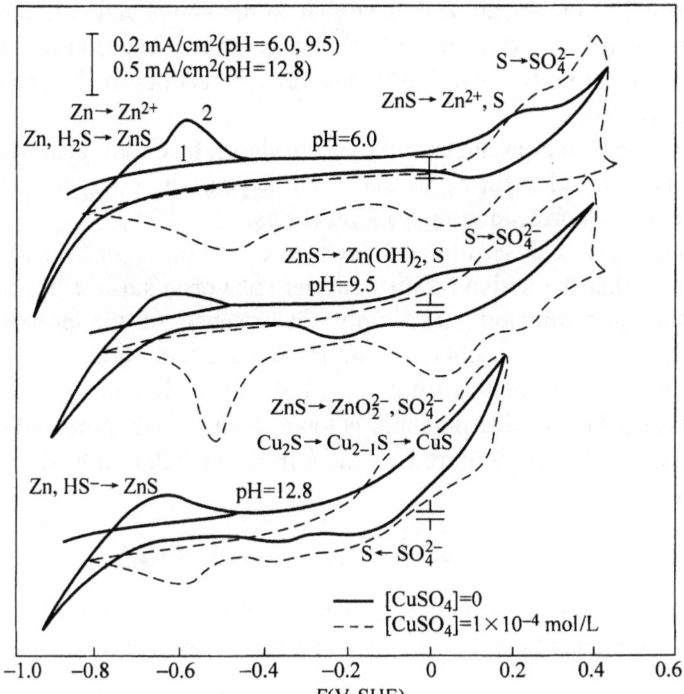

Figure 6.2 Cyclic voltammetric curves for sphalerite compound electrode in buffer solution with different pH at 293K, KNO_3 concentration of 0.5 mol/L and sweep rate of 20 mV/s

At pH = 9.5, the cyclic voltammogram of the sphalerite activated by $CuSO_4$ is relatively complex. At the upper limit of the scan potential shown in Fig. 6.1, a series of anode peaks appearing in forward scan sequentially correspond to reaction (6-9) and other reactions as follows:

$$Cu_2S + HS^- = 2CuS + H^+ + 2e^- \quad (6\text{-}10)$$

$$E^0 = -0.126(V)$$

$$CuS + HS^- = S - CuS + H^+ + 2e^- \quad (6\text{-}11)$$

$$E^0 = -0.066(V)$$

$$CuS + 2H_2O = Cu(OH)_2 + S + 2H^+ + 2e^- \quad (6\text{-}12)$$

$$E^0 = -0.863(V)$$

It can be seen that $Cu(OH)_2$ may exist on the electrode surface as well as Cu_2S and CuS. The existence of element S depends on the oxidation potential. Anode peaks appearing in inverse scan are sequentially associated with reaction (6-9), the back reaction of reaction (6-12) and reduction reaction of element S (HS^- is the reduction product).

In high alkaline media (pH = 12.8), where the upper limit of the scan potential is carried to 0.2 V, the anode peak is similar to the above and corresponds to the reaction that Cu_2S is oxidized to CuS and S – CuS. The anode peaks at the potential varying from 0.1 V to 0.2 V may contain reaction (6-12), reaction (6-7) and reaction (6-10).

At that moment, the products at the electrode surface may comprise of Cu_2S, CuS, $Cu(OH)_2$ and element S, so that there appears a series of cathode peaks which indicate the corresponding back reactions.

According to the analyses of the above cyclic voltammogram results, it can not be determined that the active products at the sphalerite surface is Cu_2S or CuS. However this phenomenon may imply the essence of the process itself. It suggests that in the ZnS – $CuSO_4$ system the Cu concentration in active products keeps equilibrium with that in solution, which is strictly controlled by chemical potential. When the working potential is lower than the rest potential of CuS, the active products will absorb more Cu^{2+} ions from the solution to form high Cu/S ratio compound.

$$Cu_xS_x + yCu^{2+} + 2ye^- \longrightarrow Cu_x + yS_x \qquad (6\text{-}13)$$

When the working potential is higher than the rest potential of CuS, sulphide will partly lose element Cu to form low Cu/S ratio compound.

$$Cu_xS_x \longrightarrow Cu_{x-2}S_x + 2Cu^{2+} + 2e^- \qquad (6\text{-}14)$$

In the actual flotation system, the activators on the sphalerite surface are some copper sulphide compounds containing different amounts of copper, namely some non-chemometric cupric sulphide. Based on the electrochemical theory of chalcocite in geologically minerogenic process, they may be chalcocite (Cu_2S), diulerite($Cu_{1.96}S$), $Cu_{1.71-1.82}S$, anilite($Cu_{1.75}S$), geerite($Cu_{1.60}S$), spionkopite ($Cu_{1.40}S$), Cu1.12S and covellite (CuS).

6.2 Electrochemical Mechanism of Copper Activating Zinc-Iron Sulphide Minerals

6.2.1 Activation Flotation

The result of activation flotation of marmatite is given in Fig. 6.3 with 10^{-4} mol/L ethyl xanthate as a collector and 2×10^{-4} mol/L $CuSO_4$ as an activator. It follows that marmatite is floatable only at pH < 6 in the absence of cupric ion and the recovery is about 90%. At pH > 6, the recovery of marmatite decreases sharply to below 30%. However, marmatite exhibits very good flotation response at wider pH range in the presence of cupric ion showing the stronger activation of copper ion on marmatite. The recovery of marmatite increases to above 90% in the range of pH = 2 – 12.

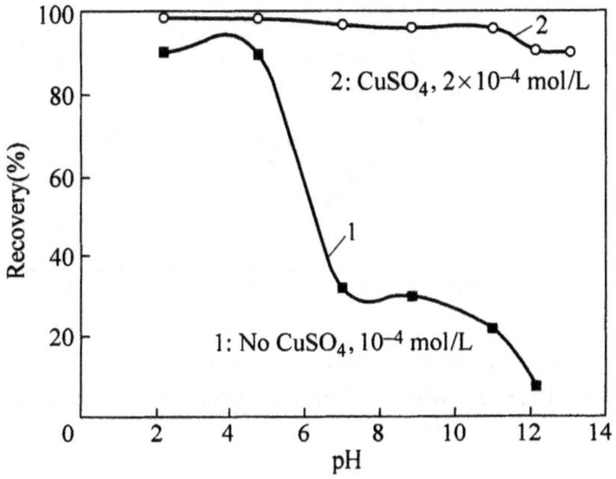

Figure 6.3 Flotation recovery of marmatite as a function of pH with ethyl xanthate (10^{-4} mol/L) as a collector

6.2.2 Effect of Pulp Potential on Activation Flotation of Zinc-Iron Sulphide Minerals

Figure 6.4 to Fig. 6.6 show the effect of pulp potential on the copper activation flotation of marmatite, arsenopyrite and pyrrhotite in the presence of 10^{-4} mol/L Cu^{2+} with 10^{-4} mol/L butyl xanthate as a collector. It is obvious that marmatite, arsenopyrite and pyrrhotite appear to have better flotation response at certain potential range at different pH conditions.

Figure 6.4 shows that the initial potential of marmatite flotation is around 0.26 V and is almost independent of pH, but the upper limit potential of flotation varies with pH, which is higher at acidic pH value. The recovery of marmatite can be above 90% only at certain pulp potential ranges at given pH. The optimal flotation potential range is 0.35 – 0.6 V at pH = 4.5, 0.28 – 0.5 V at pH = 6.5 and 0.25 – 0.3 V at pH = 9.2. It shows the stronger activation of copper ion on marmatite flotation.

Figure 6.5 demonstrates that arsenopyrite starts flotation at a potential around 0.18 V also independent of pH and ceases flotation at different potential values dependent on pH. The recovery is about 80% in the pH range 4.5 – 6.5 and potential range of 300 – 600 mV, indicating some activation of copper ion on arsenopyrite in weak acidic media. At pH = 9.2, the floatability of arsenopyrite is poor and the activation of copper ion is weak.

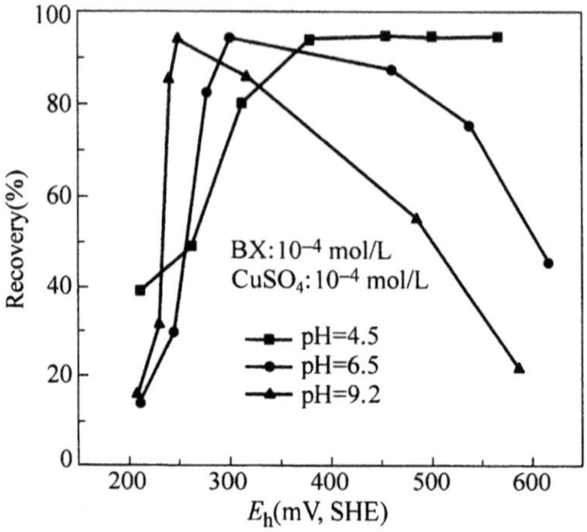

Figure 6.4 Flotation recovery of marmatite as a function of pulp potential in the presence of 10^{-4} mol/L Cu^{2+} with 10^{-4} mol/L butyl xanthate as a collector

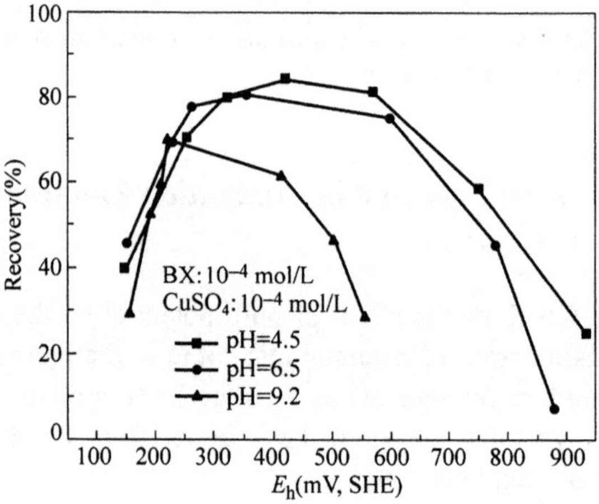

Figure 6.5 Flotation recovery of arsenopyrite as a function of pulp potential in the presence of 10^{-4} mol/L Cu^{2+} with 10^{-4} mol/L butyl xanthate as a collector

The lower and upper limit flotation potential of pyrrhotite activated by copper ion change with pH as shown in Fig. 6.6. At pH = 4.5, pyrrhotite exhibits better flotation response with recovery about 80% in the potential range of 0.25 – 0.65 V indicating the stronger activation of copper ion. At pH = 6.5, the recovery of pyrrhotite reaches 80% around potential 0.3V. At pH = 9.2, the floatability of pyrrhotite is low.

Taking the flotation recovery 50% as a criterion, above which the mineral is

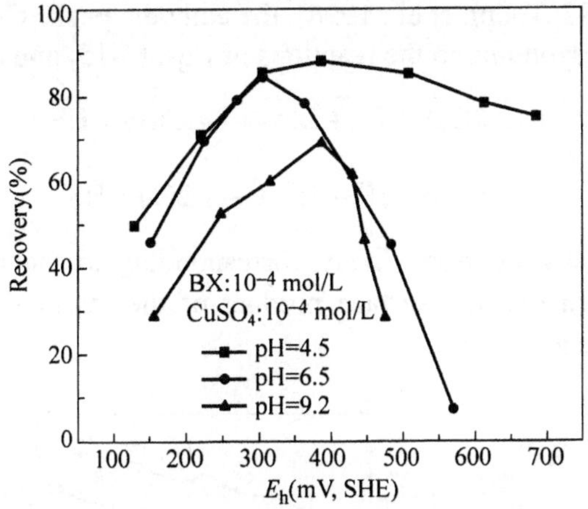

Figure 6.6 Flotation recovery of pyrrhotite as a function of pulp potential in the presence of 10^{-4} mol/L Cu^{2+} with 10^{-4} mol/L butyl xanthate as a collector

considered to be floatable or otherwise nonfloatable, the upper and lower potential limits of flotation of marmatite, arsenopyrite and pyrrhotite at different pH are presented in Table 6.1. It follows that in weak acidic media these minerals have wider potential area of activation flotation by copper ion, which is narrowed in alkaline media. Marmatite and arsenopyrite show larger upper limit potential and pyrrhotite exhibits smaller upper limit potential.

Table 6.1 Eh-pH area of flotation of Zinc-Iron sulphide minerals in the presence of 10^{-4} mol/L Cu^{2+} with 10^{-4} mol/L butyl xanthate as a collector

Minerals	Upper and Lower Limit Potential E_h(mV, SHE)	pH		
		4.5	6.5	9.2
Marmatite	E_h(u)	>800	590	500
	E_h(l)	260	260	220
Arsenopyrite	E_h(u)	800	750	500
	E_h(l)	180	170	180
Pyrrhotite	E_h(u)	>700	480	460
	E_h(l)	110	150	260

6.2.3 Electrochemical Mechanism of Copper Activating Marmatite

Electrochemical mechanism of copper activating marmatite is investigated by using voltammetric method. The voltammogram of the marmatite electrode in the presence of 10^{-4} mol/L Cu^{2+} is presented in Fig. 6.7. It can be seen that in the presence of cupric ion marmatite surface exhibits the electrochemical character of activation products. In the light of E_h-pH diagram of the Cu-S-H_2O system

(Woods et al., 1987; Young et al., 1998), the cathodic peaks C_1, C_2 in Fig. 6.7 are respectively corresponding to the reactions of Eqs. (6-15) and (6-16).

$$C_1\text{: } 2CuS + H^+ + 2e^- \longrightarrow Cu_2S + HS^- \quad (6\text{-}15)$$

$$C_2\text{: } Cu_2S + H^+ + 2e^- \longrightarrow 2Cu + HS^- \quad (6\text{-}16)$$

The anodic peaks are respectively corresponding to the reactions (6-10) to (6-12). It indicated that the surface product of the marmatite electrode in the presence of cupric ion is CuS.

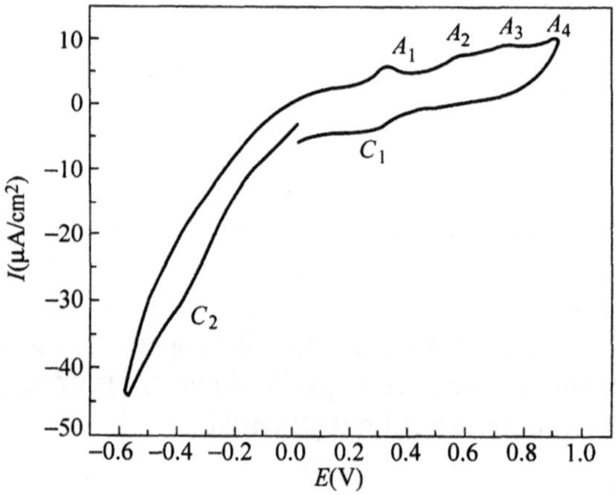

Figure 6.7 Voltammogram of marmatite electrode activated by Cu^{2+} in 0.1 mol/L KNO_3 buffer solution of pH = 9.18 (scanning rate: 20 mV/s, from cathode to anode)

The activating mechanism of cupric ion on marmatite is further investigated by holding marmatite electrode for 2 min at different potential and then scanning from cathode to anode or from anode to cathode. The results are shown in Fig. 6.8. It can be seen from Fig. 6.8 that the less the holding potential is, the less the current of cathodic peak and the greater the current of anodic peak are. At − 278 mV, the cathode shows no sign of the reduction of CuS to Cu_2S. With the increase of the holding potential, the cathodic peak of reduction of CuS to Cu_2S appears. When scanning from anode to cathode as seen from Fig. 6.8(b), the current of anode and cathode peaks is higher at the holding potential − 78 mV than at +122 mV and +322 mV. The formation of CuS is prominent. It suggests that the surface product of marmatite electrode in the presence of Cu^{2+} is CuS at higher potential and Cu_2S at lower potential.

6.2.4 Surface Analysis of Mechanism of Copper Activating Marmatite

The FTIR spectra of ethyl xanthate adsorption on marmatite in the presence of

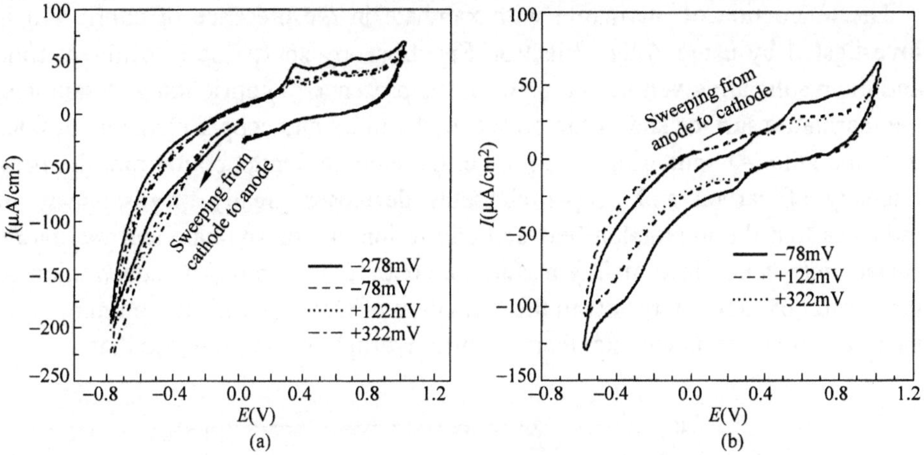

Figure 6.8 Voltammograms of marmatite electrode activated by Cu^{2+} in 0.1 mol/L KNO_3 borate buffer solution of pH = 9.18 by holding 2 minutes at a definite potential (scanning rate: 50 mV/s). (a) from cathode to anode; (b) from anode to cathode

cupric ion is illustrated in Fig. 6.9. The characteristic bands of xanthate at 1195, 1125, 1035 cm^{-1} is observed indicating the formation of CuEX. At pH = 8.8 and pH = 12.1, the intensity of reflection peak is almost the same. The interaction of marmatite with ethyl xanthate in the presence of cupric ion can take place in wide pH region even in strong pH, which accounts for the activation action of copper ion on marmatite flotation in wide pH range as shown in Fig. 6.3.

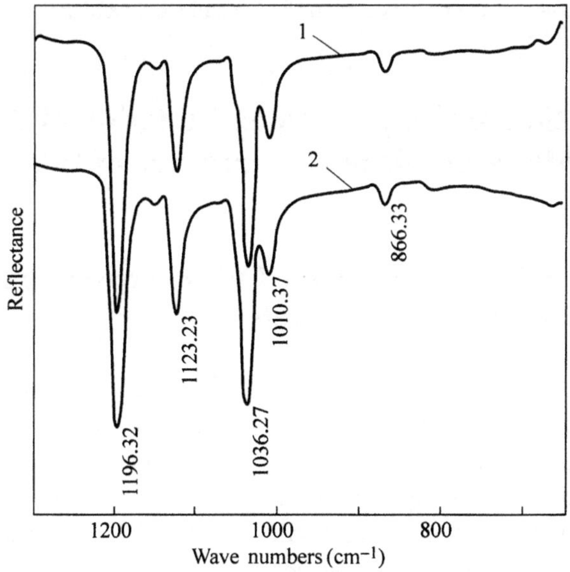

Figure 6.9 FTIR spectra of marmatite in the presence of cupric ion and ethyl xanthate ($CuSO_4$: 5×10^{-3} mol/L; KEX: 5×10^{-3} mol/L; 1: pH = 8.8; 2: pH = 12.1)

The interaction of marmatite with xanthate in the presence of cupric ion is investigsted by using Auger Electron Spectroscopy sputtering at different time and the results are given in Fig. 6.10. In the presence of cupric ion and xanthate, the marmatite surface shows the presence of carbon and copper elements besides iron and zinc. At sputtering 0.4 min, i.e. the etching depth about 1 nm, the peak intensity of carbon and copper elements decreases greatly or disappears. It indicates that the interaction between cupric ion and/or xanthate and marmatite surface may take place mainly at surface depth about 1 nm. The substitution of copper ion on zinc ion results in the formation of CuS or Cu_2S film on marmatite surface, which reacts with xanthate to form hydrophobic cuprous xanthate.

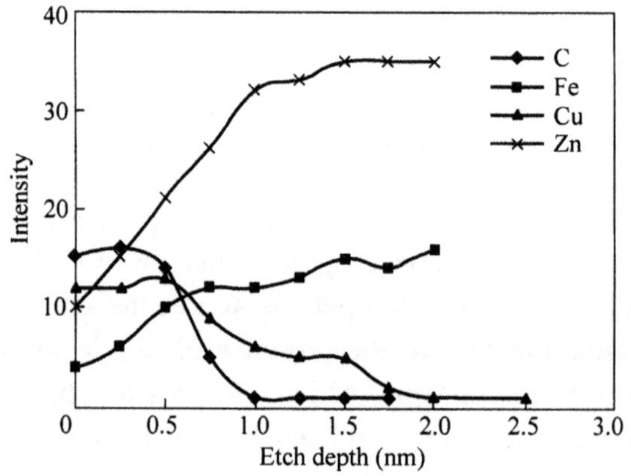

Figure 6.10 Depth profile of Auger Electron Spectroscopy (AES) sputtering of marmatite surface in the presence of cupric ion and ethyl xanthate

6.3 Activation of Copper Ion on Flotation of Zinc-Iron Sulphide Minerals in the Presence of Depressants

6.3.1 Effect of Depressant on the $CuSO_4$ Activating Flotation of Zinc-Iron Sulphide Minerals

The influence of copper ion on the flotation of zinc-iron sulphide minerals in the presence of depressant with butyl xanthate 1.0×10^{-4} mol/L as a collector is presented in Fig. 6.11 to Fig. 6.14. It can be seen from Fig. 6.11 and Fig. 6.12 that in the presence of 120 mg/L 2-hydroxyl ethyl dithio carbonic sodium (GX1) and 2,3 dihydroxyl propyl dithio carbonic sodium (GX2), marmatite is activated by copper ion and exhibits very good flotation with a recovery above 90% in the pH range of 4 – 8. The flotation of arsenopyrite and pyrrhotite is poor with a

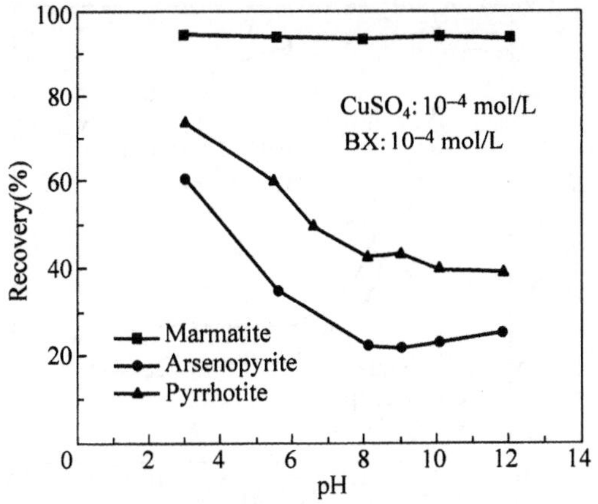

Figure 6.11 Flotation response of zinc-iron sulphide minerals as a function of pH in the presence of 120 mg/L GX1 and 1.0×10^{-4} mol/L $CuSO_4$ with butyl xanthate as a collector

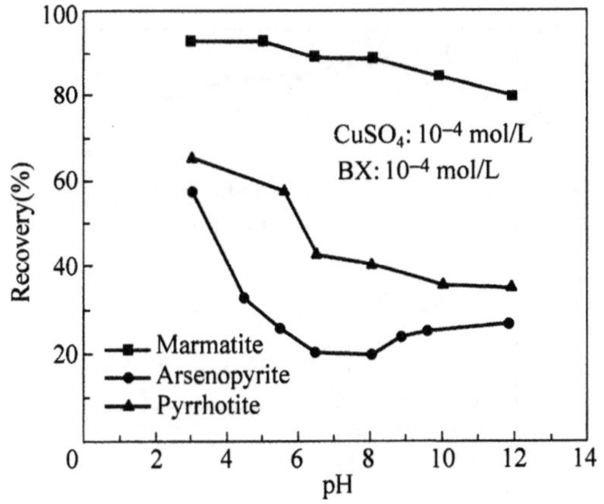

Figure 6.12 Flotation response of zinc-iron sulphide minerals as a function of pH in the presence of 120 mg/L GX2 and 1.0×10^{-4} mol/L $CuSO_4$ with butyl xanthate as a collector

recovery of below 50% at pH > 6. It indicates that the GX1 and GX2 have stronger depressing action on xanthate flotation of pyrrhotite and arsenopyrite even in the presence of copper ion and does not affect the flotation of marmatite, showing the possibility of flotation separation of marmatite from arsenopyrite and pyrrhotite in the presence of copper ion and depressant GX1 or GX2.

Figure 6.13 and Fig. 6.14 demonstrate the flotation results of zinc-iron sulphide minerals with 1.0×10^{-4} mol/L butyl xanthate as a collector in the presence of (1-carbonic sodium-2-hydroxyl) sodium propronate dithio carbonic sodium (TX3) or (1-carbonic sodium-2-sodium acetate) sodium propronate dithio carbonic

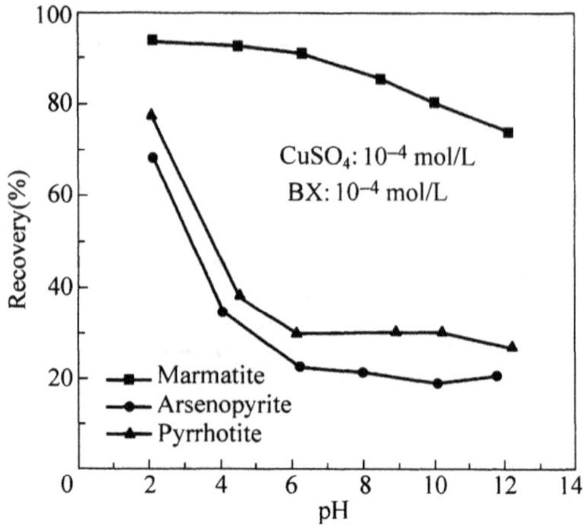

Figure 6.13 Flotation response of zinc-iron sulphide minerals as a function of pH in the presence of 120 mg/L TX3 and 1.0×10^{-4} mol/L $CuSO_4$ with butyl xanthate as a collector

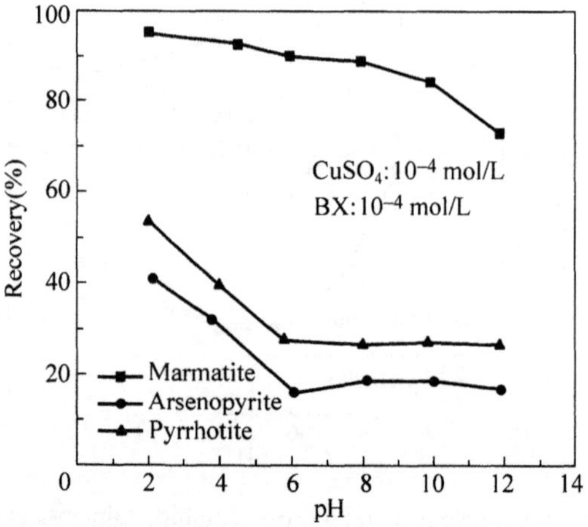

Figure 6.14 Flotation response of zinc-iron sulphide minerals as a function of pH in the presence of 120 mg/L TX4 and 1.0×10^{-4} mol/L $CuSO_4$ with butyl xanthate as a collector

sodium (TX4) and 1.0×10^{-4} mol/L $CuSO_4$. It is obvious that the flotation of marmatite activated by copper ion is not affected by the addition of depressants TX3 and TX4 and the recovery is above 80% in the pH region of 2 – 10. However, TX3 and TX4 exhibit strong depressing action on arsenopyrite and pyrrhotite even in the presence of $CuSO_4$ and the flotation recovery of these two minerals are below 30% at pH > 6. It demonstrates that the flotation separation of marmatite from arsenopyrite and pyrrhotite may be completed in the presence of the depressant TX3 or TX4 and $CuSO_4$ with butyl xanthate as a collector.

6.3.2 Influence of Pulp Potential on the Copper Ion Activating Flotation of Zinc-Iron Sulphide Minerals in the Presence of Depressant

Figure 6.15 to Fig. 6.18 show the influence of pulp potential on the flotation of zinc-iron sulphide minerals in the presence of depressant and $CuSO_4$ with butyl xanthate as a collector. It can be seen from Fig. 6.15 that in the presence of depressant 120 mg/L GX2 copper ion still activates the xanthate flotation of marmatite. The initial potential of marmatite flotation is around 0.3 V and the recovery can reach to 90%. The upper limit potential of marmatite flotation can be high to 0.8 V at pH = 4.5 and around 0.6 V at pH = 6.5 or 9.2. Figure 6.16 shows that the flotation of arsenopyrite is very poor at various pH and wide potential region. The recovery is below 20% in most potential ranges and the maximum recovery is 35%. It further indicates the possibility of flotation separation of marmatite from arsenopyrite and pyrrhotite using xanthate as a collector in the presence of copper ion and depressant GX1 or GX2 by modifying the pulp potential.

Similarly, in the presence of 150 mg/L TX4, copper ion still activates the xanthate flotation of marmatite as shown in Fig. 6.17. Marmatite starts flotation at a potential around 0.3 V with recovery of above 80%. At pH = 4.5, the upper limit potential of flotation of marmatite can be high to above 0.6 V with a recovery about 90%. At pH = 6.5 and 9.2, the upper limit potential of flotation of marmatite decreases to about 0.5 V. Arsenopyrite can not be floated in the same conditions with a recovery of below 30% as seen from Fig. 6.18. This result also suggests that the flotation separation of marmatite from arsenopyrite may be accomplished by using TX4 as a depressant and xanthate as a collector in the presence of copper ion through the control of pulp potential and pH.

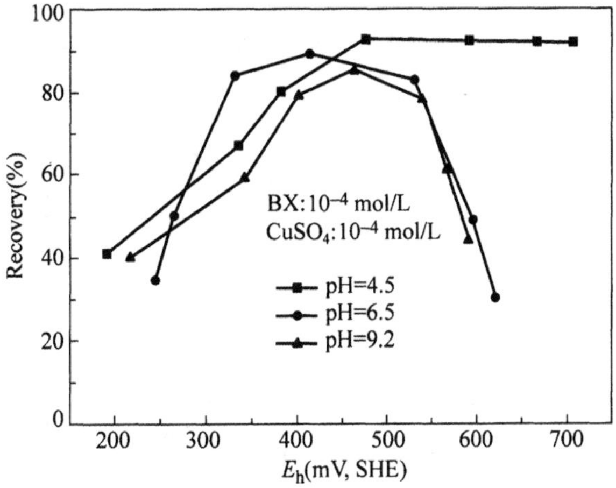

Figure 6.15 Flotation recovery of marmatite as a function of pulp potential in the presence of 120 mg/L GX2 and 10^{-4} mol/L $CuSO_4$

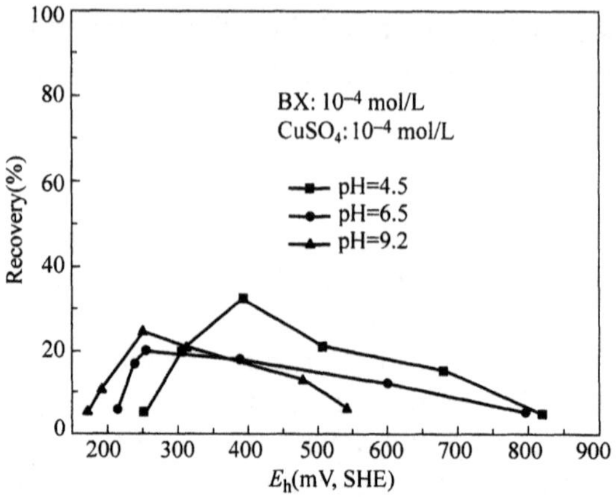

Figure 6.16 Flotation recovery of arsenopyrite as a function of pulp potential in the presence of 120 mg/L GX2 and 10^{-4} mol/L $CuSO_4$

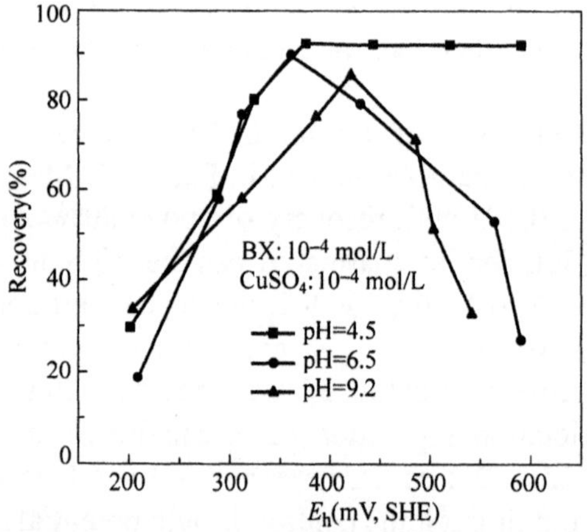

Figure 6.17 Flotation recovery of marmatite as a function of pulp potential in the presence of TX4 and $CuSO_4$

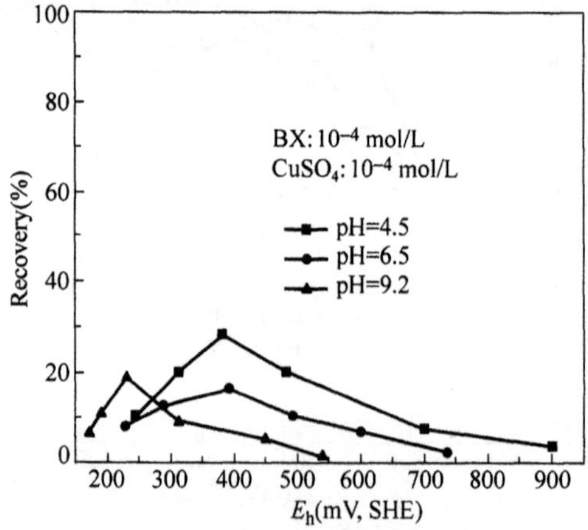

Figure 6.18 Flotation recovery of arsenopyrite as a function of pulp potential in the presence of TX4 and $CuSO_4$

6.3.3 Zeta Potential of Zinc-Iron Sulphide Minerals in the Presence of Flotation Reagents

The zeta potential of zinc-iron sulphide minerals in various flotation reagents are presented in Fig. 6.19 to Fig. 6.22 as a function of pH. It can be seen from Figs. 6.19 and 6.20 that marmatite and arsenopyrite are negatively charged in wide pH range. By addition of 1.0×10^{-4} mol/L butyl xanthate, the zeta potential of marmatite and arsenopyrite are more negative indicating the adsorption of xanthate on these two minerals. Depressants 2,3-dihydroxyl propyl dithio carbonic sodium (GX2) evidently affect the zeta potential of marmatite and arsenopyrite, which became much more negative in the presence of 120 mg/L GX2, indicating the adsorption of GX2 on these two minerals. The negative zeta potential of arsenopyrite is increased more than that of marmatite in the presence of GX2,

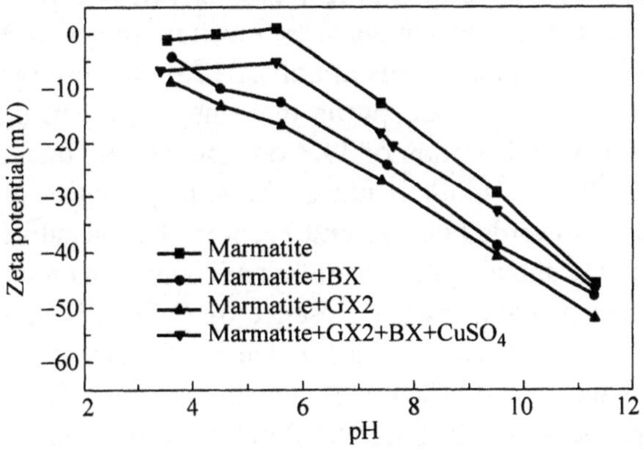

Figure 6.19 Zeta potential of marmatite as a function of pH in butyl xanthate/GX2/$CuSO_4$ system

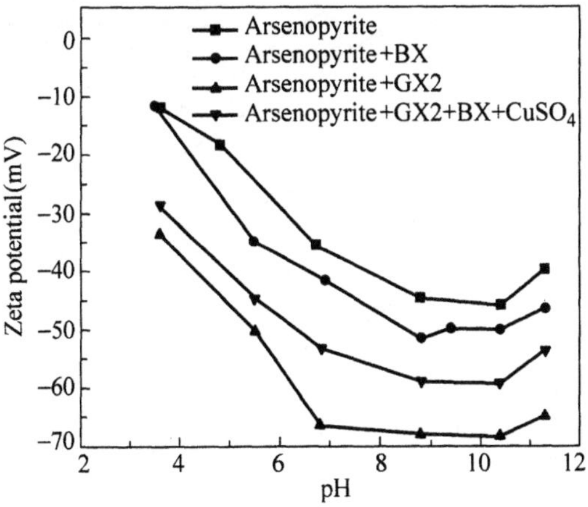

Figure 6.20 Zeta potential of arsenopyrite as a function of pH in butyl xanthate/GX2 /$CuSO_4$ system

suggesting the stronger adsorption of GX2 on arsenopyrite than on marmatite. Further by the addition of 1.0×10^{-4} mol/L $CuSO_4$, the zeta potential of marmatite and arsenopyrite moves towards positive and becomes less negative compared with the absence of $CuSO_4$, showing the adsorption of copper ion on these two minerals. The change of zeta potential in the presence of $CuSO_4$, is more for marmatite than arsenopyrite. It demonstrates that in butyl xanthate/depressant/ $CuSO_4$ system copper ion may be adsorbed more on marmatite and the GX2 may be adsorbed more on arsenopyrite accounting for the flotation results from Fig. 15 to Fig. 16, i.e. the depressants GX2 exhibits stronger depressing action on arsenopyrite even in the presence of copper ion and affect little on the copper ion activation flotation of marmatite.

Similarly, Figs. 6.21 and 6.22 show that by the addition of 1.0×10^{-4} mol/L butyl xanthate, the zeta potential of marmatite and arsenopyrite are more negative. The addition of depressants (1-carbonic sodium-2-sodium acetate) sodium propronate dithio carbonic sodium (TX4) evidently affects the zeta potential of marmatite and arsenopyrite, which becomes much more negative in the presence of 150 mg/L TX4, indicating the adsorption of TX4 on these two minerals. TX4 makes the zeta potential of arsenopyrite more negative than that of marmatite, suggesting the stronger adsorption of TX4 on arsenopyrite than on marmatite. In the presence of 1.0×10^{-4} mol/L $CuSO_4$, the zeta potential of marmatite and arsenopyrite moves towards positive and becomes less negative compared with the absence of $CuSO_4$, showing the adsorption of copper ion on these two minerals. The change of zeta potential in the presence of $CuSO_4$, is more for marmatite than arsenopyrite. It demonstrates that in the butyl xanthate/depressant/$CuSO_4$ system copper ion may be adsorbed more on marmatite and TX4 may be adsorbed more on arsenopyrite accounting for the flotation results that the TX4 exhibites

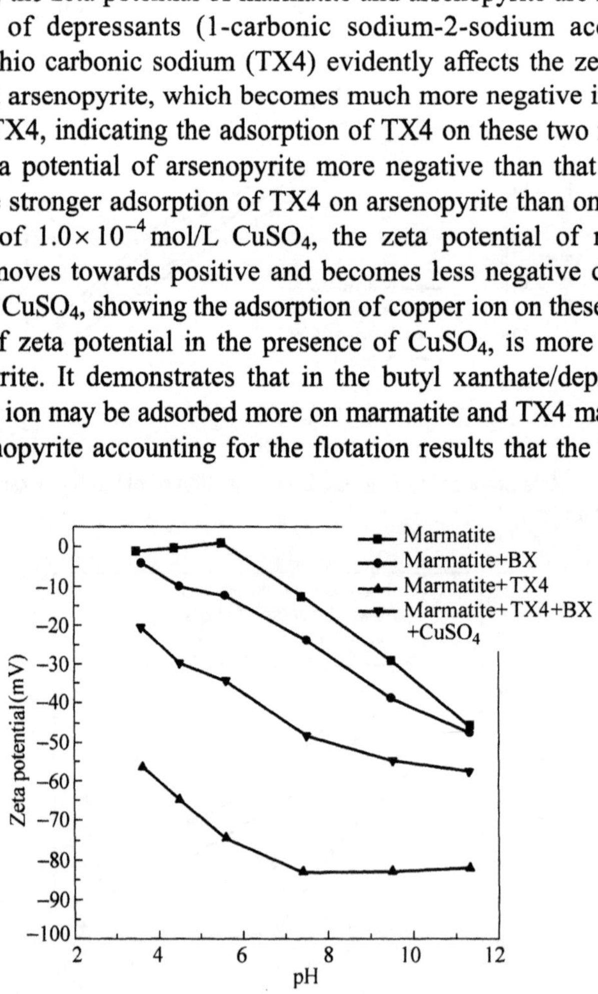

Figure 6.21 Zeta potential of marmatite as a function of pH in butyl xanthate/ TX4 /$CuSO_4$ system

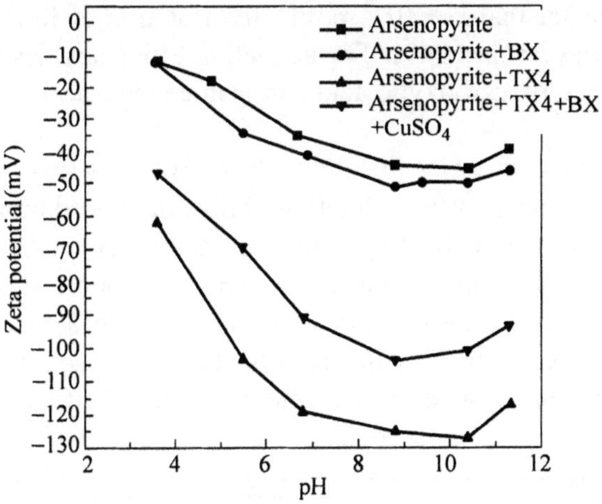

Figure 6.22 Zeta potential of arsenopyrite as a function of pH in butyl xanthate/ $TX4/CuSO_4$ system

stronger depressing action on arsenopyrite even in the presence of copper ion and does not influence the copper ion activation flotation of marmatite.

6.4 Surface Chemistry of Activation of Lime-Depressed Pyrite

In the flotation of poly-metallic sulphide minerals, especially complex sulphide ores, the pyrite can only be depressed at very high additions of lime CaO (at pH > 12) such is the case at the Fankou Mine in China where the pyrite is very floatable. After the flotation of galena and sphalerite, the activation and flotation of pyrite caused a predicament due to the extremely high pH. Sulfuric acid is the main activator for the flotation of lime-depressed pyrite, which causes environmental and calcification problems due to the large additions of sulfuric acid required. As suggested from Fig. 5.5 the formation of $Ca(OH)_2$ and $CaSO_4$ takes place at the pyrite surface because of the high pH and the oxidizing atmosphere. At high pH, the formation of $Fe(OH)_3$ is also expected .These hydrophilic species account for the complete depression of pyrite by lime treatment. The activation flotation of the lime-depressed pyrite is to remove these hydrophilic coatings.

6.4.1 Activation Flotation of Lime-Depressed Pyrite

A series of organic acids, inorganic acids and their salts have been tested as possible activators for the flotation of lime depressed pyrite. The results are presented in Fig. 6.23 and Fig. 6.24. It can be seen from Fig. 6.23 that oxalic acid is the strongest activator in the organic acid series. When the concentration of oxalic acid is greater than 5×10^{-3} mol/L, the flotation of lime depressed pyrite can be restored and reaches 94%. The activating efficiency descends in the order of oxalic acid > butyl bicarboxylic acid > propanic acid > acetic acid > poly amino phosphoric acid.

For the inorganic acid series, it follows from Fig. 6.24 that sulfuric acid is the strongest activator. The activation flotation of lime depressed pyrite can reach 80% with 5×10^{-3} mol/L. The activating efficiency decreases in the order of sulfuric acid > poly-phosphoric acid > phosphoric acid > carbonic acid > hydrochloric acid. At a high dosage, phosphoric acid and carbonic acid also exhibit good activation ability. According to the results above, the activating efficiency of oxalic acid and sulphuric acid on the lime depressed pyrite ore (flotation tailing

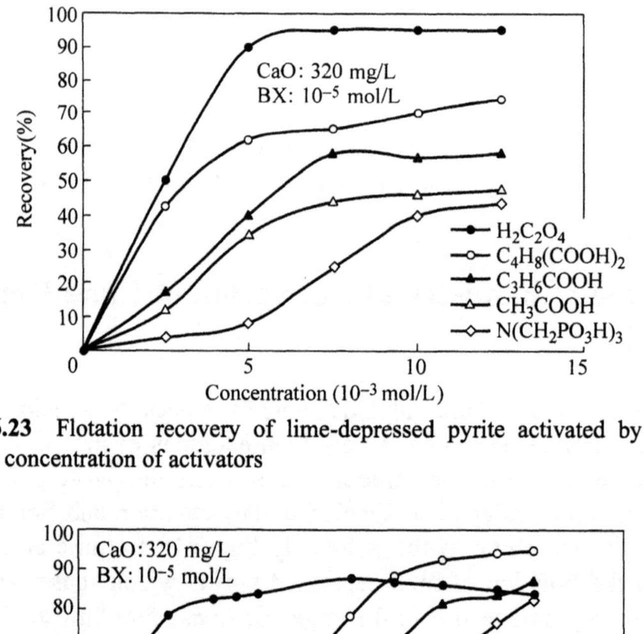

Figure 6.23 Flotation recovery of lime-depressed pyrite activated by organic acids vs. concentration of activators

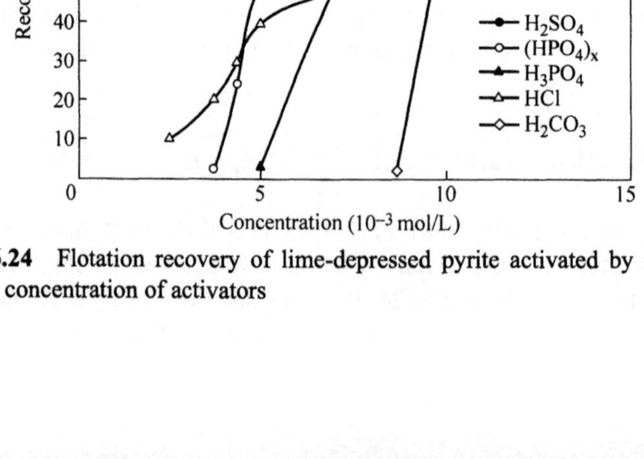

Figure 6.24 Flotation recovery of lime-depressed pyrite activated by inorganic acids vs. concentration of activators

from the Fankou Pb-Zn circuit) has been tested. The results are compared in Table 6.2. It can be seen that both oxalic acid and sulphuric acid can be used as effective activators for the flotation of lime-depressed pyrite. A pyrite concentrate with a high sulphur grade and recovery can be obtained. Table 6.2 also shows that, however, the dosage of oxalic acid is half that of sulphuric acid, which suggests that organic acids compete with sulphuric acid and reduce the environmental impact as well as calcification of pipelines.

Table 6.2 The results of activation flotation of lime-depressed pyrite ore from the tailing of flotation of Pb-Zn sulphide ores

Experiments	Products	Yield (%)	Grade (%)	Recovery (%)
Oxalic acid 8 kg/t	S conc.	52.23	48.92	94.84
Ethyl xanthate 220 g/t, pine oil 130 g/t, pH = 7.8	S tailing	47.77	2.91	5.16
	S feed ore	100.00	26.94	100.00
Sulphuric acid 16 kg/t, ethyl xanthate 220 g/t, pine oil 130 g/t, pH = 7.2	S conc.	46.88	50.67	93.74
	S tailing	53.12	3.00	6.26
	S feed ore	100.00	25.34	100.00

6.4.2 Solution Chemistry Studies on Activation Flotation of Lime-Depressed Pyrite

Under certain conditions, oxalic acid (let L^{2-} represent the oxalate anion) can form soluble complexes with calcium and iron ions.

$$Fe^{3+} + L^{2-} = FeL^{+} \tag{6-17}$$

$$K_1 = 10^{7.56}$$

$$Fe^{3+} + 2L^{2-} = FeL_2^{-} \tag{6-18}$$

$$K_2 = 10^{13.64}$$

$$Fe^{3+} + 3L^{2-} = FeL_3^{3-} \tag{6-19}$$

$$K_3 = 10^{18.49}$$

$$Me^{2+} + L^{2-} = FeL \tag{6-20}$$

$$\beta_1 : Ca^{2+}\ \ 10^{1.6};\quad Fe^{2+}\ 10^{3.05}$$

$$Me^{2+} + 2L^{2-} = FeL_2^{2-} \tag{6-21}$$

$$\beta_2 : Ca^{2+}\ \ 10^{2.69};\quad Fe^{2+}\ 10^{5.15}$$

Let $[Fe^{3+}]$ express the total concentration of species containing Fe^{3+}, then

$$[Fe^{3+}] = [Fe^{3+}] + [FeL^{+}] + [FeL_2^{-}] + [FeL_3^{3-}] \tag{6-22}$$

Let Φ_0, Φ_1, Φ_2, Φ_3, represent the percentage of various complexing species Fe^{3+}, FeL^{+}, FeL_2^{-} and FeL_3^{3-}, then

$$\Phi_0 = \frac{[Fe^{3+}]}{[Fe^{3+}]_T} = \frac{1}{1+K_1[L^{2-}]+K_2[L^{2-}]^2+K_3[L^{2-}]^3} \tag{6-23}$$

$$\Phi_1 = K_1[L^{2-}]\Phi_0 \tag{6-24}$$

$$\Phi_2 = K_2[L^{2-}]^2\Phi_0 \tag{6-25}$$

$$\Phi_3 = K_3[L^{2-}]^3\Phi_0 \tag{6-26}$$

If c_L represents the total concentration of oxalate anion, then

$$[L^{2-}] = \frac{c_L}{\alpha_L} \tag{6-27}$$

$$\alpha_L = 1+\beta_1^H[H^+]+\beta_2^H[H^+]^2, \quad \beta_1^H = 10^{4.27}, \beta_2^H = 10^{5.52} \tag{6-28}$$

where α_L is the coefficient for the protonation of oxalate.

The percentage distribution of various complexes in such systems can be calculated using the equations above and the results are presented in Fig. 6.25. It may be seen that in the presence of oxalic acid, the dominant ferric complex is FeL_3^{3-}, even at lower dosage. The dominant complexes of Fe^{2+} are FeL and FeL_2^{2-},

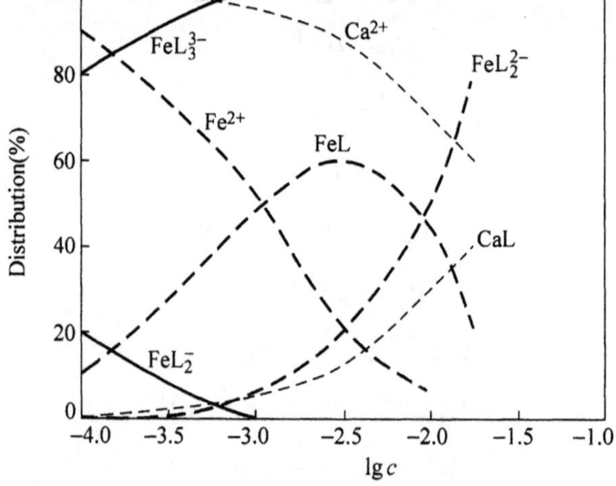

Figure 6.25 The percentage distribution of complexing species formed between oxalate and metal ions vs. the concentration of oxalate

respectively, at a concentration of oxalate greater than 10^{-3} mol/L and 10^{-2} mol/L. The formation of CaL occurs only at a concentration of oxalate greater than 10^{-2} mol/L.

Similarly, for the system of iron/calcium/phosphate, the percentage distribution of various complexes can also be calculated using solution equilibrium calculations as shown in Fig. 6.26. It follows that depending on solution pH, the dominant complexes is $CaPO_4^-$ at pH = 10, whereas $CaHPO_{4(aq)}$ and $CaH_2PO_4^+$ are dominant at pH = 8.

The soluble complexes formed by activators will desorb cation from the lime depressed pyrite surface, which will expose a fresh pyrite surface and activate pyrite flotation. Therefore, the moderately strong acids such as oxalic acid and phosphoric acid exhibit a strong activation action on lime-depressed pyrite because of their ability to decrease pulp pH and to form soluble complexes with hydrated surface cations.

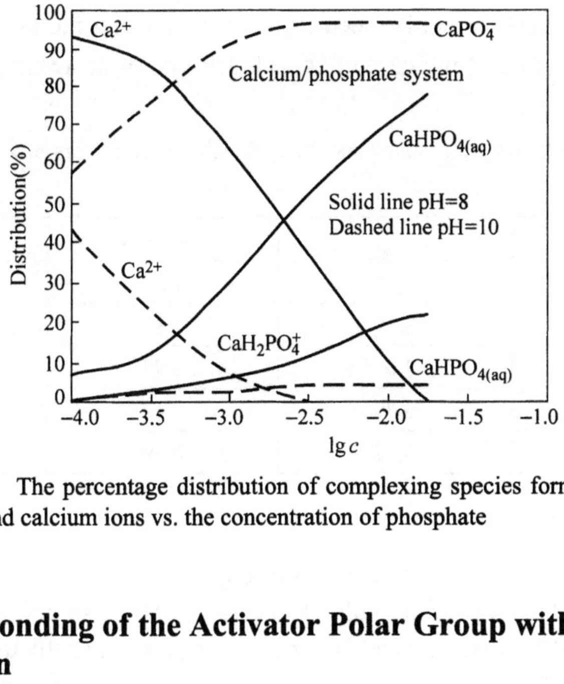

Figure 6.26 The percentage distribution of complexing species formed between phosphate and calcium ions vs. the concentration of phosphate

6.4.3 The Bonding of the Activator Polar Group with Surface Cation

In order to form stable complexes with calcium and iron ions, the activator must have a strong affinity for the cations. The bonding strength can be determined by the group electronegativity as given by the following equation:

$$X_g = 0.31\left(\frac{n^* + 1}{r}\right) + 0.5 \tag{6-29}$$

$$n^{*}=(N-P)+\sum 2m\frac{X_{A}}{X_{A}+X_{B}}+\sum s\left(\frac{X_{B}-X_{A}}{X_{A}+X_{B}}\right)$$
$$+\sum\frac{2m_{i}+s_{i}}{\alpha_{i}}\left(\frac{X_{i+1}-X_{i}}{X_{i+1}+X_{i}}\right) \quad (6\text{-}30)$$

where n^{*} is the effective valence electron numbers; r is the covalent radius of the bonding atom; N is the valence electron numbers of bonding; P is the electron numbers of bonding atom bonded by neighboring atom B;X is electronegativity of element; i is the order number of the group; m is the number of electron pairs for the bond between A and B; s is the number of electron pairs unbonded between A and B; α_i is the separated constant (usually taken as 2.7). The X_g value reflects the bonding ability of the activator with a particular cation. A better activator should have a greater X_g values for some activators which are calculated and given in Table 6.3. Also shown in Table 6.3 are the pK_a values for the activators of interest. It can be seen from Table 6.3 that the activating ability is increased in the order of increase of X_g values, and of decrease of pK_a values.

Table 6.3 The pK_a, X_g values and calcium salts of activators and their activity index A_i

Activator	X_g	pK_a (20)	$A_i = X_g/\text{p}K_a$
$H_2C_2O_4$	3.74	1.27	2.94
H_2SO_4	3.69	1.9*	1.94
$(HPO_4)_n$	3.6	1.94	1.86
H_3PO_4	3.59	2.12	1.69
H_2CO_3	3.62	6.39	0.57
HCl	2.69		

Therefore, the efficiency of an activator depends on its acidity which decreases pulp pH, and its bonding ability with cations which forms soluble stable complexes with cations at the pyrite surface. In other words, a better activator should have a lower pK_a value(stronger acidity) and a greater X_g value (stronger affinity). A combination of these factors defines the activity index A_i as follows:

$$A_i=\frac{X_g}{\text{p}K_a} \quad (6\text{-}31)$$

The greater the A_i value, the stronger the activating action of an activator. Table 6.3 demonstrates that the activating efficiency of activators is in the same order as the A_i value.

6.4.4 Surface Analysis of Lime-Depressed Pyrite in the Presence of Activator

Figure 6.27 shows the XPS spectra of the pyrite surface under different conditions. The characteristic peaks Ca(2p),Ca(2s) and Fe^{3+} disappear by the addition of oxalic acid when compared to spectra in the absence of activator. In the XPS extended spectra for oxygen and carbon (presenting since the system is open to atmosphere) at the pyrite surface, no change is found after the addition of oxalic acid, indicating that no adsorption of oxalate occurs at the pyrite surface. The results in Fig. 6.27 also indicate that the soluble complexes formed by activators with cation hydroxides of $Ca(OH)_2$ and $Fe(OH)_3$ desorb from the pyrite surface.

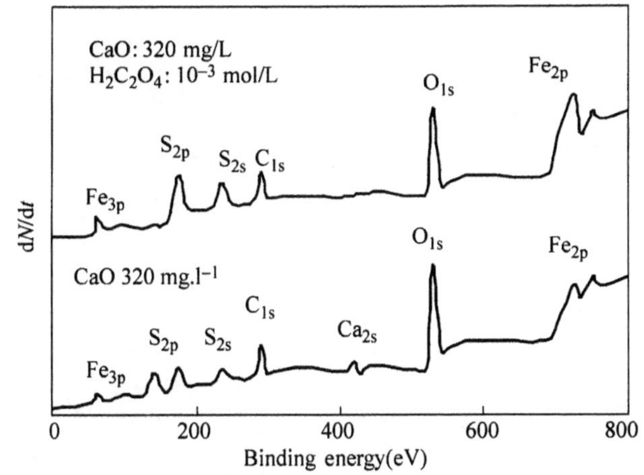

Figure 6.27 XPS spectra of lime depressed pyrite in the absence and presence of activator

Figure 6.28 presents the anodic current of the pyrite electrode in different solutions. The results show that in the absence of activator, the anodic current evidently increases at a potential above 0.1 V and the strong oxidation of pyrite surface takes place at a high dosage of CaO. In the presence of activator, the potential at which the anodic current begins to rise is greatly increased, in the order of hydrochloric acid, sulfuric acid, oxalic acid and phosphoric acid, indicating that activators will inhibit the surface oxidation of pyrite by the high dosage of CaO. Therefore, the mechanism that the organic or inorganic acid activates the lime-depressed pyrite flotation may be attributed to that these activators prevent the pyrite surface oxidation and expose the bared surface to recover pyrite floatability through their complexing action.

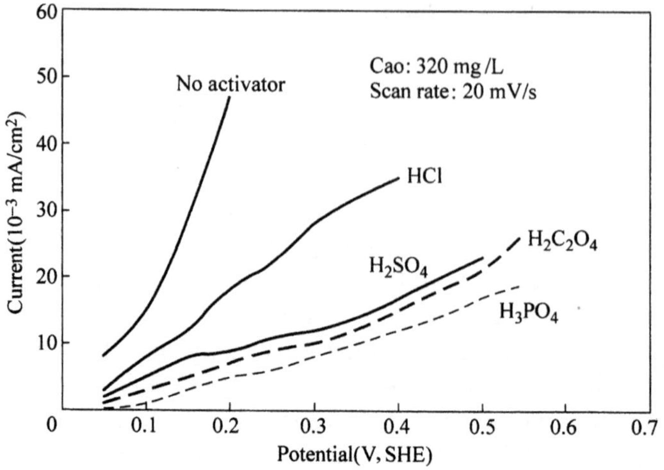

Figure 6.28 The anodic current of pyrite electrode in different solutions

Chapter 7 Corrosive Electrochemistry of Oxidation-Reduction of Sulphide Minerals

Abstract The flotation mechanism is discussed in the terms of corrosive electrochemistry in this chapter. In corrosion the disolution of minerals is called self-corrosion. And the reaction between reagents and minerals is treated as inhibition of corrosion. The stronger the ability of inhibiting the corrosion of minerals, the stronger the reagents react with minerals. The two major tools implied in the research of electrochemical corrosion are polarization curves and EIS (electrochemistry impedance spectrum). With these tools, pyrite, galena and sphalerite are discussed under different conditions respectively, including interactions between collector with them and the difference of oxidation of minerals in NaOH solution and in lime. And the results obtained from this research are in accordance with those from other conventional research. With this research some new information can be obtained while it is impossible for other methods.

Keywords corrosive electrochemistry; corrosive potential; corrosion inhibition; polarization curves; Electrochemistry Impedance Spectrum

7.1 Corrosive Electrochemistry

As is known to all, the flotation mechanism of sulphide minerals can be explained based on electrochemistry because sulphide minerals have the semiconductor character and a series of electrochemistry reaction occurring in solution. After these reactions, the surface of sulphide minerals changes and forms a new phase. We called it as self-corrosion of sulphide minerals. As before, the essence of the reaction between the collector and the minerals is the formation of the hydrophobic entity on the mineral surface, and then minerals can be floated. We can find that the reaction between the collector and the minerals is similar to the depression on mineral self-corrosion. In the corrosion, we called this effect as inhibition, and this kind of reagent is an inhibiting reagent. There are many studies on corrosion, especially its research method and theory. Thus, we can get some new information on the mechanism of sulphide flotation from corrosive electrochemistry.

7.1.1 Concept and Significance of Mixed Potential, Corrosive Potential and Static Potential

Corrosion can be divided into two parts, chemical corrosion and electrochemical corrosion. Chemical corrosion is a direct reaction between the medium and the materials, while electrochemical corrosion is an electron transformation on different surfaces of the metal because of conductivity. In this chapter, all corrosions are referred as electrochemical corrosions.

There existed oxidation-reduction reactions with the same reaction speed on the sulphide mineral surface in water. One is the self-corrosion of sulphide mineral. Another is the reduction of oxygen. If the equilibrium potential for the anodic reaction and the cathodic reaction are, respectively, E_a^ϕ and E_c^ϕ, and the mineral electrode potential is E, the relationship among them is as follows:

$$E = E_a^\phi + \eta_1 = E_c^\phi - \eta_2 \tag{7-1}$$

where η_1 and η_2 are the over-potential for the anodic reaction and cathodic reaction, respectively. Mineral electrode potential is unequal potential, not only for the anode but also for the coupling cathode. Electrode potential E is named as the mixed potential of the whole electrode reactions. It is also named as the corrosion potential in corrosion or static potential for mineral in flotation electrochemistry.

It can react only when $E_c - E_a > 0$. In other words, sulphide minerals are corrupted only when the reduction equilibrium potential for an oxidant is higher than the oxidation equilibrium potential in sulphide mineral solution. Generally speaking, the oxidant is oxygen. Oxygen is the essential condition for the electrochemistry flotation of sulphide mineral.

The concept of mixed potential can be extended to multi-electrode reactions on one electrode. If N electrode reactions happen on a single electrode simultaneously, the outside electrode current is up to zero. When $N > 2$, all these electrode reactions constitute the multi-electrode reaction coupled system, in which a part of the electrode reactions belong to the anodic reaction and the others are cathodic reaction. Given that the current of the anodic reaction is positive value and that of the cathodic reaction is negative value, the following formula can be given.

$$I = \sum_{j=1}^{N} I_j = 0 \tag{7-2}$$

I_j is the current density of reaction j. Given that the over-potential of the anodic reaction is positive value and that of the cathodic reaction is negative value, and all N reactions happen at the same mixed potential, the following formula can be given.

$$E = E^{\phi}_{e_1} + \eta_1 = E^{\phi}_{e_2} + \eta_2 = \cdots = E^{\phi}_{e_j} + \eta_i = \cdots = E^{\phi}_{e_N} + \eta_N \tag{7-3}$$

E_{e_j} is the equilibrium potential of reaction j, η_i is the over-potential of reaction i.

The formation of the multi-electrode coupled system must accord with the two above-mentioned formulas. Mixed potential E can be gained by factual measurements, and have the following characteristics generally.

(1) Total current of the reaction is zero.

(2) In the multi-electrode reaction coupled system, the mixed potential E always locates between the highest and the lowest equilibrium potential. That is to say, at least there is an electrode equilibrium potential lower or higher than the mixed potential.

(3) When the equilibrium potential is higher than the mixed potential E, the reactions occur towards cathode reaction.

Mineral static potential or mineral electrode potential is an important parameter. It plays an important role in flotation. It has been demonstrated that the floatability of minerals has direct relation with electrode potential because the grinding-flotation system is similar to the multi-electrode reaction coupled system. Therefore, it is important to study the variance rules of the mineral electrode potential under different conditions.

7.1.2 The Concept of Corrosive Current and Corrosive Speed

$E_{e,a}$ and $E_{e,c}$ represent the equilibrium potential of mineral anodic dissolution and cathode reduction of oxygen, respectively. E_{corr} represents the mineral mixed potential in certain system. $I_{0,a}$ and $I_{0,c}$ are current density of anodic and cathode reaction, respectively. When the discharge is the controlled step of electrode reaction, according to electrochemistry theory, the equation can be described as following:

$$I_a = I_{0,a}\left[\exp\left(\frac{E_K - E_{e,a}}{\frac{RT}{n\alpha_a F}}\right) - \exp\left(\frac{E_K - E_{e,a}}{\frac{RT}{n(1-\alpha_a)F}}\right)\right] \tag{7-4}$$

$$|I_c| = I_{0,c}\left[\exp\left(\frac{E_K - E_{e,c}}{\frac{RT}{n(1-\alpha_a)F}}\right) - \exp\left(\frac{E_K - E_{e,c}}{\frac{RT}{n\alpha_a F}}\right)\right] \tag{7-5}$$

where α is transmission factor, when potential is E_{corr}

$$I_a = |I_c| = I_{corr} \tag{7-6}$$

I_{corr} is corrosive current density, used to describe the corrosive speed of minerals. Corrosive current value indicates the speed of electrochemistry reaction on the mineral surface.

7.1.3 The Corrosion Inhibitor, Inhibiting Corrosive Efficiency and Its Relationship with Collector Action

Corrosion inhibitor is a chemical which prevents metal from corrosion in a medium at certain concentration. For sulphide minerals, due to the addition of collector and pH regulator, the hydrophobic film of a collector or hydrophilic film of hydroxyl forms on the surface of minerals in the process of flotation, which prevents sulphide mineral from being corrosive further because of the poor film conductivity. The study of inhibiting corrosive efficiency may be helpful to understand the mechanism of sulphide minerals-collector-pH regulator system. Generally speaking, inhibiting corrosive efficiency can be expressed by the following formula:

$$\eta = \frac{R_{p'} - R_p}{R_{p'}} \tag{7-7}$$

where R_p and $R_{p'}$ are polarization resistance in water system and in corrosion inhibitor system, respectively. Polarization resistance is resistance which inhibits electron transfer on the surface of sulphide mineral. The more the resistance is, the more difficult the reaction is. The higher the inhibiting corrosive efficiency is, the stronger the interaction between sulphide mineral and collector is.

7.2 Self-Corrosion of Sulphide Minerals

Figure 7.1 is the polarization curves for pyrite, galena and sphalerite electrode in 0.1 mol/L KNO_3 at natural pH. Electrochemistry parameters obtained by the computer PARcal are listed in Table 7.1. The results indicate that the corrosive potential and the corrosive current of pyrite are the highest. The corrosive potential of galena is the lowest, while its corrosive current is lower than that of pyrite and higher than that of sphalerite. The corrosive potential of sphalerite is in between that of pyrite and galena and its corrosive current is the lowest. These results suggest that the oxidation to be easier in the order of pyrite > sphalerite > galena at given potential. The oxidation of pyrite would be the fastest.

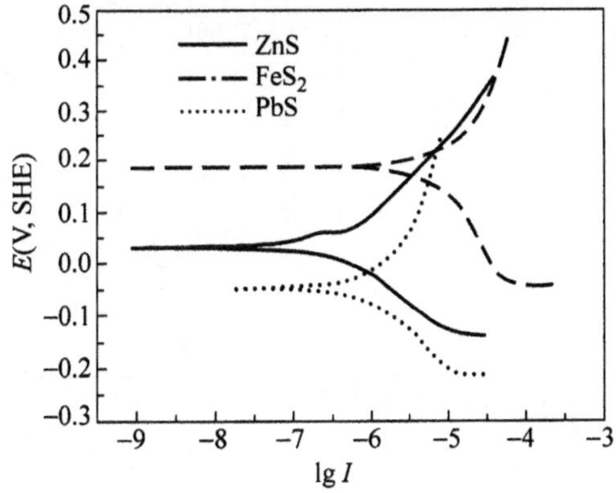

Figure 7.1 Polarization curves of sulphide minerals electrode at natural pH (KNO_3: 0.1 mol/L; unit of I: A/cm^2)

Table 7.1 Tafel parameters of sulphide minerals electrode at natural pH

Minerals	I_{corr} (μA/cm^2)	E_{corr} (mV, SHE)	b_a (mV)	b_c (mV)	R_p (kΩ)
Pyrite	10.78	187	124	125	6.2
Galena	3.45	−48	122	117	13.3
Sphalerite	0.13	42	129	120	48.6

Figure 7.2 shows the electrochemistry impedance spectrum (EIS) of sulphide minerals electrode at natural pH under the corrosive potential. The ESI are in the x-y quadrant, x represents the resistance and y represents capacitance. It can be seen from Fig. 7.2 that three sulphide minerals appear as a single capacitive loop, suggesting that the electron transfer circuit on a mineral surface is equivalent to a parallel circuit of a capacitance and resistance as shown in Fig. 7.3. In Fig. 7.3, R_e represents the solution resistance by comparing the electrode to a working electrode. C represents the double layer interface capacitance of sulphide minerals. R_F represents the Faraday transfer resistance, which is generally equal to polarization resistance of a mineral. The polarization resistance reflected the difficult degree of electron transfer on a mineral surface. The polarization resistance can be determined according to the intersection point between capacitive loop and x-axis so that the longer the radius of capacitive reactance loops, the bigger of resistance of a mineral surface. It can be seen from Fig. 7.2 that the radius of capacitive reactance loop decreases and hence surface resistance descends in the order of pyrite, sphalerite and galena, showing that the oxidation products are formed on the mineral surface with more on pyrite. i.e. the oxidation of the three minerals descends in the order of pyrite, sphalerite and galena.

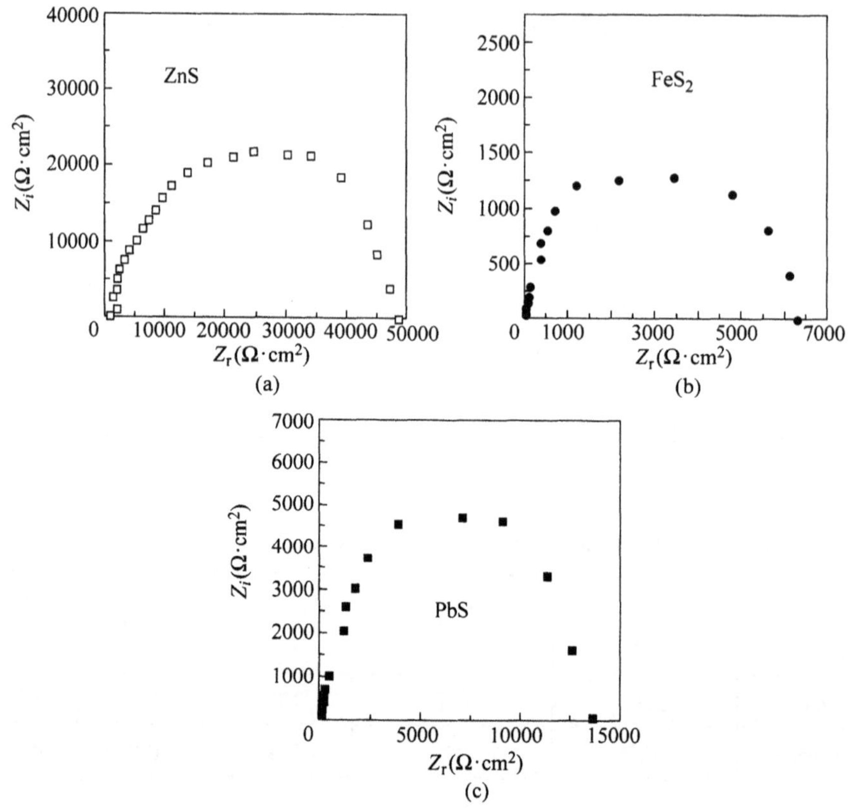

Figure 7.2 EIS of sulphide minerals electrode at natural pH (KNO_3: 0.1 mol/L)

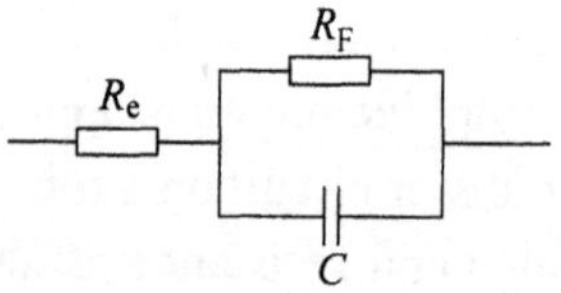

Figure 7.3 The equivalent circuit for measure impedance of mineral

7.3 Corrosive Electrochemistry on Surface Redox Reaction of Pyrite under Different Conditions

7.3.1 The Oxidation of Pyrite in NaOH Medium

Figures 7.4 and 7.5 are the polarization curves and EIS of pyrite under different concentration of the NaOH media. The results show that its corrosive potential move towards negatively and the corrosive current density decreases with the increase of NaOH concentration. The capacitance loop radius increases with the increase of NaOH concentration. The reaction resistance increases from 6000 in the absence of NaOH to 8500 in the presence of 10^{-2} mol/L NaOH. EIS exhibits passivation characteristic. The results indicate the formation of surface oxidation products which prevents from transferring of electron, giving rise to the increase of surface resistance and the descent of corrosive current. With the increase of NaOH concentration, the surface oxidation is stronger and the corrosive current density decreases more.

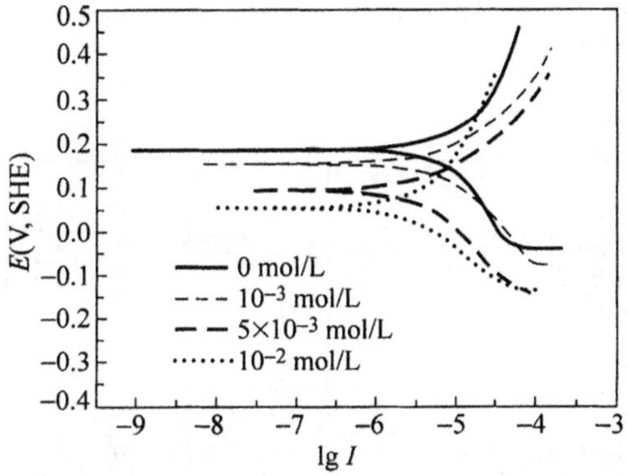

Figure 7.4 Polarization curves of pyrite electrode at natural pH and in NaOH medium (KNO_3: 10^{-1} mol/L; unit of I: A/cm^2)

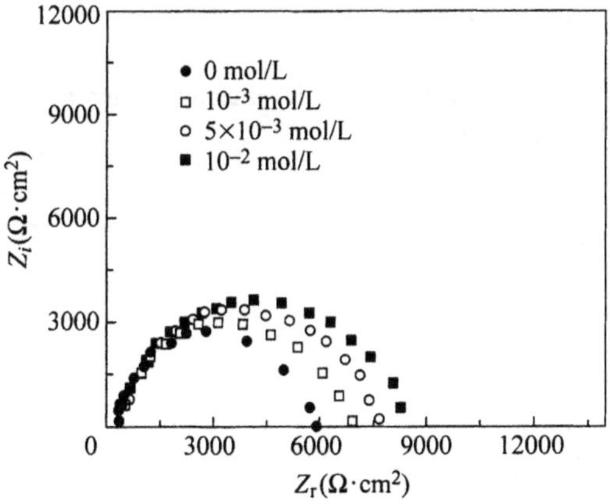

Figure 7.5 EIS of pyrite electrode under different NaOH concentration (KNO_3: 0.1 mol/L)

Figure 7.6 is the EIS of pyrite under different potential conditions in NaOH solution. The relationship between polarization resistance and potential can be further demonstrated by Fig. 7.7. It can be seen from Fig. 7.6 and Fig. 7.7 that when the anodic polarization potential is between 50 and 330 mV, all the curves appear as a single capacitive reactance loop. But when the potential is between 50 and 250 mV, the capacitive reactance loop radius increased with the enhancement of the electrode potential. This may be due to the formation of hydroxyl iron precipitate on the mineral surface. The surface resistance increases from 8700 to 11600 with the increase of potential from 50 to 250 mV and the growth of surface oxide film is the controlled step. When the potential is between 270 and 330 mV, the capacitive reactance loop radius decreases from 1100 to 6000 at potential from 270 to 330 mV, indicating that the dissolution of hydroxyl precipitate is related on dissolution speed. At potential > 270 mV, the polarization resistance decreases with the decrease of electrode potential, indicating that the dissolution of passivation film may be the controlled step of the electrode process.

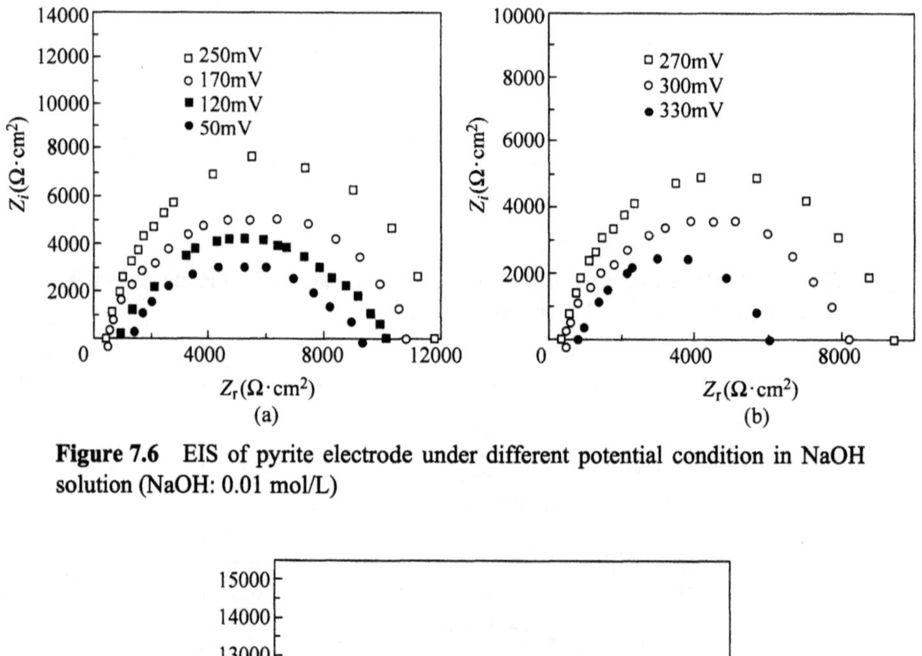

Figure 7.6 EIS of pyrite electrode under different potential condition in NaOH solution (NaOH: 0.01 mol/L)

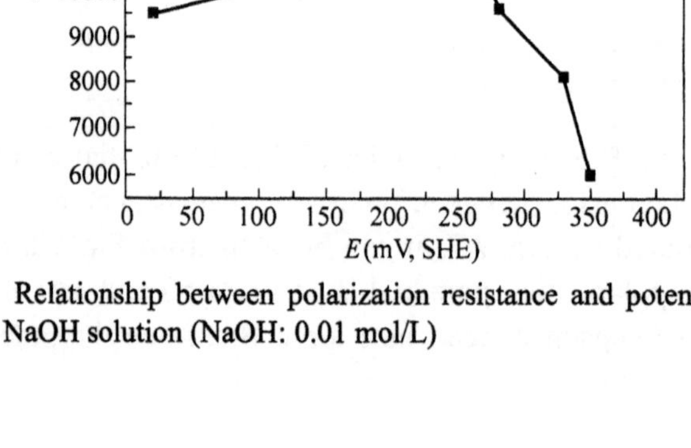

Figure 7.7 Relationship between polarization resistance and potential of pyrite electrode in NaOH solution (NaOH: 0.01 mol/L)

The oxidation of FeS_2 presents passivation characteristic in sodium hydroxide solution due to the following reactions:

$$FeS_2 + 2OH^- \longrightarrow Fe(OH)_2 + 2S + 2e^- \tag{7-8}$$

$$FeS_2 + 8OH^- \longrightarrow Fe(OH)_2 + 2SO_3^{2-} + 6H^+ + 10e^- \tag{7-9}$$

$$FeS_2 + 5OH^- \longrightarrow Fe(OH)_2 + S_2O_3^{2-} + 3H^+ + 6e^- \tag{7-10}$$

$$FeS_2 + 10OH^- \longrightarrow Fe(OH)_2 + 2SO_4^{2-} + 8H^+ + 14e^- \tag{7-11}$$

The formation of hydroxyl precipitate prevents from the transfer of electron especially between oxygen and the mineral surface. As a result of all these processes, EIS represents passivation characteristic. And the corrosive potential moves towards negatively, the surface resistance increases, and the corrosive current decreases. The formation of surface hydroxyl iron precipitates makes the pyrite surface very hydrophilic.

7.3.2 Oxidation of Pyrite in Lime Medium

Figures 7.8 and 7.9 are the polarization curves and EIS for pyrite at natural pH and in the lime medium, respectively. Obviously, after adding lime, the corrosive potential of pyrite electrode moves towards negatively about 150 mV and the corrosive current density decreases from 10.7 μA/cm^2 to 6.2 μA/cm^2. The anodic and cathodic slope has almost no change. Whereas, the capacitive reactance

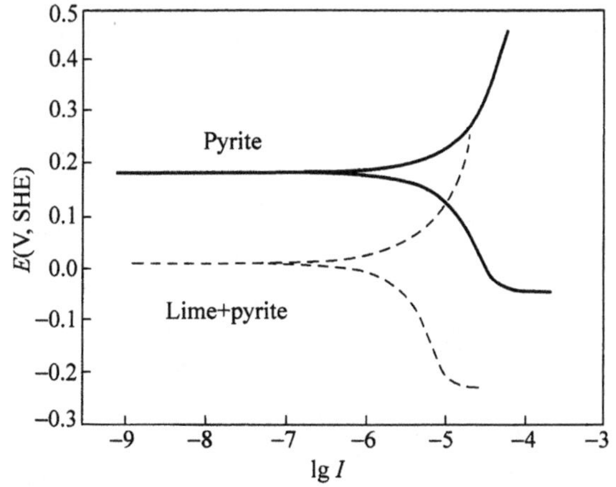

Figure 7.8 Polarization curves of pyrite electrode at natural pH and at pH = 12 modified by lime (KNO_3: 0.1 mol/L; unit of I: A/cm^2)

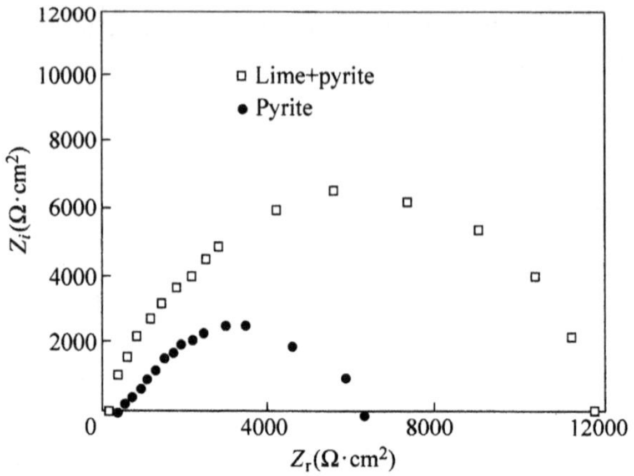

Figure 7.9 EIS of pyrite electrode at natural pH and at pH = 12 modified by lime (KNO_3: 10^{-1} mol/L)

loop radius increases evidently in the presence of lime. The electrode resistance increases about 5600 Ω, i.e. from 6200 Ω to 11800 Ω. The EIS exhibits passivation characteristic indicated the formation of surface oxidation products. And the electrode surface resistance about 11800 Ω in the lime medium is much higher than that of about 8500 Ω in NaOH solution comparing with Fig. 7.5. It demonstrates that pyrite surface is strongly oxidized in the lime medium.

Further, the EIS of pyrite under different polarization potential in the lime medium are measured and shown in Fig. 7.10. The relationship between polarization resistance and potential can be demonstrated in Fig. 7.11. It can be seen from Fig. 7.10 and Fig. 7.11 that when the anodic polarization potential is between 20 and 350 mV, all the curves appear as a single capacitive reactance loop. But surface process of pyrite electrode is controlled by a quite different

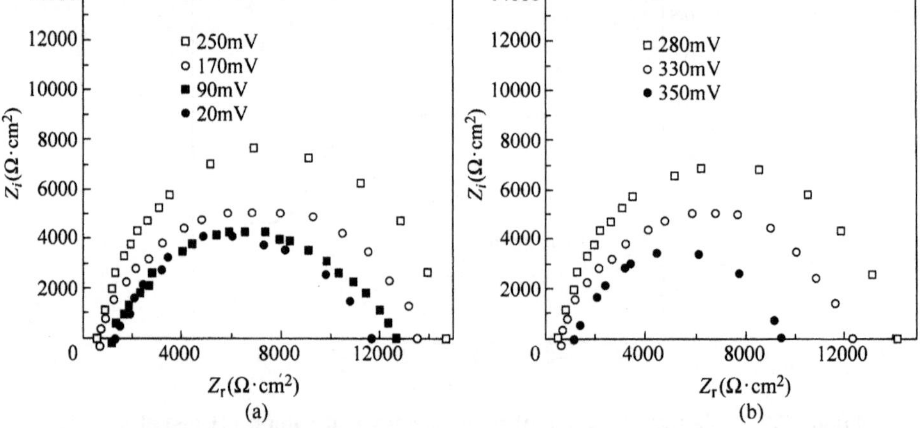

Figure 7.10 EIS of pyrite at different potential in the lime medium (pH = 12; KNO_3: 0.1 mol/L)

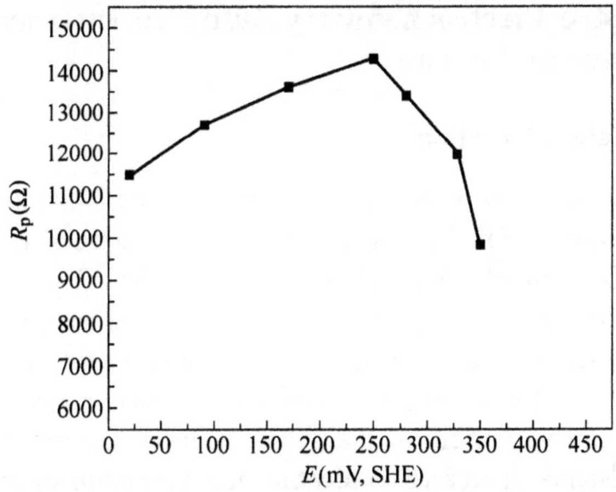

Figure 7.11 Relationship between polarization resistance and potential of pyrite in the lime medium

process. When the potential is between 20 and 250 mV, the capacitive reactance loop radius increases with the enhancement of the electrode potential. This indicates the formation of hydroxyl iron precipitate on the mineral surface. The surface resistance is increased from 11500 Ω to 14600 Ω with the increase of potential and the growth of surface oxide film is the controlled step. When the potential is between 280 and 350 mV, the capacitive reactance loop radius decreased with the enhancement of electrode potential indicates that the dissolution of hydroxyl precipitate occurs and capacitive reactance loop radius is dependent on the dissolution speed. Polarization resistance decreases from 14400 Ω to 9500 Ω with the decrease of the electrode potential from 280 mV to 350 mV. Here, the dissolution of the passivation film is the controlled step of the electrode process.

Comparing EIS for pyrite in lime medium and different polarization potential (Figs. 7.10 and 7.11) with those in sodium hydroxide solution (Fig. 7.6 and Fig. 7.7), it can be seen that at a potential of about 50 mV the surface resistance in the NaOH solution (about 8700 Ω) is lower than that in the lime medium (about 12000 Ω). It means that besides the formation of hydroxyl precipitate on the surface of pyrite in lime media similar to that in NaOH media, there may be other surface products being formed, which may be due to the following reactions:

$$Ca^{2+} + SO_4^{2-} = CaSO_4 \quad (7\text{-}12)$$

$$CaOH^{+} + SO_4^{2-} = CaSO_4 + OH^{-} \quad (7\text{-}13)$$

The formation of hydroxyl iron and calcium sulphate precipitates makes pyrite surface very hydrophilic and inhibits other electrochemistry reaction on pyrite like collector giving rise to the depression of pyrite. The depression effect of lime on pyrite may be stronger than that of NaOH.

7.3.3 Corrosive Electrochemistry Study on Interactions between Collector and Pyrite

1. Pyrite/Xanthate Interaction

Figure 7.12 is the polarization curves of pyrite electrode in xanthate solution with different concentration for dipping for 48 hours. Electrochemistry parameters determined by the computer PARcal are listed in Table 7.2. Inhibiting efficiency can be calculated by Eq. (7-7), $R_{p'}$ is the polarization resistance after adding collector, R_p is the polarization resistance without collector. It can be seen from Fig. 7.12 and Table 7.2 that, with the increase of xanthate concentration, corrosive potential and corrosive current of the pyrite electrode decrease gradually while polarization resistance increases, indicating the formation of surface oxidation products.

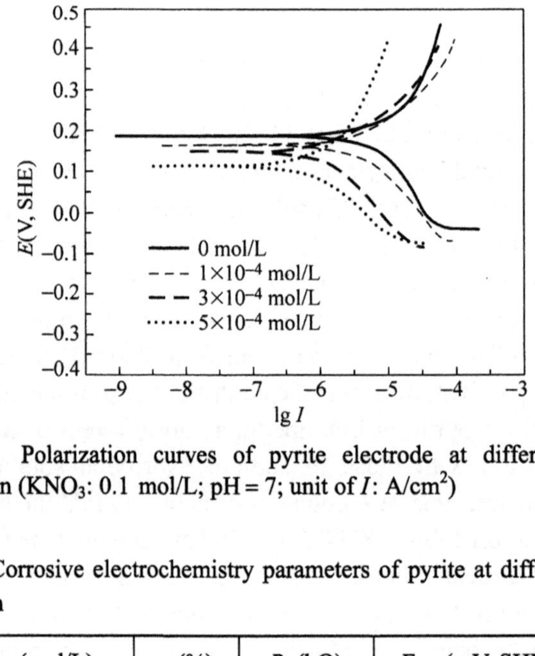

Figure 7.12 Polarization curves of pyrite electrode at different xanthate concentration (KNO_3: 0.1 mol/L; pH = 7; unit of I: A/cm^2)

Table 7.2 Corrosive electrochemistry parameters of pyrite at different xanthate concentration

Collector	c (mol/L)	η (%)	R_p (kΩ)	E_{corr} (mV, SHE)	I_{corr} (μA/cm^2)
Xanthate	0	–	6.2	187	10.78
	10^{-4}	46.09	11.5	175	5.96
	2×10^{-4}	53.73	13.4	153	5.2
	5×10^{-4}	57.53	15.6	122	5.99

I_{corr}—corrosive current; E_{corr}—corrosive potential; R_p—polarization resistance; η—inhibit efficiency.

Figure 7.12 is the EIS of pyrite under different xanthate concentration. Figure 7.12 exhibits two capacitive reactance loops existing in the pyrite-xanthate system different from the self-oxidation of pyrite. The capacitive reactance loop with high resistance value represents the process of charging or discharging between electrode surface and double layer in solution. Low value capacitive reactance loop is caused by specific adsorption of species on the electrode surface.

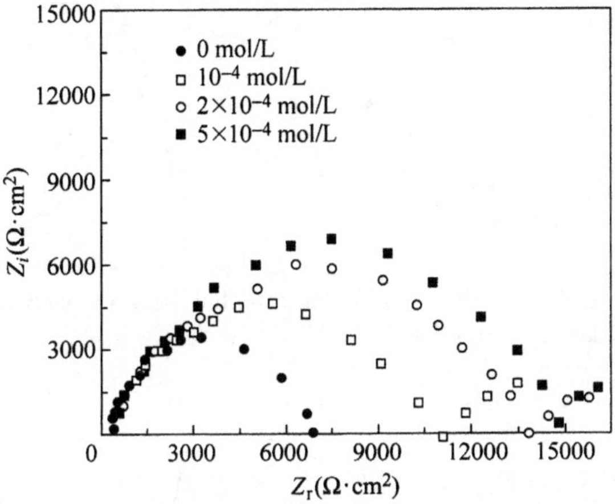

Figure 7.13 EIS of pyrite electrode at different xanthate concentration (KNO_3: 10^{-1} mol/L, pH = 7)

In xanthate solution, FeS_2 have reactions as follows:

$$FeS_2 + X^- = FeS_2 - X^-_{(ab)} + e^- \tag{7-14}$$

$$FeS_2 - X^-_{(ab)} + X^- = FeS_2 - X_2 \tag{7-15}$$

$$FeS_2 - X_2 = FeS_2 + X_2 \tag{7-16}$$

It may be derived that high value capacitive reactance loops are induced by electron transfer from Eq. (7-14), i.e. the chemisorption of xanthate ion on the pyrite surface, and low value capacitive reactance loops are induced by the adsorption and desorption of X_2 from Eqs. (7-15) and (7-16). The capacitive reactance loop radius increases with the increase of xanthate concentration, indicating the thickening of the collector film on pyrite surface, increasing conduction resistance and weakening dissolution of pyrite. It suggests that the adsorption of xanthate on the pyrite surface may undergo several steps such as the adsorption of xanthate ion, the formation of dixanthogen due to the oxidation of the adsorbed ion and the increase of dixanthogen film with the increase of xanthate concentration.

Figure 7.14 is the EIS of pyrite at different polarization potential with xanthate as a collector. The relationship between polarization resistance and potential can be further demonstrated by Fig. 7.15.

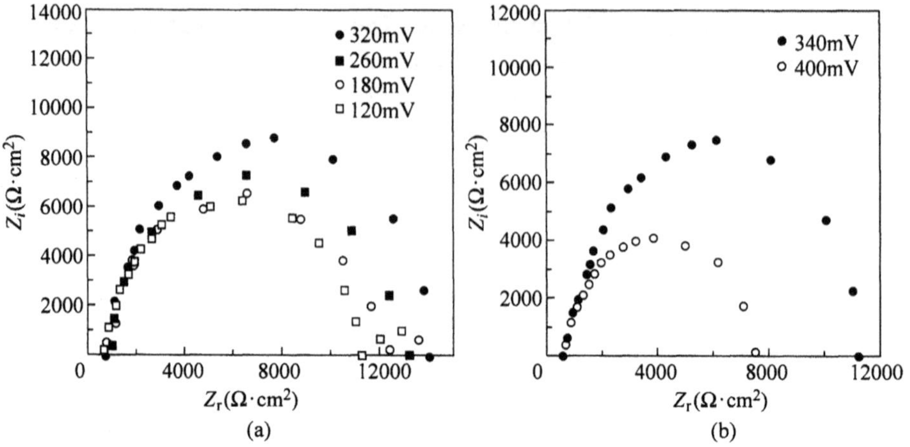

Figure 7.14 EIS of pyrite at different potential with xanthate as the collector (pH = 7; xanthate: 10^{-4} mol/L)

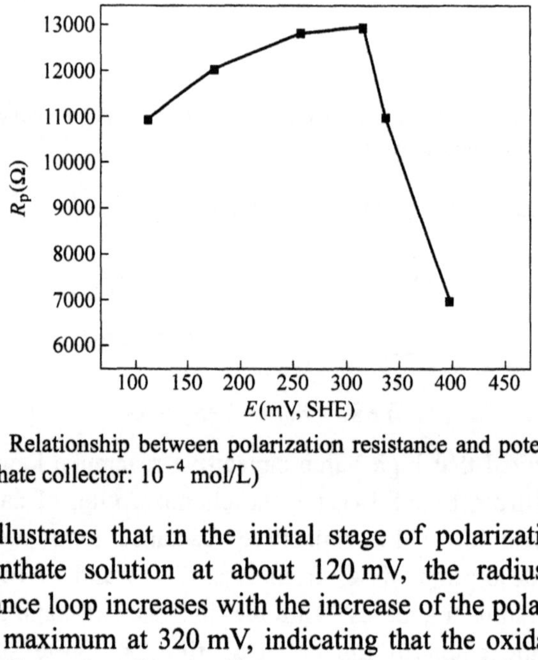

Figure 7.15 Relationship between polarization resistance and potential of pyrite (pH = 7; xanthate collector: 10^{-4} mol/L)

Figure 7.14 illustrates that in the initial stage of polarization of the pyrite electrode in xanthate solution at about 120 mV, the radius of high value capacitive reactance loop increases with the increase of the polarization potential and reaches the maximum at 320 mV, indicating that the oxidation of xanthate increases gradually and collector film on pyrite surface becomes thicker. It increases the conduction resistance and the growth of collector film is the controlled step resulting in pyrite surface hydrophobic. When the polarization potential increases from 320 mV to 400 mV, the capacitive reactance loop radius decreases, indicating the decrease of transferring conduction resistance as can be seen in Fig. 7.15. It belongs to the step of film dissolution. Capacitive reactance loop radius decreases obviously when the potential is larger than 400 mV, at where the collector film falls off and the anodic dissolution of pyrite occurs. The controlled step is the anodic dissolution of pyrite and the surface becomes hydrophilic. The results in Figs. 7.14 and 7.15 account for the flotation potential range of pyrite and voltammetric results with xanthate as a collector in Figs. 4.23 and 4.24 where the formation of dixanthogen and the growth of the collector film takes place.

2. Pyrite/Dithiocarbamate Interaction

Figure 7.16 is the polarization curves of the pyrite electrode in dithiocarbamate solution at different concentration for dipping for 48 hours. Electrochemistry parameters determined by the computer PARcal are listed in Table 7.3. It can be seen from Fig. 7.16 and Table 7.3 that the corrosive potential of pyrite electrode decreases gradually from 187 to 160 mV and the corrosive current decreases from 10.78 to 6.01 μA/cm^2 without or with the DDTC addition of 5×10^{-4} mol/L, while polarization resistance increases from 6.2 to 10.1 kΩ with the increase of dithiocarbamate concentration. It indicates the formation of surface oxidation products. Comparing with xanthate, DDTC has less effect on corrosive potential, current and polarization resistance. It indicates that collector function of DDTC on pyrite is less than that of xanthate.

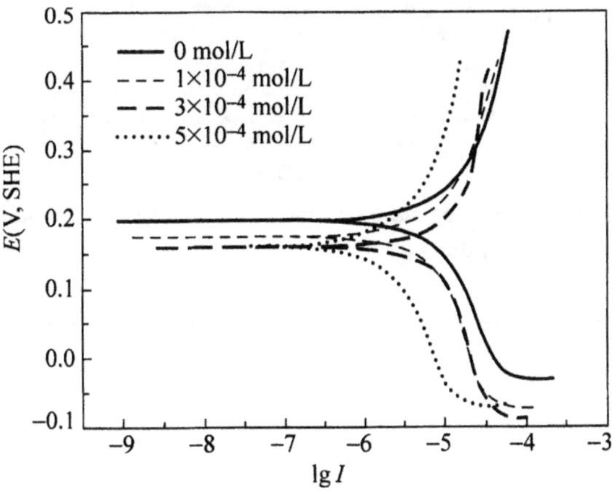

Figure 7.16 Polarization curves of pyrite electrode at different dithiocarbamate concentration (KNO_3: 0.1 mol/L; pH = 7; unit of I: A/cm^2)

Table 7.3 Corrosive electrochemistry parameters of pyrite at different collector concentration

Collector	c (mol/L)	η (%)	R_p (kΩ)	E_{corr} (mV, SHE)	I_{corr} (μA/cm^2)
DDTC	0	–	6.2	187	10.78
	10^{-4}	35.74	9.5	179	6.34
	2×10^{-4}	36.08	9.7	164	6.13
	5×10^{-4}	38.61	10.1	160	6.01

Figure 7.17 is the EIS of pyrite electrode under different DDTC concentration. If only according to the measured rest potential (reversible potential for oxidation to thiouram disulphide (Finkelstein and Goold, 1972)), the products of the interaction of pyrite with sodium dimethyl dithio carbamate are predicted to be thiouram disulphide. However, the EIS of the pyrite electrode in the Fig. 7.17 shows a single capacitive reactance loop characteristic, indicating that the interaction between DDTC and pyrite may be dominated by the adsorption of DDTC on the pyrite surface similar to Eq. (7-14) and the oxidation of DDTC into disulphide $(DDTC)_2$ is relatively difficult comparing with xanthate, indicating weak collecting ability of dithiocarbamate on pyrite.

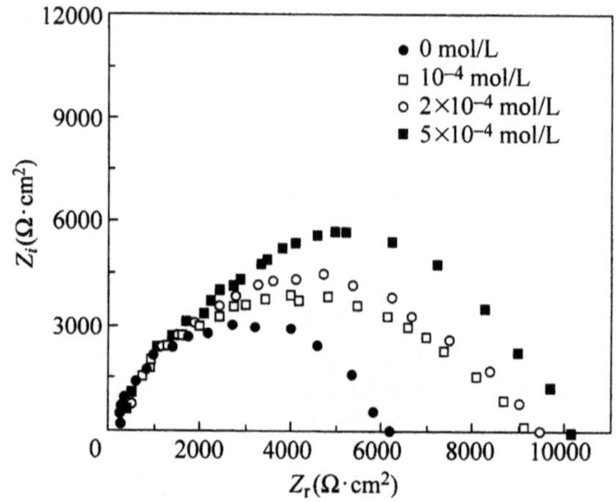

Figure 7.17 EIS of pyrite electrode in different DDTC concentration (KNO_3: 0.1 mol/L; pH = 7)

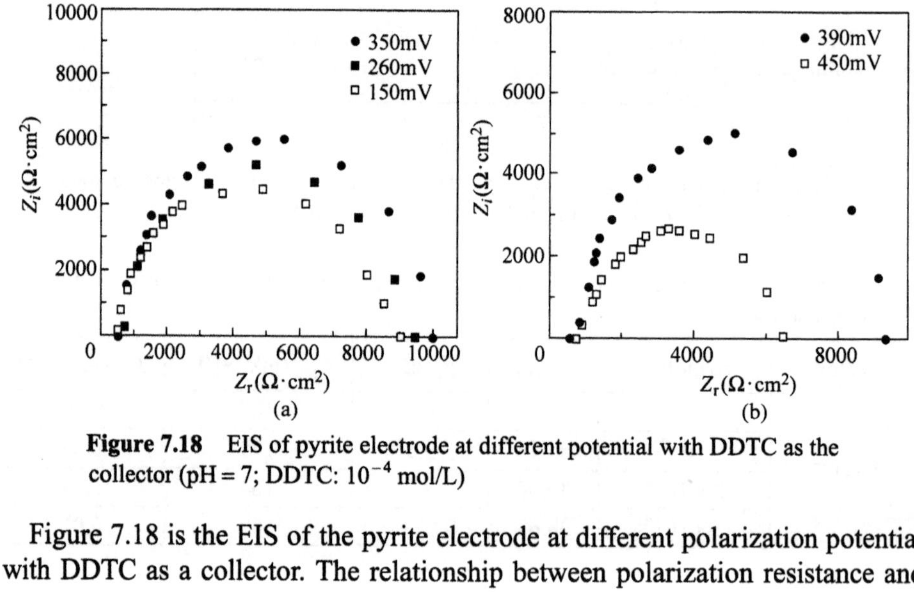

Figure 7.18 EIS of pyrite electrode at different potential with DDTC as the collector (pH = 7; DDTC: 10^{-4} mol/L)

Figure 7.18 is the EIS of the pyrite electrode at different polarization potential with DDTC as a collector. The relationship between polarization resistance and potential can be further demonstrated by Fig. 7.19. It can be observed from Fig. 7.18 and Fig. 7.19 that in DDTC solution, the capacitive reactance loop radius increases with the increase of polarization potential and reaches the maximum at 350 mV, indicating that the adsorption of DDTC increases gradually and the adsorbed collector film on pyrite surface becomes thicker. When the polarization potential increases to 390 mV, capacitive reactance loop radius and the transferring of conduction resistance decrease, indicating the desorption of adsorbed collector. Capacitive reactance loop radius decreases obviously when the potential is larger than 450 mV, suggesting that the collector film falls off and the anodic dissolution of pyrite occurs. This potential is higher than that in the xanthate system.

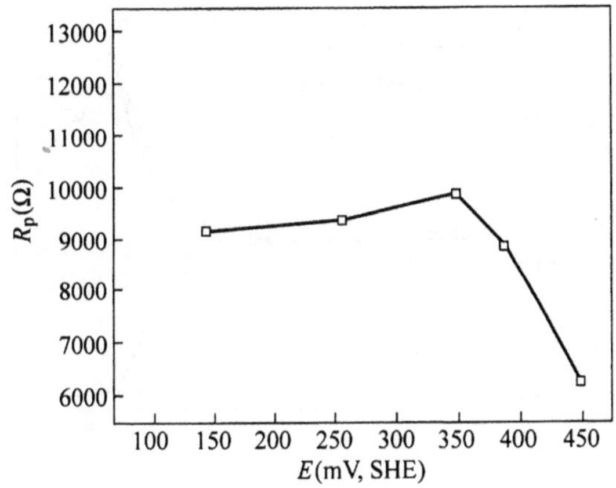

Figure 7.19 Relationship between polarization resistance and potential of pyrite (pH = 7; DDTC collector: 10^{-4} mol/L)

7.3.4 Interaction between Collector and Pyrite in High Alkaline Media

Figures 7.20 and 7.21 are the polarization curves of pyrite interaction with collector at pH = 12 modified by NaOH and lime, respectively. It can be seen from Fig. 7.20 that the corrosive electrochemistry parameters do not change too much whether collector is added or not at pH = 12 modified by NaOH. In the presence of xanthate, corrosive current decreases slightly. Xanthate shows certain inhibiting corrosive function, i.e. adsorption on pyrite even at high pH media. Hence, xanthate still shows some collecting ability on FeS_2 at pH = 12 modified by NaOH. Figure 7.21 shows that the DDTC has no effect on the corrosive current and corrosive potential at pH = 12 modified by lime. The inhibiting corrosion action of DDTC is much lower than that of xanthate, i.e. no adsorption of DDTC on pyrite at high pH modified by lime. It demonstrates that DDTC has no collecting action on pyrite at pH = 12 modified by lime, accounting for that

DDTC has better selective action than xanthate in flotation separation of Pb-Zn-Fe sulphide ores in high lime medium.

Figures 7.20 and 7.21 also show that in the presence of collectors, the decreasing degree of the corrosive current of pyrite electrode in the NaOH system is slightly bigger than that in the lime system. It demonstrates that the inhibiting corrosive function and the collecting ability of collector on pyrite in the NaOH system is stronger than that in the lime system. That is to say, xanthate has certain collecting function on pyrite at pH = 12 modified by NaOH, while the collecting function decreases greatly in the lime system. Lime may be the effective depressant for pyrite with xanthate and DDTC as collectors.

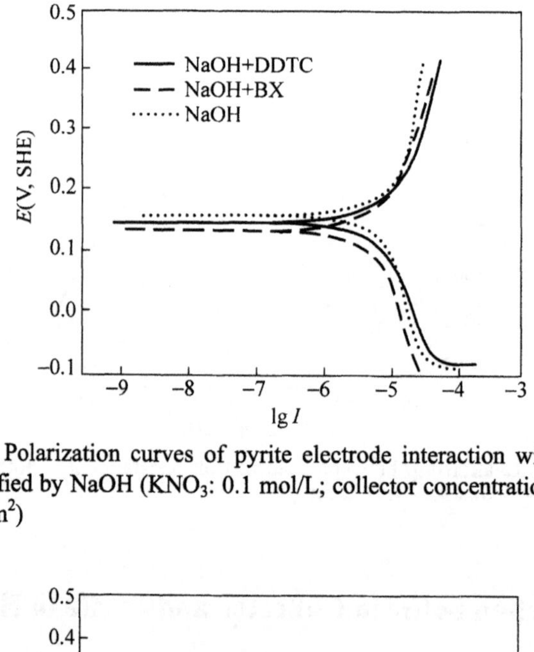

Figure 7.20 Polarization curves of pyrite electrode interaction with collector at pH = 12 modified by NaOH (KNO_3: 0.1 mol/L; collector concentration: 10^{-4} mol/L; unit of I: A/cm^2)

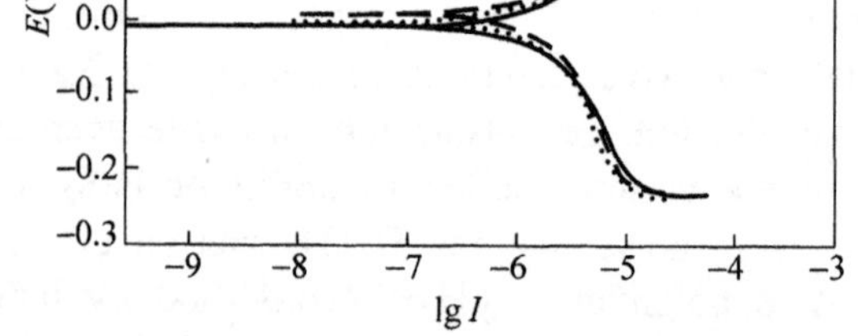

Figure 7.21 Polarization curves of pyrite electrode interaction with collector at pH = 12 modified by lime (KNO_3: 0.1 mol/L; collector concentration: 10^{-4} mol/L; unit of I: A/cm^2)

Figures 7.22 and 7.23 are the EIS of pyrite interaction with collector at pH = 12 modified by NaOH and lime respectively. Figure 7.22 shows that the radius of the capacitive reactance loop is, respectively, 8700 Ω without collector, 9400 Ω by the addition of xanthate and 10000 Ω by the addition of dithiocarbamate. The small increase of the radius of the capacitive reactance loop at pH = 12 modified by NaOH when adding xanthate and dithiocarbamate, shows that the two collectors still have certain inhibiting corrosion action and hence adsorption on pyrite. The action of xanthate is some stronger than that of DDTC because the radius of the capacitive reactance loop in the presence of xanthate is slightly larger than that in the presence of DDTC. However, Fig. 7.23 demonstrates that at pH = 12 modified

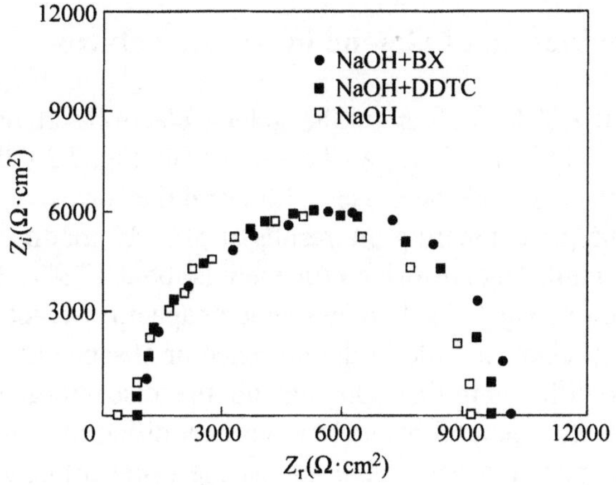

Figure 7.22 EIS of pyrite interaction with collector at pH = 12 modified by NaOH (KNO_3: 0.1 mol/L; collector concentration: 10^{-4} mol/L)

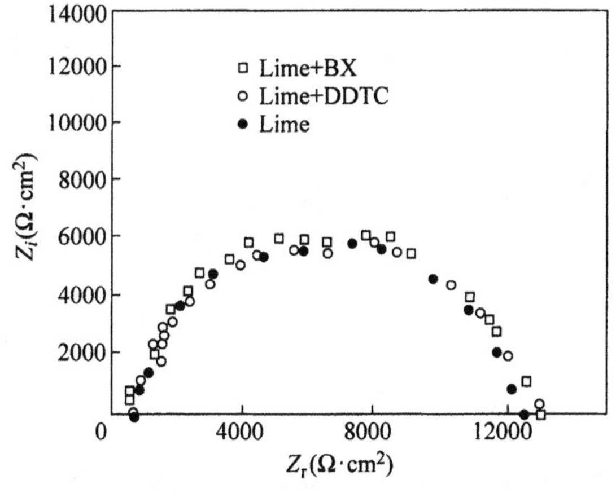

Figure 7.23 EIS of pyrite interaction with collector at pH = 12 modified by lime (KNO_3: 0.1 mol/L; collector concentration: 10^{-4} mol/L)

by lime, the radius of capacitive reactance loops changes little in the presence of xanthate and dithiocarbamate, indicating that at pH = 12 modified by lime, the oxidation products on pyrite surface may be almost the same whether in the absence or presence of collectors. It demonstrates that xanthate and dithiocarbamate exhibit very weak inhibiting corrosion action and hence collecting ability on pyrite at high pH modified by lime.

7.4 Corrosive Electrochemistry on Surface Redox Reaction of Galena under Different Conditions

7.4.1 The Oxidation of Galena in NaOH Solution

Figure 7.24 is the Tafel curves of the galena electrode at natural pH and at pH = 12 modified by NaOH. It can be seen from Fig. 7.24 that the corrosive potential of galena electrode decreases a little and the corrosive current decreases from 3.45 to 3.05 μA comparing the results at pH = 12 modified by NaOH with natural pH. The inhibiting corrosive efficiency is about 17.9%. The EIS of galena electrode is given in Fig. 7.25. It shows single capacitive reactance loop and the radius of the loop changes little in the presence or absence of NaOH, indicating that NaOH has little inhibiting function on the dissolution of galena. These results suggest that, at natural pH and high pH modified by NaOH, the oxidation products on the galena surface have almost the same influence on the surface conductivity.

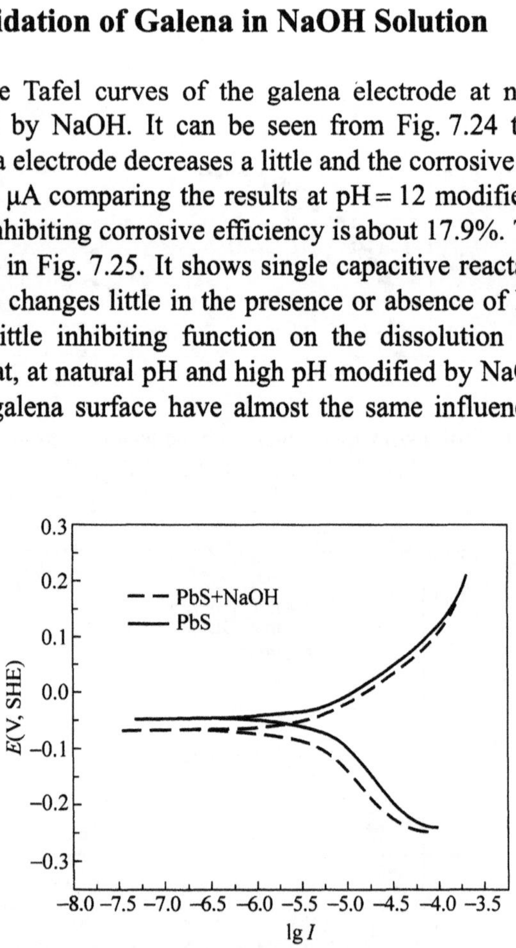

Figure 7.24 Polarization curves of galena electrode at natural pH and pH = 12 modified by NaOH (KNO_3: 0.1 mol/L; unit of I: A/cm^2)

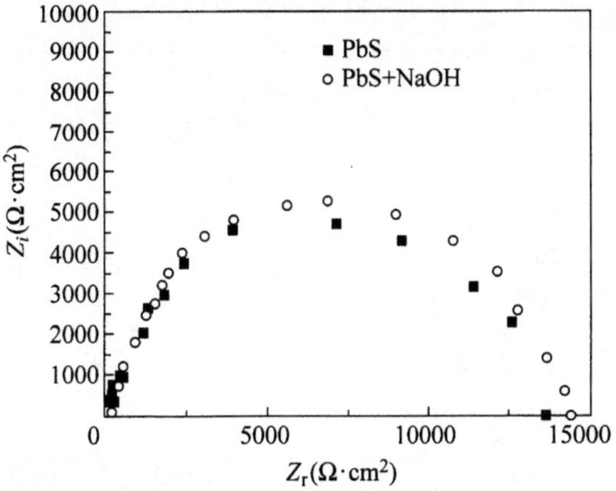

Figure 7.25 EIS of PbS at natural pH in NaOH medium (KNO_3: 0.1 mol/L)

7.4.2 The Effect of Lime on the Oxidation of Galena

The polarization curves at different pH and the EIS of galena electrode are, respectively, given in Fig. 7.26 and Fig. 7.27. It follows from Fig. 7.26 that the corrosive potential reduces about 30 mV and corrosive current decrease from 3.45 to 2.14 μA when pH increases from natural pH to pH = 11.8 by adding lime. The inhibiting corrosive efficiency of lime is about 42.05% higher than that of Na(OH) about 17.9%, suggesting the formation of surface products and inhibiting corrosive action of lime on galena, which is stronger than NaOH. The surface oxidation products are some different at natural pH in lime medium. According to the theory of corrosive electrochemistry, the negatively towards

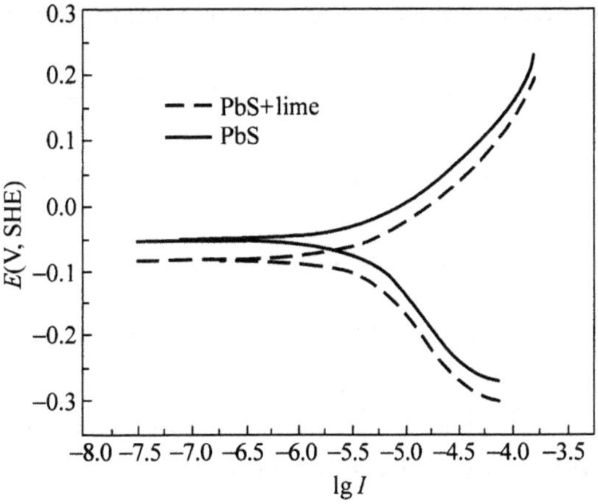

Figure 7.26 Polarization curves of galena electrode at natural pH and pH = 11.8 modified by lime (KNO_3: 0.1 mol/L; unit of I: A/cm^2)

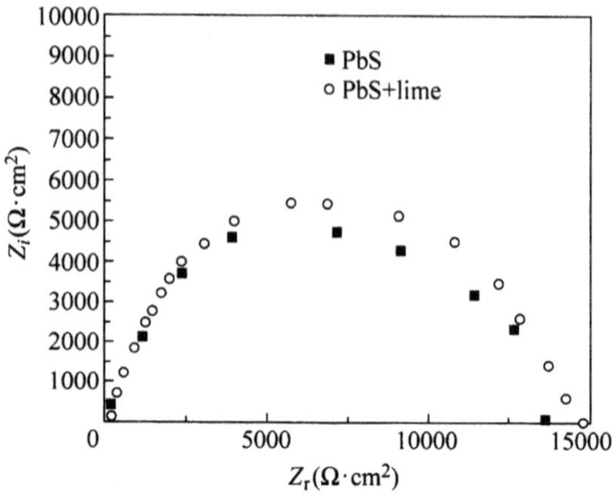

Figure 7.27 EIS of galena electrode at natural pH and pH = 11.8 modified by lime (KNO_3: 0.1 mol/L)

shift of corrosion potential by adding lime suggests that the cathode reaction is retarded by some non-electroactive materials. It makes O_2 not contact effectively with the electrode surface, giving rise to the decrease in the reduction efficiency of O_2, further inhibiting the corrosion rate of galena.

In Fig. 7.27, there appears a single capacitive reactance loop in different pH media. The capacitive reactance loop radius is bigger in the lime medium than that without lime. The reaction resistance increases from 14000 in the absence of lime to 15000 in the presence of lime. EIS exhibits passivation characteristic. The results indicate the formation of surface oxidation products which prevent the transferring of electron, giving rise to the increase of surface resistance and descending of corrosive current. It may be mainly because of the following reactions on the surface of galena in the lime medium besides Eq. (7-12):

$$PbS + 2OH^- \longrightarrow Pb(OH)_2 + S + 2e^- \tag{7-17}$$

$$PbS + 6OH^- \longrightarrow Pb(OH)_2 + SO_4^{2-} + 4H^+ + 8e^- \tag{7-18}$$

$$Pb^{2+} + SO_4^{2-} = PbSO_4 \tag{7-19}$$

Figure 7.28 is the EIS of the galena electrode at different potential in the lime medium. The relationship between polarization resistance and potential is presented in Fig. 7.9. The EIS of the galena electrode can be divided into three stages according to the different characters of the surface oxidation film. When the potential is between −70 and 300 mV, capacitive reactance loop radius and polarization resistance increases slowly due to the formation of surface oxidation products, and the growth of surface oxidation film is the controlled step of the electrode process. From 300 to 400 mV, polarization resistance decreases gradually and the dissolution of oxidation film is the controlled step of electrode process. Polarization resistance decreases sharply after 400 mV, the oxidation film falls off and the dissolution of galena is the controlled step of the electrode process, which may be corresponding to the formation of $HPbO_4^-$.

$$Pb(OH)_2 + OH^- = HPbO_2^- + H_2O \tag{7-20}$$

$$2PbS + 7H_2O = 2HPbO_2^- + S_2O_3^{2-} + 12H^+ + 8e^- \tag{7-21}$$

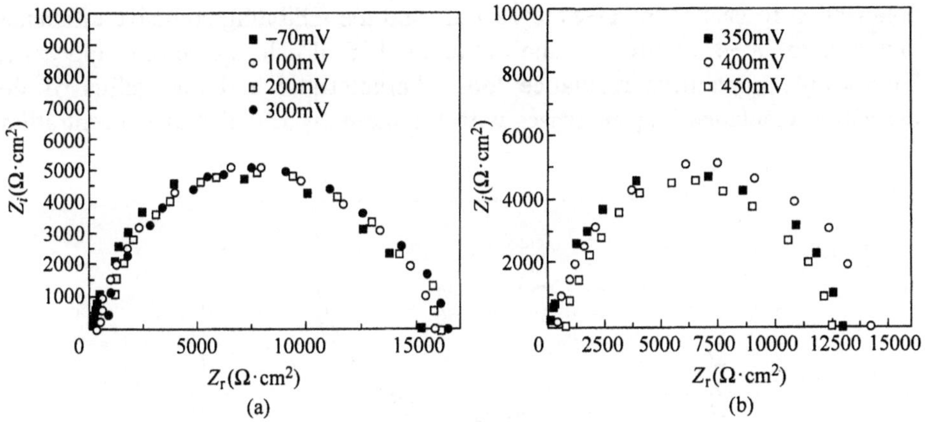

Figure 7.28 EIS of galena electrode at different potential in lime medium (KNO_3: 0.1 mol/L)

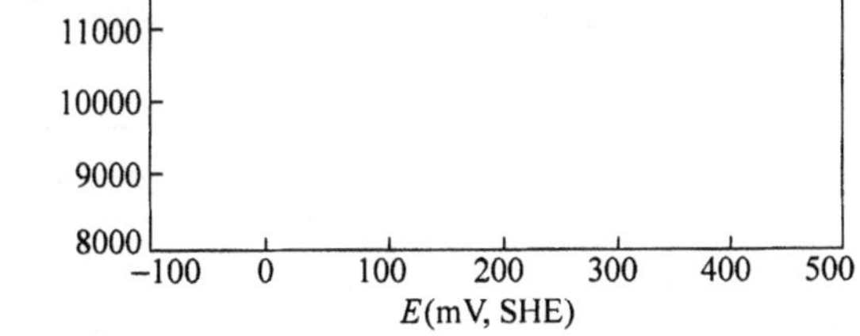

Figure 7.29 Relationship between polarization resistance and potential of galena electrode in lime medium

7.4.3 Corrosive Electrochemistry Study on Interactions between Collector and Galena

1. Galena/Xanthate Interaction

Figure 7.30 is the polarization curves of the galena electrode in xanthate solution with different concentration. Figure 7.31 is the EIS of the galena electrode in xanthate solution with different concentration.

The corrosive electrochemistry parameters are listed in Table 7.4. From Fig. 7.30 and Table 7.4, it can be seen that, after adding xanthate 5×10^{-4}mol/L, the corrosive potential of galena electrode decreases from −48 to −94 mV, the corrosive current of the galena electrode decreases from 3.45 to 0.99 μA/cm^2, the polarization resistance increases to 18.7 kΩ and the inhibiting corrosive efficiency increases to 28.34. Figure 7.31 shows that the EIS of galena electrode appears to have single capacitive reactance loop characteristic and the radius of the capacitive reactance loop increases with the increase of collector concentration.

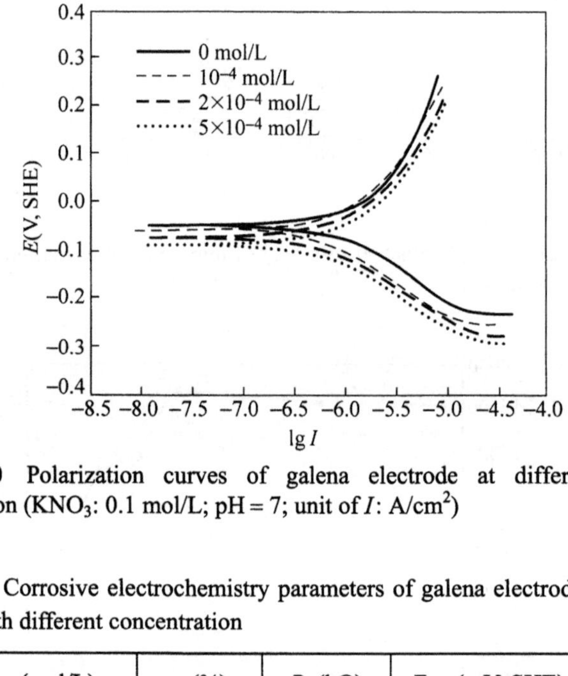

Figure 7.30 Polarization curves of galena electrode at different xanthate concentration (KNO_3: 0.1 mol/L; pH = 7; unit of I: A/cm^2)

Table 7.4 Corrosive electrochemistry parameters of galena electrode in xanthate solution with different concentration

Collectors	c (mol/L)	η (%)	R_p (kΩ)	E_{corr} (mV, SHE)	I_{corr} (μA/cm^2)
BX	0	—	13.4	−48	3.45
	10^{-4}	18.4	16.3	−68	1.46
	2×10^{-4}	27.47	18.2	−77	1.25
	5×10^{-4}	28.34	18.7	−94	0.99

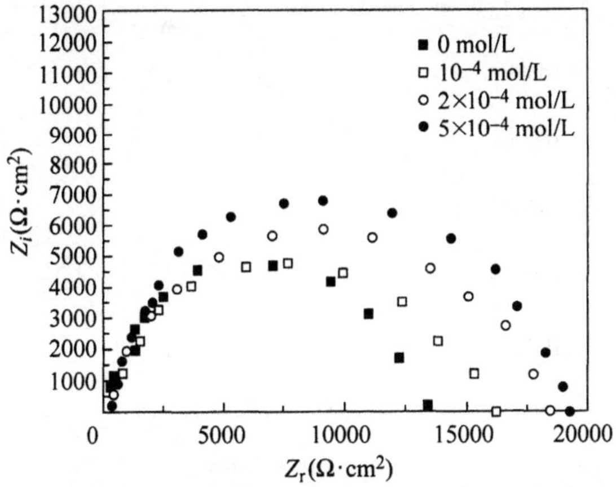

Figure 7.31 EIS of galena electrode in xanthate solution with different concentration (KNO_3: 0.1 mol/L; pH = 7)

The above results demonstrate that the formation of surface products of the interaction between xanthate and galena increases with the increase of xanthate concentration and xanthate has stronger inhibiting corrosive action on galena. The single capacitive reactance loop characteristic indicates that the lead xanthate may be the main oxidation product on galena surface, which is responsible for surface hydrophobicity. The amount of formation of lead xanthate will be enhanced with the increase of xanthate concentration based on the change of corrosive potential and current.

Figure 7.32 is the EIS of the galena electrode at different polarization potential in xanthate solution. The relationship between polarization resistance and potential can be further demonstrated by Fig. 7.33. Performance of galena interaction with xanthate is different at different potential. From −50 to 300 mV potential ranges, the radius of capacitive reactance loop is enhanced with the polarization potential,

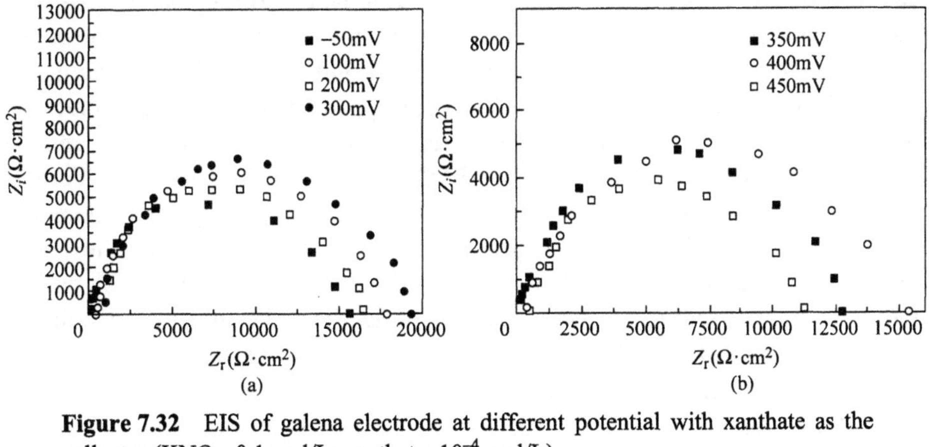

Figure 7.32 EIS of galena electrode at different potential with xanthate as the collector (KNO_3: 0.1 mol/L; xanthate: 10^{-4} mol/L)

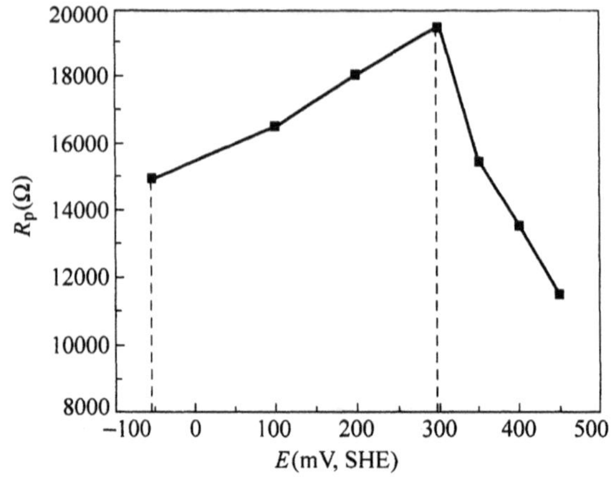

Figure 7.33 Relationship between polarization resistance and potential of galena electrode

the growth of the collector film PbX_2 is the controlled step of the electrode process. Within this potential range, xanthate may exhibit stable action and good collecting property on galena. Above this potential range (> 350 mV), the polarization resistance and capacitive reactance loop radius decrease evidently with the increase of potential, which may be attributed to the film dissolution of PbX_2. The controlled step of electrode process is determined by the dissolution of the collector film and the further oxidation of galena, which may result in the falling off of the collector film and galena loses its floatability. The EIS of the galena electrode may explain the results in Fig. 4.6.

2. Galena/Dithiocarbamate Interaction

The polarization curves and the EIS of the galena electrode in dithiocarbamate solution with different concentration are, respectively, presented in Fig. 7.34 and Fig. 7.35. The corrosive electrochemistry parameters are listed in Table 7.5. It can be seen from Fig. 7.34 and Table 7.5 that, in the absence of dithiocarbamate, the corrosive potential, corrosive current and polarization resistance of the galena electrode are, respectively, −48 mV, 3.45 μA/cm^2 and 13.4 kΩ. By the addition of 5×10^{-4} mol/L dithiocarbamate, the corrosive potential of the galena electrode decreases to −99 mV, the corrosive current decreases to 1.21 μA/cm^2 and the polarization resistance increases to 18.3 kΩ. The inhibiting corrosive efficiency is 27.32. It indicates the formation of surface products through the interaction between galena and collector. Figure 7.35 shows that the EIS of the galena electrode appears to have single capacitive reactance loop characteristic and the radius of the capacitive reactance loop increases with the increase of collector concentration. It demonstrates that the formation of surface products of the interaction between dithiocarbamate and galena increases with the increase of DDTC concentration. The single capacitive reactance loop characteristic indicates that the lead dithiocarbamate salt may be the dominant oxidation product on the

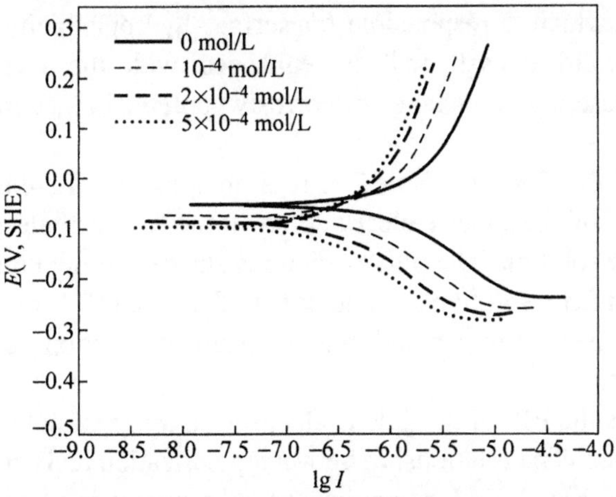

Figure 7.34 Polarization curves of galena electrode in DDTC solution with different concentration (KNO_3: 0.1 mol/L; pH = 7; unit of I: A/cm^2)

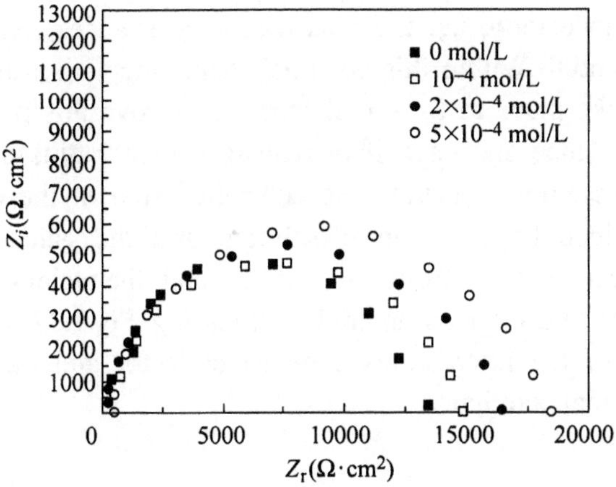

Figure 7.35 EIS of PbS at different DDTC concentration (KNO_3: 0.1 mol/L; pH = 7)

Table 7.5 Corrosive electrochemistry parameters of galena electrode in DDTC solution with different concentration

Collector	c (mol/L)	η (%)	R_p (kΩ)	E_{corr} (mV, SHE)	I_{corr} (μA/cm²)
DDTC	0	—	13.4	−48	3.45
	10^{-4}	13.15	15.2	−71	1.67
	2×10^{-4}	18.79	16.5	−81	1.42
	5×10^{-4}	27.32	18.3	−99	1.21

galena surface, which is responsible for surface hydrophobicity. The amount of formation of collector salt will be enhanced with the increase of DDTC concentration based on the change of corrosive potential and current with DDTC concentration.

Comparing Table 7.4 and Table 7.5, it is obvious that the corrosive potential and the current of the galena electrode are lower in xanthate solution than in dithiocarbamate solution. The polarization resistance and the inhibiting corrosive efficiency in xanthate solution are higher than that in DDTC solution. It indicates that xanthate has stronger inhibiting corrosive action on galena than dithiocarbamate.

Figure 7.36 is the EIS of the galena electrode at different polarization potential in DDTC solution. The relationship between polarization resistance and potential can be shown in Fig. 7.37. It is obvious that the interaction between galena and dithiocarbamate also shows different behavior at different potential. When the polarization potential ranges from −50 to 350 mV, the growth of collector film PbD_2 will be the controlled step of electrode process because the radius of capacitive reactance loop, i.e. the polarization resistance is enhanced with the polarization potential. Within this potential range, the action of dithiocarbamate on galena can take place to form collector salt. Above this potential range, the polarization resistance and capacitive reactance loop radius decreased and the collector film dissolution occurs. The controlled step of the electrode process may be determined by the film dissolution and the galena oxidation. The collector film may detach from the surface and the galena surface may be oxidized to form hydroxyl compound. Comparing Fig.7.37 with Fig.7.33, the potential range of the interactions between collector and galena is wider for dithiocarbamate than xanthate.

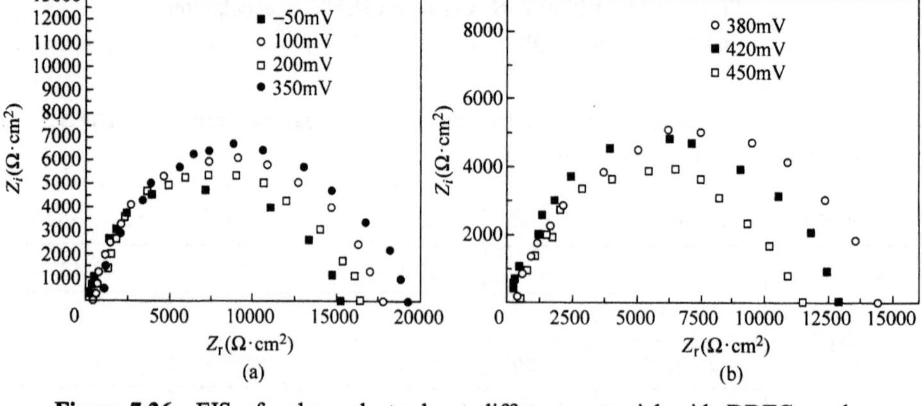

Figure 7.36 EIS of galena electrode at different potential with DDTC as the collector (KNO_3: 0.1 mol/L; DDTC: 10^{-4} mol/L)

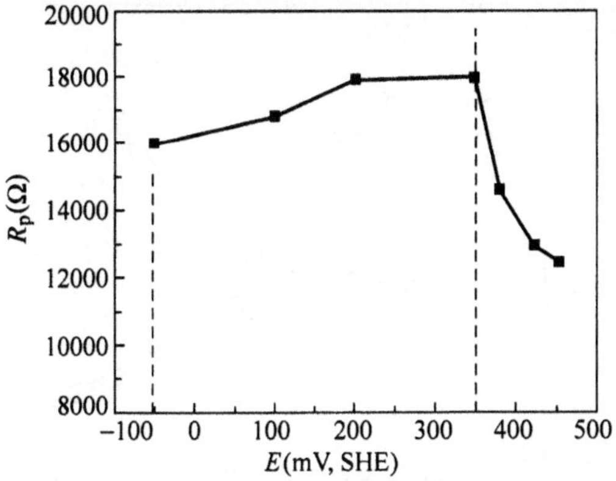

Figure 7.37 Relationship between polarization resistance and potential of galena with DDTC as the collector

7.4.4 Interactions between Collector and Galena at High pH

Figure 7.38 and Fig. 7.39 present the polarization curves of the galena electrode in collector solution at pH = 12 modified by NaOH and lime, respectively. The corrosive electrochemistry parameters are listed in Table 7.6. From Fig. 7.24 to Fig. 27, it has been found that in the absence of collector, the corrosive current and polarization resistance are, respectively 3.05 μA/cm^2 and 14 kΩ at pH = 12 modified by NaOH; 2.14 μA/cm^2 and 15 kΩ at pH = 12 modified by lime. Table 7.6 shows that the corrosive potential and corrosive current decreases and polarization resistance increases in the presence of xanthate and dithiocarbamate at high pH made by NaOH and lime. The corrosive current and polarization resistance are, respectively 1.32 μA/cm^2 and 17.5 kΩ at pH = 12 modified by NaOH; 1.29 μA/cm^2 and 17.4 kΩ at pH = 12 modified by lime. The corrosive current and polarization resistance are, respectively 1.64 μA/cm^2 and 15.3 kΩ at pH = 12 modified by NaOH; 1.59 μA/cm^2 and 15.4 kΩ at pH = 12 modified by lime. These results show that at high pH modified by NaOH or lime, the formation of collector salt on the galena surface is still possible resulting in the decrease of corrosive current and the increase of polarization resistance of the galena electrode. On the other hand, comparing with natural pH, corrosive potential and corrosive current of the galena electrode decreases very little and polarization resistance has almost not changed, indicating that the interaction between xanthate and/or dithiocarbamate and galena is almost not affected by pH change modified by lime or NaOH. Xanthate and dithiocarbamate may still exhibit good collecting capability on galena in high alkaline media modified by lime or NaOH. It is the basis for selective flotation separation of lead-iron sulphide ore by using xanthate or dithiocarbamate as a collector and lime as a depressant.

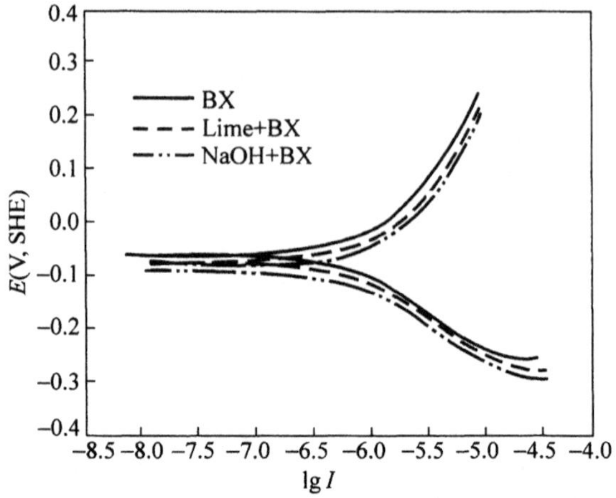

Figure 7.38 Polarization curves of galena electrode with xanthate as the collector in different pH media (KNO_3: 0.1 mol/L; xanthate: 10^{-4} mol/L; unit of I: A/cm^2)

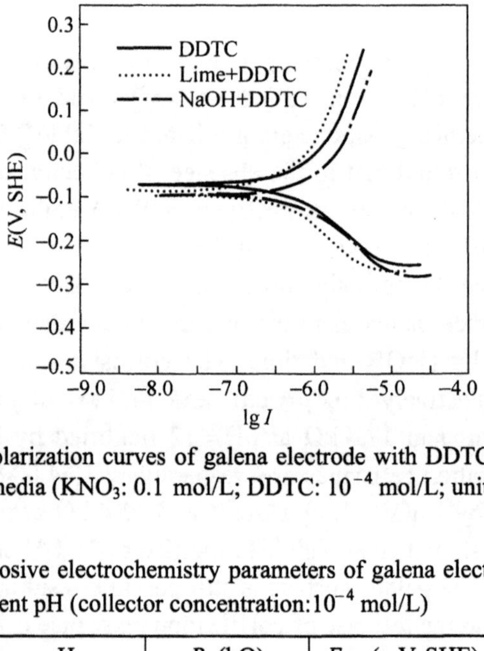

Figure 7.39 Polarization curves of galena electrode with DDTC as the collector in different pH media (KNO_3: 0.1 mol/L; DDTC: 10^{-4} mol/L; unit of I: A/cm^2)

Table 7.6 Corrosive electrochemistry parameters of galena electrode in collector solution at different pH (collector concentration: 10^{-4} mol/L)

Collectors	pH	R_p (kΩ)	E_{corr} (mV, SHE)	I_{corr} (μA/cm^2)
Xanthate	nature	16.3	−68	1.46
	12(NaOH)	17.5	−77	1.32
	12(lime)	17.4	−78	1.29
DDTC	nature	15.2	−71	1.67
	12(NaOH)	15.3	−74	1.64
	12(lime)	15.4	−76	1.59

7.5 Corrosive Electrochemistry on Surface Redox Reaction of Sphalerite in Different Media

7.5.1 Influence of Different pH Media on Sphalerite Oxidation

Because sphalerite is a poor semiconductor, the sphalerite electrode is made of carbon and mineral powder. The corrosive electrochemistry test of the sphalerite electrode is only carried out by electrode polarization. Figure 7.40 is the polarization curves of the sphalerite combination electrode in different pH media. Corrosive electrochemistry parameters are listed Table 7.7. It follows that the corrosive potential and current of sphalerite electrode are, respectively, 42 mV and 0.13 $\mu A/cm^2$ at natural pH, 34 mV and 0.1 $\mu A/cm^2$ at pH = 12 modified by NaOH, and −12 mV and 0.03 $\mu A/cm^2$ at pH = 12 modified by lime. Comparing with natural pH condition, the corrosive potential of sphalerite electrode moves towards negatively and corrosive current decreases when in high pH condition modified by NaOH or $Ca(OH)_2$, indicating the formation of new surface products like hydroxyl species, resulting in the increase of polarization resistance from 48.6 kΩ at natural pH to 49.2 kΩ at pH = 12 modified by NaOH and 68.4 kΩ at pH modified by lime. The change of corrosive potential and current is greater at pH = 12 modified by $Ca(OH)_2$ than that by NaOH, indicating that $Ca(OH)_2$ has stronger inhibiting function on the sphalerite than NaOH and more surface hydrophilic species will be formed in the lime media, which may produce more depression on sphalerite.

Besides, the polarization resistance of sphalerite electrode is the highest among lead-zinc-iron sulphide minerals, illustrating its poor conductivity of ZnS.

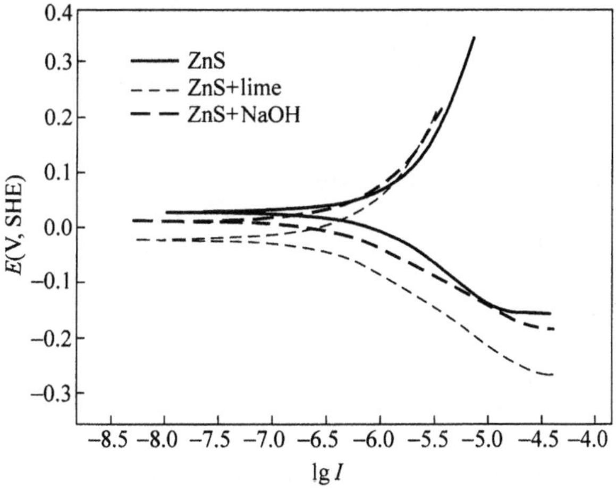

Figure 7.40 Polarization curves of sphalerite combination electrode in different pH media (KNO_3: 0.1 mol/L; NaOH: 2×10^{-4} mol/L; unit of I: A/cm^2)

Table 7.7 Corrosive electrochemistry parameters of sphalerite at different pH

Media	pH	R_p (kΩ)	E_{corr} (mV, SHE)	I_{corr} (μA/cm^2)
Water	7	48.6	42	0.13
Lime	12	68.4	−12	0.03
NaOH	12	49.2	34	0.10

NaOH concentration: 0.0002mol/L.

7.5.2 Inhibiting Corrosive Mechanism of Collector on Sphalerite Electrode

Figure 7.41 is the polarization curves of sphalerite-carbon combination electrode in different collector solution at natural pH. The corrosive electrochemistry parameters are listed in Table 7.8. These results show that xanthate and dithiocarbamate have distinctly different effects on sphalerite. The corrosive potential and current of sphalerite electrode are, respectively, 42 mV and 0.13 μA/cm^2 at natural pH in the absence of collector, −7 mV and 0.01 μA/cm^2 in the presence of xanthate, and 32 mV and 0.12 μA/cm^2 in the presence of dithiocarbamate. The corrosive potential and current decrease sharply with xanthate as a collector, indicating that the electrode surface has been totally covered by the collector film from the electrode reaction. Xanthate has big inhibiting corrosive efficiency and stronger action on sphalerite. However, the corrosive potential and current of sphalerite electrode have small change with dithiocarbamate as a collector, indicating that DDTC exhibits a weak action on sphalerite.

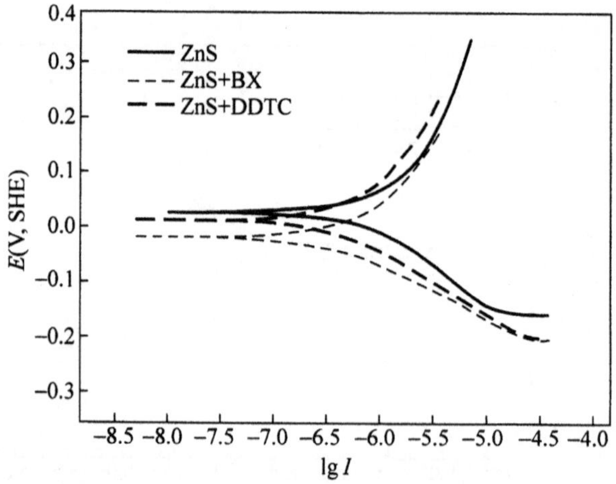

Figure 7.41 Polarization curves of sphalerite combination electrode with different collectors at natural pH (KNO_3: 0.1 mol/L; collector: 10^{-4} mol/L; unit of I: A/cm^2)

Table 7.8 Corrosive electrochemistry parameters of ZnS with different collectors

Collectors	η (%)	R_p (kΩ)	E_{corr} (mV, SHE)	I_{corr} (μA/cm^2)
No collector	7	48.6	42	0.13
Xanthate	32.02	71.5	−7	0.01
DDTC	5.7	52.3	32	0.12

NaOH: 2×10^{-4} mol/L.

Figures 7.42 and 7.43 are the polarization curves of sphalerite electrode in xanthate and DDTC solution at different pH media. It is obvious that xanthate

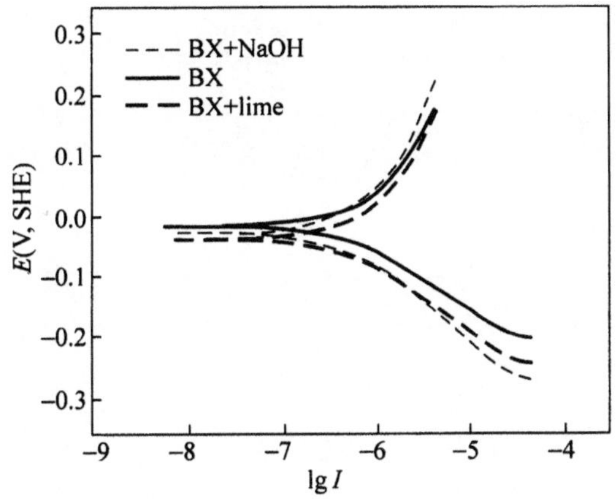

Figure 7.42 Polarization curves of sphalerite combination in xanthate solution at different pH media (KNO_3: 0.1 mol/L; xanthate: 10^{-4} mol/L; unit of I: A/cm^2)

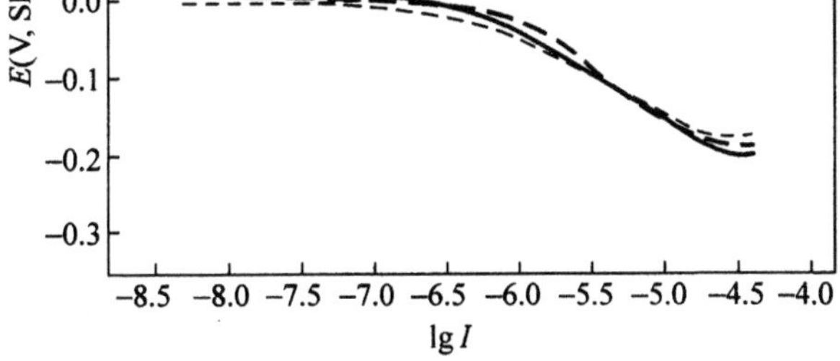

Figure 7.43 Polarization curves of sphalerite combination electrode in dithiocarbamate solution at different pH media (KNO_3: 0.1 mol/L; DDTC: 10^{-4} mol/L; unit of I: A/cm^2)

and DDTC also have different effects on sphalerite oxidation at different pH media. When using NaOH and lime as pH modifiers, the corrosive potential and current of the sphalerite electrode decrease in xanthate solution. Xanthate may still have stronger action and certain collecting capacity on sphalerite. However, the corrosive potential and the current of the sphalerite electrode change little in DDTC solution. Dithiocarbamate has not collecting effect on sphalerite at pH = 12 by using NaOH and lime as pH modifiers, which provides the way for flotation separation of galena from sphalerite using dithiocarbamate as the collector.

Chapter 8 Mechano-Electrochemical Behavior of Flotation of Sulphide Minerals

Abstract This chapter introduces the changes of electrochemical behavior of sulphide minerals leading by mechnical force exerted in the process of grinding, which is called mechano-electrochemical behavior. To investigate it, two new equipments are applied. The configuration of them is illustrated. And the mechano-electrochemical behavior of pyrite and galena is investigated in this chapter. It is found that after grinding the corrosion and oxidation of these mineral surface are inhibited and produce a reduction atmosphere, which may be useful for the interaction between galena and xanthate to form collector salt and not suitable for the formation of dixanthogen. Meanwhile, surface changes of sulphide minerals under mechanical force are also investigated with Nissan mineral phase micro-camera for observing the surface top graph. And the fact turns out to be that among three sulphide minerals the pyrite erodes the fastest and its eroding current is the largest which is consistent with the previous one.

Keywords mechano-electrochemical behavior; pyrite; sphalerite; galena

Grinding is essential for the liberation of sulphide minerals in order to achieve effective flotation. The grinding process, however, may also have a various effects on the flotation separation because of the galvanic interactions among the grinding media and the different minerals due to the high redox activity of sulphide mineral and iron media as well as thio-reagents.

Rey and Formanek (1960) investigated the effect of the grinding media on the flotation of sulphide minerals and showed that sphalerite has natural floatability after grinding with ceramic mill, while has no similar floatability after grinding with steel mill in alkaline pulp. Thornton (1973) reported that the steel media result in the decrease of the floatability of galena, but the floatability of galena had been greatly improved when ground in pyrite media mill.

Galvanic cells are set up by redox reactions taking place due to the difference in their rest potential of various species in the grinding system. One of higher potential acts as a cathode, while that of lower potential as an anode. In the grinding-flotation circuit of sulphide ores, galvanic interactions exist between mineral and mineral as well as minerals and grinding media. The galvanic reactions are also governed by the mixed potential principle.

The interfacial galvanic interactions of sulphide minerals in the grinding system are affected by hitting or erasing force resulting in the formation of crystal lack, dislocation and the nascent surface. This may be called as mechanical electrochemistry behavior, which results from mechanical and electrochemical process, affecting the semiconductor character of sulphide minerals and its electrochemical behavior. The change of the electrochemical behaviors of sulphide minerals in grinding system will be discussed in this chapter.

8.1 Experiment Equipment

A mechanical electrochemical equipment and corrosive couple equipment were designed in order to study the electrochemical behavior of sulphide mineral surface and galvanic interaction based on the method used by Kneer (1997) as shown in Fig. 8.1 and Fig. 8.2, respectively.

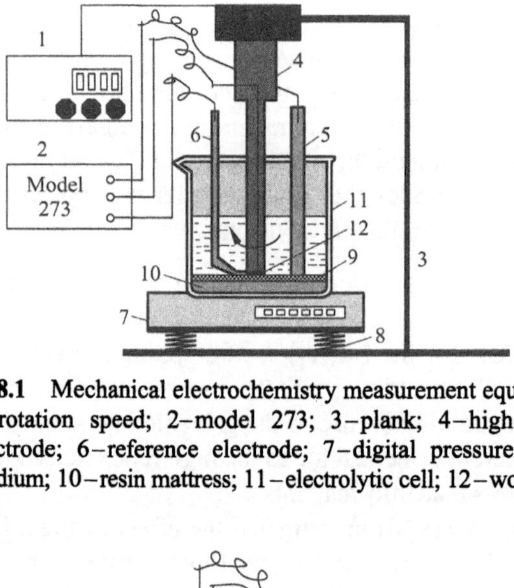

Figure 8.1 Mechanical electrochemistry measurement equipment
1–adjuster of rotation speed; 2–model 273; 3–plank; 4–high speed motor; 5–opposite electrode; 6–reference electrode; 7–digital pressure gauge; 8–lift platform; 9–medium; 10–resin mattress; 11–electrolytic cell; 12–working electrode

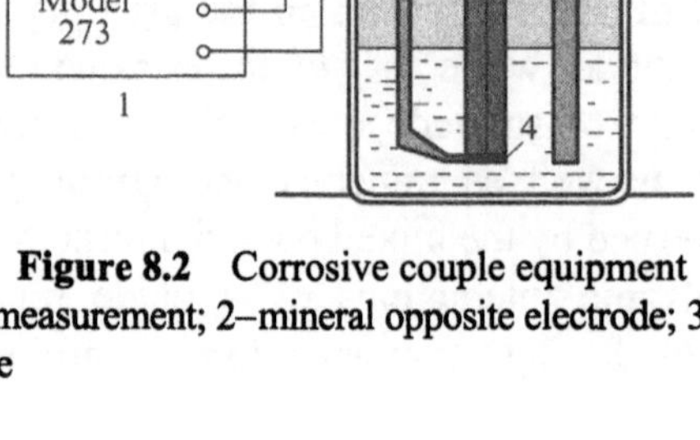

Figure 8.2 Corrosive couple equipment
1–electrochemistry measurement; 2–mineral opposite electrode; 3–reference electrode; 4–working electrode

The mechanical electrochemistry equipment is different from conventional electrochemistry equipment, which can exert pressure to the surface of the electrode by adjusting the height of the lifting platform and change the pressure of electrode (see Fig. 8.1). Before the experiment, the electrode surface was polished, and the grinding media, electrolyte and reagent were added in order.

8.2 Mechano-Electrochemical Behavior of Pyrite in Different Grinding Media

Different flexibility distortion will be produced on the surface of material due to partial mechanical force, which induce the change of corrosion potential and material surface state.

Figure 8.3 shows the variation of corrosion potential and out circuit current of pyrite at static state with Fe as the opposite electrode at natural pH in the presence of xanthate. Figure 8.3 showed that the corrosive potential and current of pyrite decreases gradually with the increase of grinding time. Pyrite appears as cathode when Fe is the opposite electrode, and corrosion potential is finally stabilized at about −100 mV giving rise to reduction atmosphere. Pyrite surface exhibits cathode current and reduction reaction may occur. In this condition, the corrosion and oxidation of pyrite surface is inhibited. According to Chapter 4 and Chapter 7, the potential range of pyrite interaction with xanthate is 100–300 mV. Therefore, dixanthogen can not be formed on pyrite surface by grinding only with iron media.

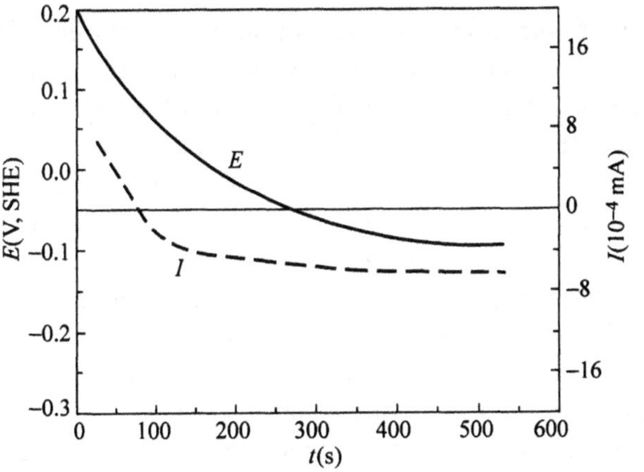

Figure 8.3 Variation of corrosion potential and out circuit current of pyrite electrode at static state with Fe as opposite electrode (pH = 7, BX: 2×10^{-4} mol/L)

Figure 8.4 is the variation of potential and the current of pyrite electrode at different pressure with Fe powder as the grinding media at natural pH in the presence of xanthate. It is obvious that the potential of the pyrite electrode decreases with the increase of mechanical pressure exerted on it, which results in the increase of the grinding action between pyrite electrode and iron media. At about 7 min and the mechanical pressure at 800 g, the potential of the pyrite electrode is only the lowest –0.25 V. The pyrite surface exhibits stronger reduction atmosphere under grinding with iron media. Here, the pyrite surface potential is far lower than the potential for the formation of dixanthogen and floatation of pyrite. Therefore, mechanical pressure due to grinding only in iron media is detrimental to electrochemical action between xanthate and pyrite.

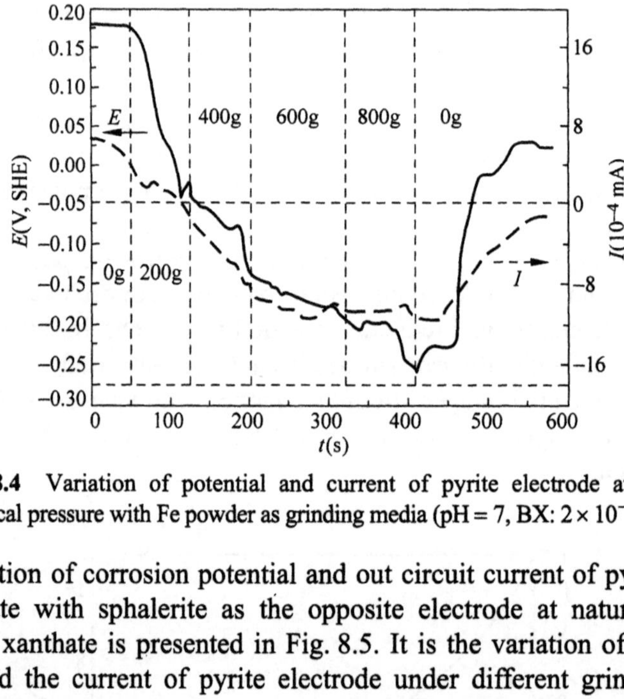

Figure 8.4 Variation of potential and current of pyrite electrode at different mechanical pressure with Fe powder as grinding media (pH = 7, BX: 2×10^{-4} mol/L)

The variation of corrosion potential and out circuit current of pyrite electrode at static state with sphalerite as the opposite electrode at natural pH in the presence of xanthate is presented in Fig. 8.5. It is the variation of the electrode potential and the current of pyrite electrode under different grinding pressure with sphalerite as the grinding media at natural pH in the presence of xanthate.

It can be seen from Fig. 8.5 that pyrite still exhibits the cathodic characteristic when sphalerite is used as the opposite electrode at static state. The corrosion potential of the pyrite electrode decreases at the beginning and is finally stabilized at about 140 mV. The pyrite electrode has not exhibited obvious cathode current. When sphalerite is used as the grinding media as seen from Fig. 8.6, the potential of pyrite electrode decreases with the increase of the mechanical pressure exerted on it and the grinding time. Pyrite exhibits cathodic characteristic, but the degree of cathode polarization is less than that in Fe grinding media. Corrosion potential of the pyrite electrode reaches to the lowest value about 145 mV at pressure of 800 g and 8 min.

Comparing Fig. 8.3 with Fig. 8.5, and Fig. 8.4 with Fig. 8.6, it can be found that the corrosive potential of the pyrite electrode is higher either by using

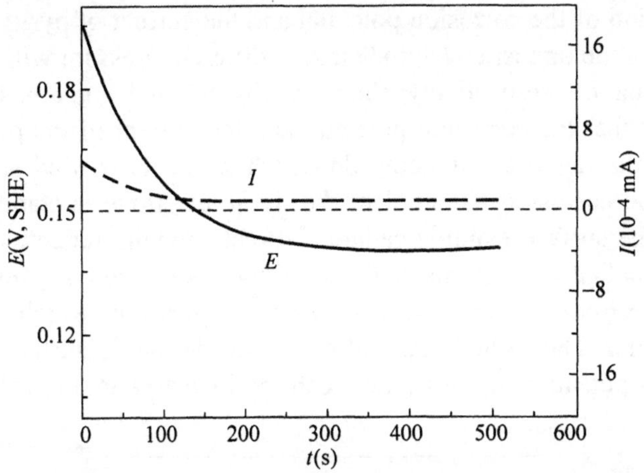

Figure 8.5 Variation of corrosion potential and out circuit current of pyrite electrode at static state with sphalerite as opposite electrode (pH = 7, BX: 2×10^{-4} mol/L)

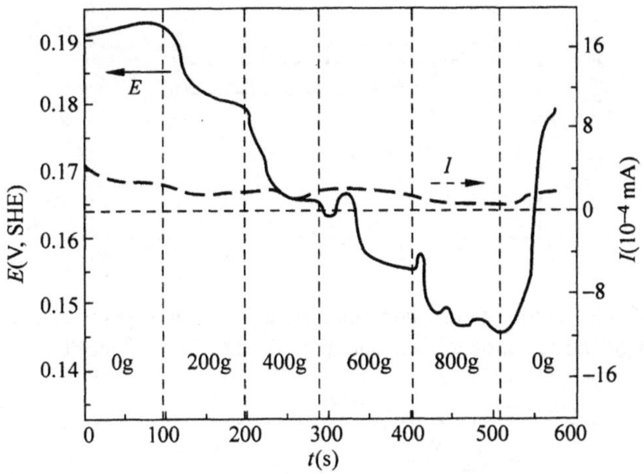

Figure 8.6 Variation of electrode potential and current of pyrite at different mechanical pressure with sphalerite as grinding media (pH = 7, BX: 2×10^{-4} mol/L)

sphalerite instead of iron powder as opposite electrode or by using sphalerite instead of iron powder as grinding media. When iron powder is used as the opposite electrode or as the grinding media, pyrite surface exhibits strong cathode current. When sphalerite is used as the opposite electrode or as the grinding media, no cathode current occurs on the pyrite surface. These results demonstrate that the strong reduction atmosphere will be induced for pyrite in iron medium grinding, which may be changed to the less reduction atmosphere when coexisted with sphalerite. For the coupling electrodes of sphalerite and pyrite, xanthate may show weak interaction with sphalerite at this potential and no action with pyrite.

The variation of the corrosion potential and the current of pyrite at static state with galena as the opposite electrode and at different pressure with galena as the grinding media are, respectively, shown in Fig. 8.7 and Fig. 8.8. It can be seen from Fig. 8.7 that the corrosion potential and the current of the pyrite electrode demonstrate the tendency of going down when galena is used as the opposite electrode. The corrosion potential of the pyrite electrode is stabilized at about 70 mV and the surface exhibits cathode current showing reduction atmosphere. Fig. 8.8 shows that with galena as the grinding media, the corrosion potential of the pyrite electrode changes with the exerting pressure, which reaches about 70 mV in 6 min. The cathode current occurs on the pyrite surface. After 8 min, the corrosion potential of the pyrite electrode increases to 150 mV and a small

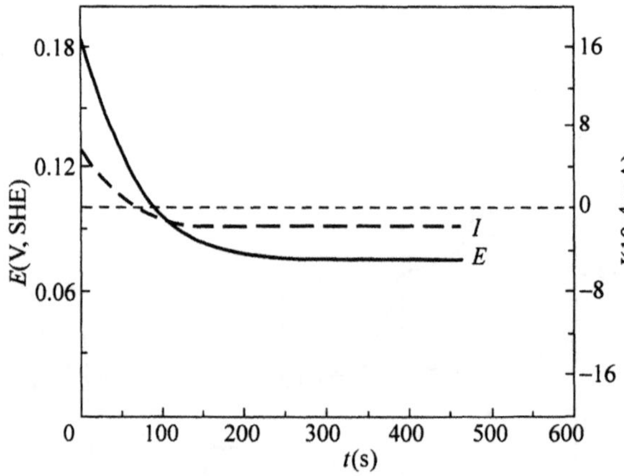

Figure 8.7 Variation of corrosion potential and current of pyrite electrode at static state with galena as opposite electrode (pH = 7, BX: 2×10^{-4} mol/L)

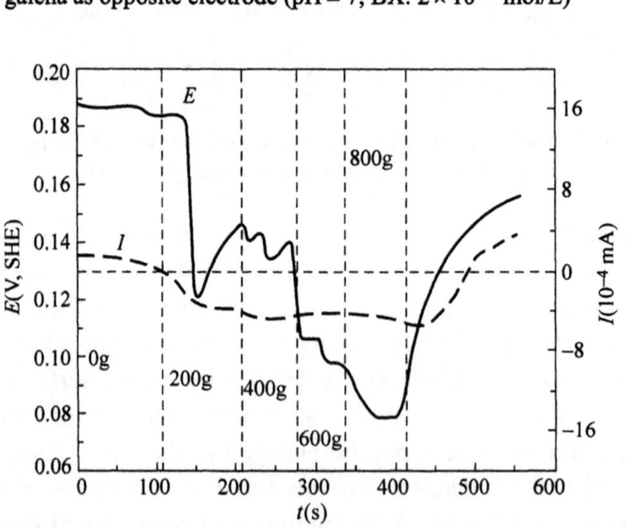

Figure 8.8 Variation of potential and current of pyrite electrode at different mechanical pressure with galena as grinding media (pH = 7, BX: 2×10^{-4} mol/L)

anode current occurs on the pyrite surface. These results suggest that for the coupling electrodes of galena and pyrite, the potential of the pyrite electrode is lower than the formation potential of dixanthogen, so pyrite can not interact with xanthate. Although with the increase of grinding time, the electrode potential of pyrite may increase to some extent, it is still outside of the range of the potential of action with xanthate.

The results from Fig. 8.3 to Fig. 8.8 demonstrate that the potential of the pyrite electrode always decreases when galena, sphalerite and iron are used as the grinding media. The reason is that a corrosion couple has been established when pyrite contacted with other media. The principle may be schematically presented in Fig. 8.9. It shows that the rest potential of pyrite, sphalerite, galena and iron are 180, 50, –50, –440 mV, respectively. Due to the high static potential of pyrite, it becomes the cathode and other media become the anode when pyrite contacts with other media. During the process of reaction, the surface static potential of pyrite reaches towards anode polarization and other media reach towards cathode polarization. When cathode potential is equal to anode potential the reaction achieves equilibrium, accounting for the decreasing tendency of the potential of pyrite electrode when using galena, sphalerite and iron powder as the opposite electrode, which is more in the grinding atmosphere.

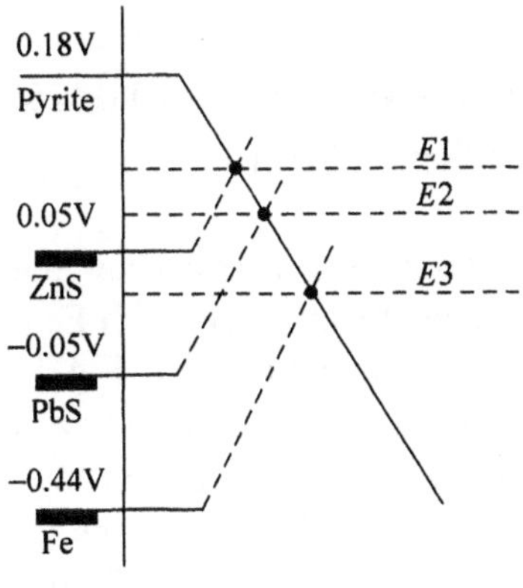

Figure 8.9 Illustration of polarization of corrosion couple between minerals and media

In flotation practice, the reducing environment produced in the grinding system may be useful for the interaction between galena and xanthate to form collector salt and not suitable for the formation of dixanthogen. It may promote the selective flotation separation of galena from pyrite.

Figure 8.10 is the variation of the potential and the current of the pyrite electrode with pyrite as the grinding media at different pressure at natural pH in the presence of xanthate. The results show that the corrosion potential decreases with the increase of mechanical pressure. The range of potential change is in 170 – 190 mV. This change may arise from the formation of nascent surface under grinding force. That is to say, the newly exposed pyrite surface may show lower potential due to grinding. Pyrite electrode surface appears as anode current, which changes little with the time and the exerting pressure.

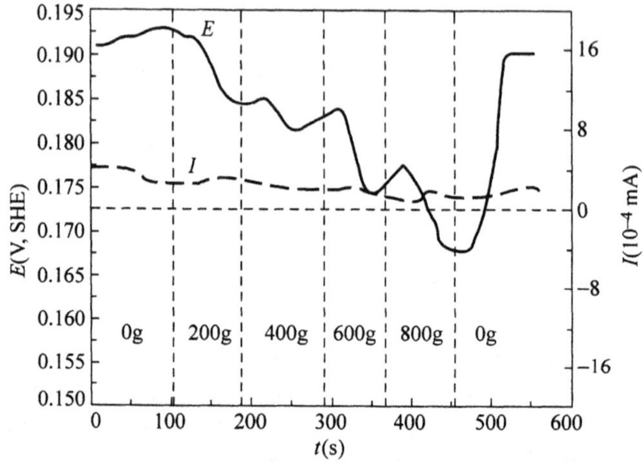

Figure 8.10 Variation of potential and current of pyrite electrode at different mechanical pressure with pyrite as the grinding media (pH = 7, BX: 2×10^{-4} mol/L)

8.3 Mechano-Electrochemistry Process of Galena in Different Grinding Media

Figure 8.11 gives the variation of the corrosion potential and the out circuit current of the galena electrode at static state with Fe as the opposite electrode in the presence of butyl xanthate at natural pH. Fig. 8.12 is the variation of the potential and the current of the galena electrode at different pressure with Fe powder as the grinding media in the presence of butyl xanthate at natural pH. It can be seen from Fig. 8.11 that the potential of the galena electrode reduces and reaches to a stable value of about –160 mV after 3 min. Galena appears as cathode when Fe is the opposite electrode and its surface exhibits cathode current, in which galena surface is protected and surface corrosion is inhibited. Figure 8.12 exhibits that when Fe powder is used as the grinding media, the corrosion potential of galena decreases with the increase of time and the exerting pressure. When the exerting pressure is 800 g, the corrosion potential of galena decreases to about –190 mV, which is lower than that when Fe is used as the opposite electrode, and the galena surface appears to have higher cathodic current. It indicates that the mechanical force by grinding may promote the reduction atmosphere. When the out exerting pressure is removed in about 500 s, the corrosion potential of galena reconciles to about –50 mV and the cathodic current of galena decreases and even a small anodic current occurs. These results suggest that the potential range after grinding in Fe media is suitable for the formation of lead xanthate and hence for the flotation of galena, which is usually corresponding to the initial potential of galena listed in Chapter 4.

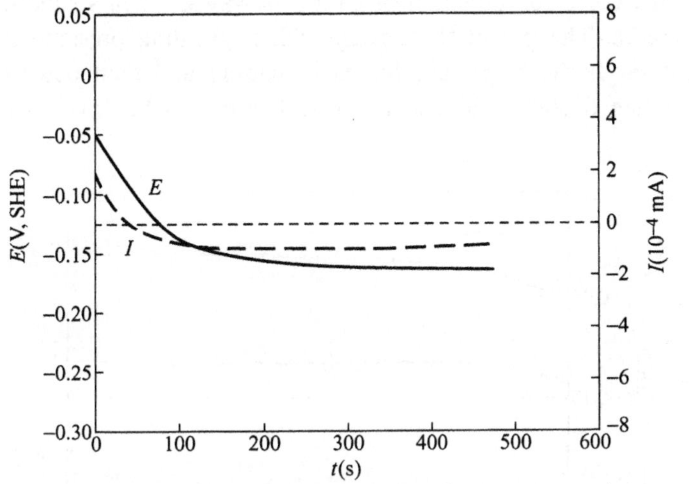

Figure 8.11 Variation of corrosion potential and current of galena electrode at static state with Fe as opposite electrode (pH = 7, BX: 2×10^{-4} mol/L)

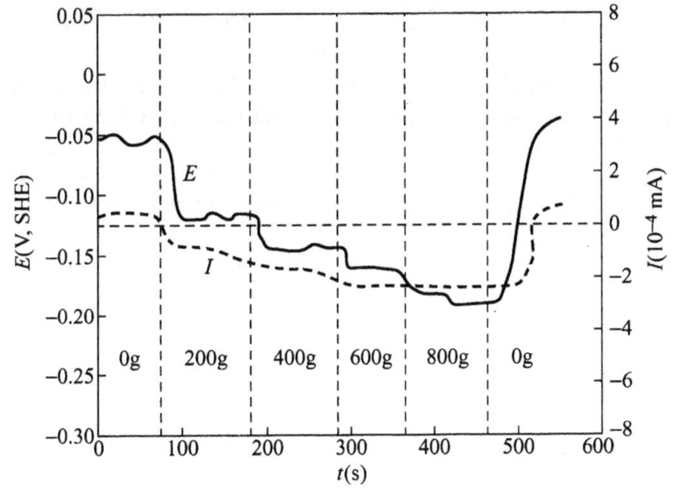

Figure 8.12 Variation of potential and current of galena electrode at different mechanical pressure with Fe powder as grinding media (pH = 7, BX: 2×10^{-4} mol/L)

In the presence of butyl xanthate and at natural pH, the variation of the corrosion potential and the out circuit current of galena electrode at static state with sphalerite as the opposite electrode is presented in Fig. 8.13. The variation of the potential and current of the galena electrode at different pressure with sphalerite as the grinding media is presented in Fig. 8.14. Figures 8.13 and 8.14

show that for the coupling electrodes of galena and sphalerite, the corrosive potential of the galena electrode reaches a stable value of about –25 mV after 3 min. Galena is anode and its surface exhibits obvious anode current. The corrosion on the surface of galena is intensified when sphalerite is the opposite electrode. The change of the corrosion potential of galena is waved when sphalerite is used as grinding media. The potential decreases when grinding pressure is put on the galena surface at the beginning for each loading and then potential increases gradually when friction achieves stability. There may be existing two kinds of

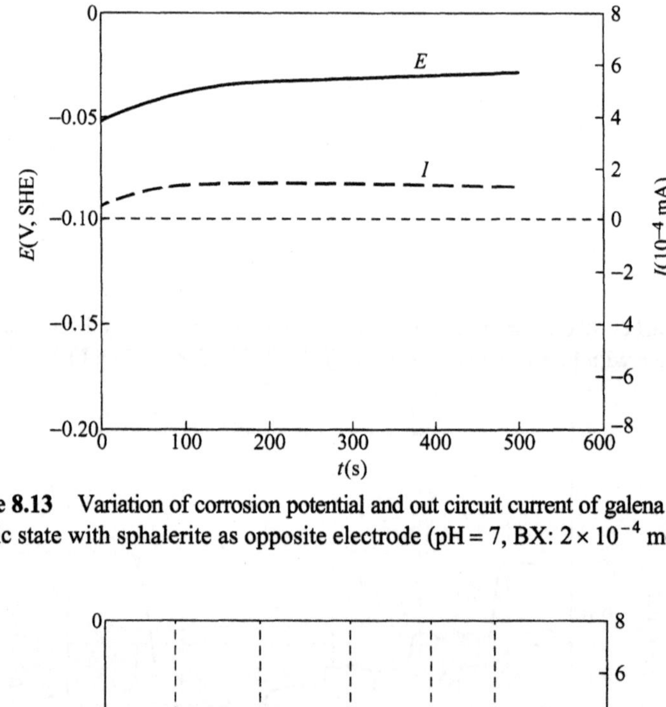

Figure 8.13 Variation of corrosion potential and out circuit current of galena electrode at static state with sphalerite as opposite electrode (pH = 7, BX: 2×10^{-4} mol/L)

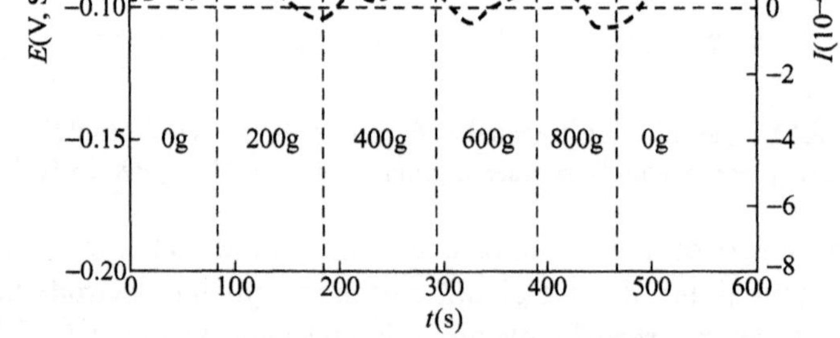

Figure 8.14 Variation of electrode potential and current of galena at different mechanical pressure with sphalerite as grinding media (pH = 7, BX: 2×10^{-4} mol/L)

factors which influence corrosion potential. One is that nascent galena surface is anode, the other is that the grinding media (sphalerite) is cathode. Nascent surface increases and the potential decreases when pressure is put on galena surface. When the increase of nascent surface achieves equilibrium, the coupling action between grinding media and galena occupies dominant position and potential starts to increase. These results suggest that in lead-zinc ore the potential of mineral changes frequently due to grinding. However, it can be found from Fig. 8.14 that the change of galena potential is around −50 mV, which may produce little effect on the interactions between the collector and galena.

Figure 8.15 shows the variation of the corrosion potential and the out circuit current of the galena electrode at static state with pyrite as the opposite electrode in the presence of butyl xanthate at natural pH. Figure 8.16 is the variation of the electrode potential and the current of the galena electrode at different pressure with pyrite as grinding media. It follows that galena appears as anode with an increase of potential and its surface exhibits obvious anode current when pyrite is used as the opposite electrode, where the anode reaction between galena surface and xanthate may be speeded up. When pyrite is used as grinding media, the same phenomena of using sphalerite as the grinding media can be observed. The potential of galena electrode changes with the exerting force and time, which decreases firstly and then increases with the grinding time increasing. This is also because two factors mutually affect corrosion potential. Nascent galena surface increases due to grinding and potential decreases when pressure is put on galena surface. When the increase of nascent surface achieves equilibrium, the coupling action between the pyrite media and the galena electrode results in the increase of potential. The change range of galena potential is from −70 mV to −40 mV. When the out exerting pressure is removed at about 500 s, the corrosion potential

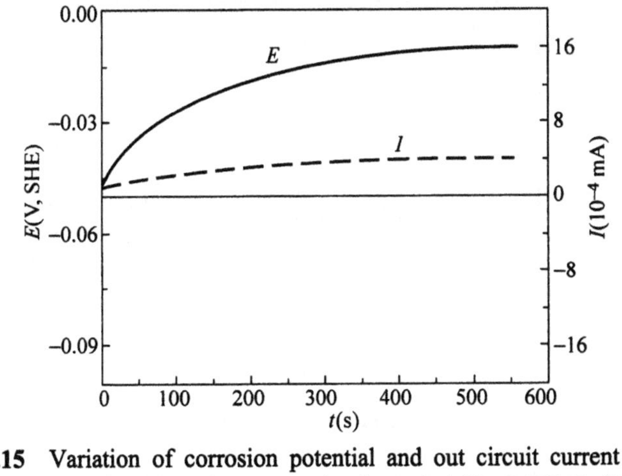

Figure 8.15 Variation of corrosion potential and out circuit current of galena electrode at static state with pyrite as opposite electrode (pH = 7, BX: 2×10^{-4} mol/L)

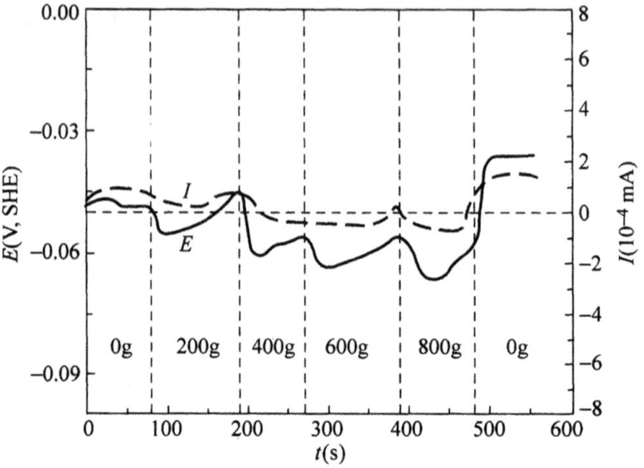

Figure 8.16 Variation of electrode potential and current of galena electrode at different mechanical pressure with pyrite as grinding media (pH = 7, BX: 2×10^{-4} mol/L)

of galena electrode reconciles to about −35 mV and a small anodic current occurs, suggesting that the potential range after grinding in sphalerite media is suitable for the formation of lead xanthate and hence for the flotation of galena.

Figure 8.17 reflects the variation of the electrode potential and the current of galena electrode at different pressure with galena as grinding media in the presence of butyl xanthate at natural pH. The results show that corrosion potential decreases with the increase of mechanical pressure. This is because the nascent galena surface has strong anode character. More and more nascent surface may be produced due to grinding giving rise to the descending of potential.

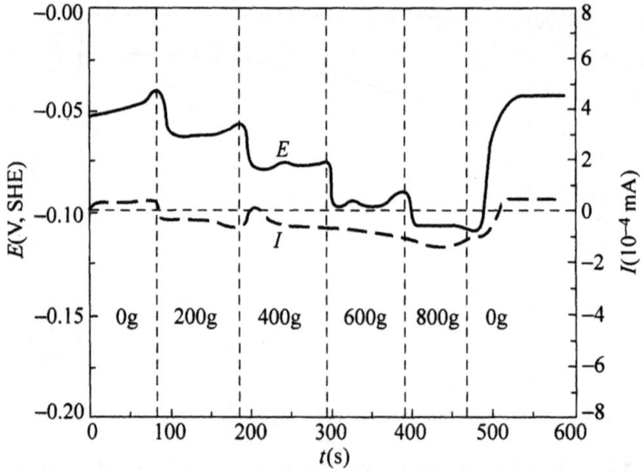

Figure 8.17 Variation of electrode potential and current of galena at different mechanical pressure with PbS as grinding media (pH = 7, BX: 2×10^{-4} mol/L)

Figure 8.18 illustrates the polarization of the corrosion couple between galena and other minerals. Galena has low static potential in the three sulphide minerals. When it contacts with other media, they form the corrosion cell. Galena appears anode and cathode polarization occurs. When galena contacts with Fe medium, which has lower potential than galena, the result is contrary (see Fig. 8.9).

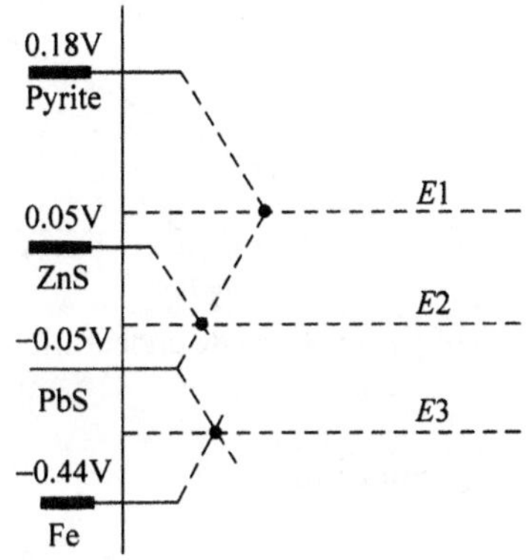

Figure 8.18 Schematical diagram of polarization of corrosion couple between galena and other minerals

8.4 Influence of Mechanical Force on the Electrode Process between Xanthate and Sulphide Minerals

As indicated by the previous research, the electrode process between collector and sulphide mineral on its surface is closely related to the surface activity of the mineral, which is greatly influenced by grinding in flotation process. It is very important to investigate the influence of mechanical force on the mineral electrode process so as to reveal the surface process of the sulphide minerals.

Figure 8.19 is the Tafel curves of the pyrite electrode in xanthate solution under the conditions of the mechanical power and non-mechanical powers. From Fig. 8.19 it can be seen that under non-mechanical powers the Tafel slope of the pyrite electrode is 120 mV and the $\lg I_0$ is –5.45, indicating that the reaction $FeS_2 + X^- = FeS_2 - X^-_{(ad)} + e^-$ is the crucial step, i.e. the adsorption of xanthate is the controlled step. Under the action of mechanical power, the Tafel slope of the pyrite electrode is 40 mV and the $\lg I_0$ is –5.35. The reaction $FeS_2 - X^-_{(ad)} + X^- = FeS_2 - X_2$ is the crucial step, i.e. the formation of dixanthogen due to the adsorbed xanthate is the controlled step. In this case the mechanical power through the increasing the number of the active points on the surface leads to the rise of the reaction rates of $FeS_2 + X^- = FeS_2 - X^-_{(ad)} + e^-$ and affects the equilibrium constant K.

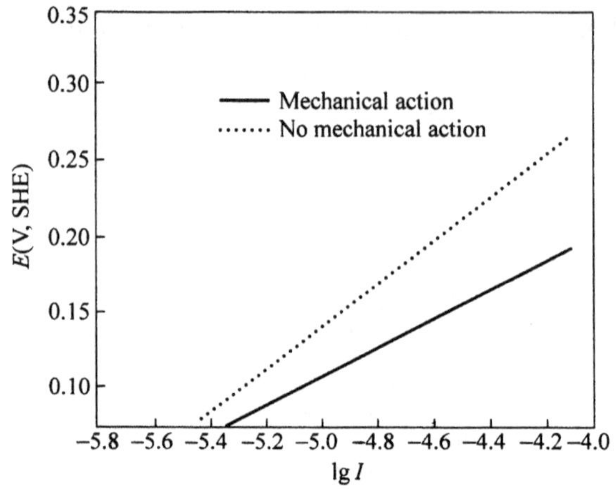

Figure 8.19 Polarization of pyrite electrode under different mechanical pressure (unit of I: A/cm^2)

According to the following equation

$$\lg I = \lg(2Fk_B K^2) + 2\lg c_X - \frac{2F}{2.3RT}E$$

The equilibrium constant K influences $\lg I_0$ value and hence affects the Tafel slope and the electrode process.

The Tafel curves of the galena electrode in xanthate solution under the conditions of the mechanical power and non-mechanical powers are given in Fig. 8.20. It follows from Fig. 8.20 that $\lg I_0$ of the anode action on the surface of galena raises from −6.5 to −5.9 with the increase of Tafel slope from −40 mV to −28 mV under the mechanical action. It shows that the reaction is favored in dynamics due to the grinding.

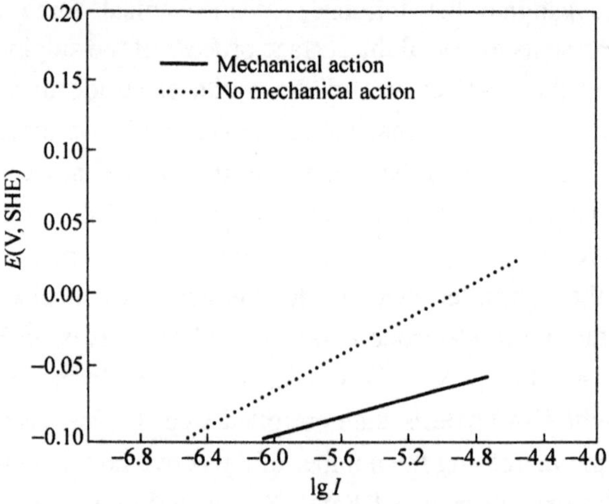

Figure 8.20 Polarization of galena electrode under different mechanical pressure (unit of I: A/cm^2)

8.5 Surface Change of Sulphide Minerals under Mechanical Force

Different mechanical force and grinding media induce the change of surface character, which causes the change of potential of mineral surfaces. The surface changing can be investigated by observing the surface top graph which results from mechanical rubbing of the mineral surfaces with the media of different sizes. Iron media sizes are as follows: for coarse grinding (+30—−70 μm), for middling grinding (+10—−30 μm), for fine grinding (−10 μm). The pictures of the top graph are taken by Nissan Mineral Phase Micro-camera with the amplification of 1000 times.

8.5.1 Surface Change of the Pyrite under Mechanical Force

The surface appearances of the pyrite ground by the Fe media of different size are presented in Fig. 8.21 to Fig. 8.24. It can be seen from Fig. 8.21 that the original surface of pyrite is smooth and undamaged. When the mechanical action is imposed to pyrite, its surface turns uneven due to the friction. Since there are height differences on the uneven surface area, the whole viewing area can not be concentrated at a focus and some parts get obscure which is most obvious in the case of the coarse grinding. At this time, the surface is the most uneven. When pyrite is abraded in the middling size grinding condition, the specific surface area increases further and some colored species exist in the viewing area, demonstrating that the minerals have experienced some reactions. Under the fine grinding media, pyrite surface appears rougher and more colorful species are observed, indicating more intensive surface reaction.

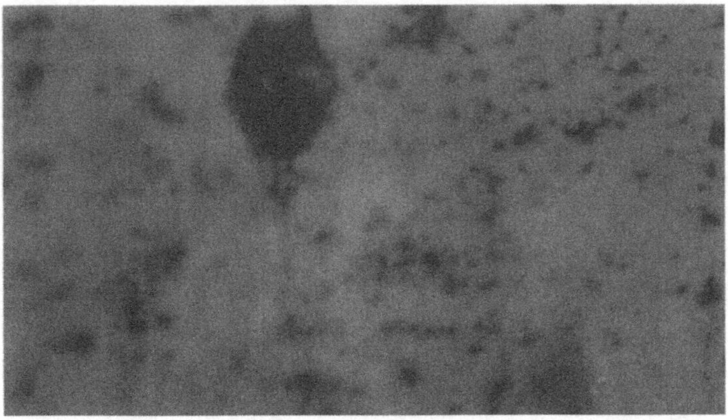

Figure 8.21 Original pyrite surface

Figure 8.22 Pyrite surface in coarse particle media

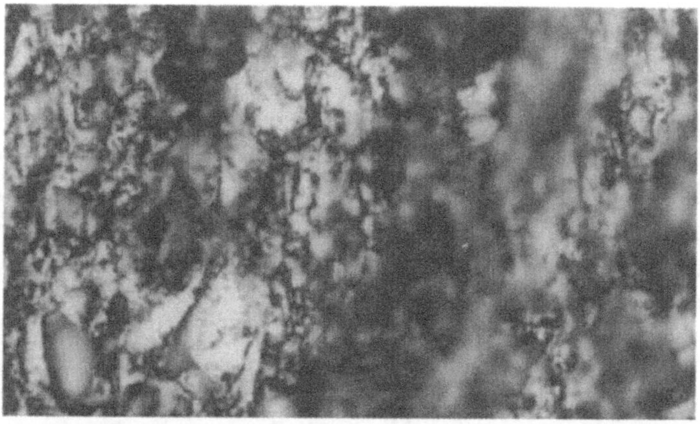

Figure 8.23 Pyrite surface in middling particle media

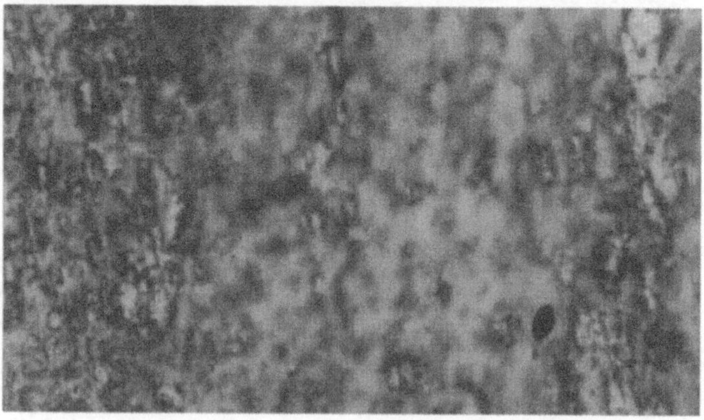

Figure 8.24 Pyrite surface in fine particle media

8.5.2 Surface Change of Sphalerite in Mechanical Force

Figures 8.25, 8.26, 8.27 and 8.28 are surface appearance of sphalerite at original state, and ground by coarse media, middling media and fine media, respectively. The figures above show that sphalerite has similar changes to those of pyrite under the four different conditions except that the colorization of sphalerite surface is not as evident as that of pyrite under the fine grinding condition. It indicates that there may be more active species to be formed on the surface of pyrite than sphalerite. This is consistent with the findings from the previous researches that among the three sulphide minerals the pyrite erodes are the fastest and its eroding current is the largest.

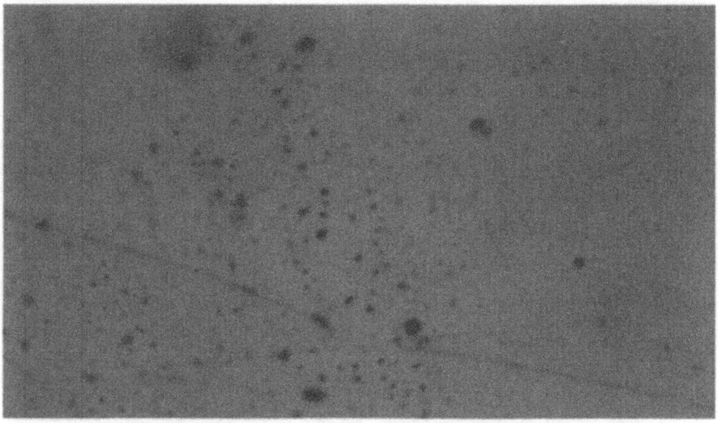

Figure 8.25 Original sphalerite surface

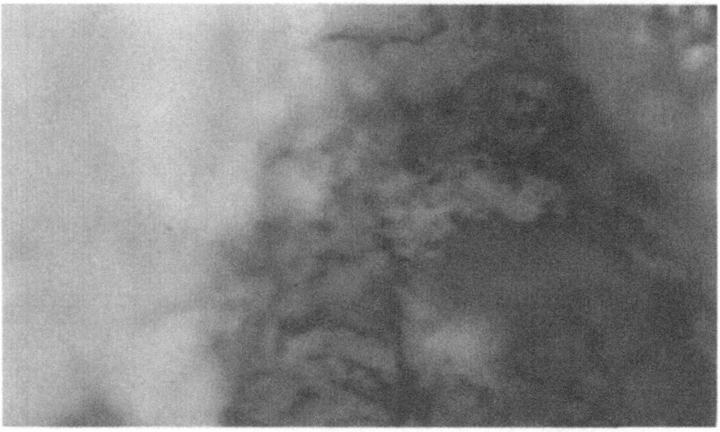

Figure 8.26 Sphalerite surface in coarse particle media

Figure 8.27 Sphalerite surface in middling particle media

Figure 8.28 Sphalerite surface in fine particle media

Chapter 9 Molecular Orbital and Energy Band Theory Approach of Electrochemical Flotation of Sulphide Minerals

Abstract In the light of quantum chemistry, sulphide minerals and the interaction of them with reagents are investigated in this chapter. With the density functional theory pyrite is first researched including its bulk properties about energy band and frontier orbitals and the property of FeS_2 (100) surface. It is found that surface Fe and S atom have more ionic properties. And both Fe^{2+} and S^{2-} have high electrochemistry reduction activity, as is the base of oxygen adsorption. Thereafter, the configuration of oxygen adsorping on the surface of pyrite is studied. And from the viewpoint of adsorption energy, the parallel form of oxygen adsorption is in preference to the perpendicular form. Similarly, some properties of ganlena are investigated. Furthermore, the activation of sphalerite is also researched. And the conclusion is that the forbidden gap of doped ZnS will decrease and the electrochemistry activity of ZnS is in return enhanced. The reaction between reagents and minerals is also discussed in the terms of molecular orbital and energy band, which provides a new method to investigate the mechanism of flotation.

Keywords molecular orbital; energy band; density functional theory(DFT)

It is well known that the flotation of sulphides is an electrochemical process, and the adsorption of collectors on the surface of mineral results from the electrons transfer between the mineral surface and the oxidation-reduction composition in the pulp. According to the electrochemical principles and the semiconductor energy band theories, we know that this kind of electron transfer process is decided by electronic structure of the mineral surface and oxidation-reduction activity of the reagent. In this chapter, the flotation mechanism and electron transferring mechanism between a mineral and a reagent will be discussed in the light of the quantum chemistry calculation and the density function theory (DFT) as tools.

9.1 Qualitative Molecular Orbital and Band Models

The frontier molecular orbital, including the HOMO (highest occupied molecular orbital) and the LUMO (lowest unoccupied molecular orbital), plays a key important role in a chemical reaction.

When a chemical reaction takes place, the electron transfers from the HOMO of a molecule to LUMO of another one, in which the HOMO and LUMO must be σ or π symmetrical and the energy of both orbitals is close.

When a group of atoms is brought together to form a solid, the individual electronic energy levels of the separate atoms overlap to form bands of closely spaced energy levels. The lower bands are completely filled with electrons, which are called filled bands or valence bands. The electrons in a filled band cannot normally move, but they can be forced up into unoccupied band by applying large amount of energy. The completely unoccupied or empty band is called an empty band, which becomes the conduction band when electrons are excited into the band from the filled band. The top of valence band and the bottom of the conduction band are separated by a forbidden energy gap or band gap. If the separation is large, i.e. an energy of 5 – 10 eV is required to promote an electron from the valence band to the conduction band then the solid would be an insulator. No forbidden energy gap exists for metal solid. If the forbidden energy gap is considerably smaller (perhaps 0.2 – 0.3 eV) then the material would be an intrinsic semiconductor, which can be divided into *p*-type which contains "holes" through which conduction may occur and *n*-type which contains electrons through which conduction may occur. The holes and electrons through which conduction occurs are called the conduction carrier of semiconductor. The conduction electrons are in the bottom of the conduction band and the conduction holes are in the top of valence band. Most sulphide minerals have intrinsic semiconductor structure. The interactions across the mineral-solution interface in the single presence or coexistence of oxygen, collector, HS^- etc. may take place by two paths, one of which is that the electrons are transferred from the bottom of the conduction band into the LUMO of molecule of the reactants and the other is from HOMO into the holes of the top of valence band, in both of which energy must be close.

9.2 Density Functional Theory Research on Oxygen Adsorption on Pyrite (100) Surface

It has been shown in the previous chapters that the product of the electrochemical reactions in the sulphide flotation system is determined by the mixed potential of the flotation pulp. The value of the potential is dependent on the equilibrium of anodic and cathodic process existing in the pulp. In general, the most important cathodic reaction existing in the pulp is the oxygen reduction. To rewrite Eq. (1-1) as the following:

$$O_2 + 2H_2O + 4e^- = 4OH^- \qquad (9\text{-}1)$$

If the access of O_2 is limited, the cathodic reaction and hence the coupled electrochemical reactions will be confined. Therefore, the oxygen reduction on the sulphide mineral surface plays a pivotal role in the flotation of the sulphide minerals. Oxygen reduction has been extensively studied in catalyze, fuel cell and corrosion (Pattabi et al., 2001; King et al., 1995). The reaction was often described as a multiple-step process, which is illustrated in Eq. (9-2):

$$O_2 \xrightarrow{\text{adsorption}} O_2^* \rightleftharpoons H_2O_2 \longrightarrow 2OH^- \tag{9-2}$$

It is reported that oxygen reduction is an electro-catalytic process (Fierro, et al., 1988; Gupta, et al., 1989) and that the rate is largely dependent on the properties of the surface. Chung's works (1998) showed that different surfaces have different affinity to oxygen, which has an important effect on oxygen reduction.

Ahlberg and Broo (1996a,b,c) studied the oxygen reduction on pyrite and galena with a rotating ring disc electrode. Their results suggest that the first electron transfer is the rate-determining step for the oxygen reduction at both galena and pyrite and hydrogen peroxide is found to be an intermediate. While galena is a poor catalyst for oxygen reduction, with minor formation of hydrogen peroxide, pyrite is a relatively good catalyst. This kind of difference may lead to different flotation behaviors.

One approach to a better understanding of the properties of the solid surface is to model the electron structure with quantum mechanical methods, which is a useful complement to experimental techniques. It allows direct observation of atomic-scale phenomena in complete isolation, which cannot be achieved in current experimental studies.

A lot of work (Opahle et al., 2000; Andrew et al., 2002a,b; Muscat et al., 2002; Edelbro et al., 2003) has been performed on bulk and surface properties of FeS_2 using various kinds of density functional theory. These works have shown that such methods are capable of producing calculated bulk properties such as lattice constants which agree well with experiment, and has provided a reference for the study of pyrite surfaces.

9.2.1 Computation Methods

The geometry optimization of FeS_2—O_2 and electric-structure-calculation of pyrite and its surface presented here is performed using ab intio simulation program Cambridge serial total energy package (CASTEP) through CERIUS2 graphical user interface.

The wave functions are expended in a plane wave basis set, and the effective potential of ions is described by ultrasoft pseudo potential. The generalized gradient approximation (GGA)-PW91, and local gradient-corrected exchange-correlation functional (LDA)-CAPZ are used for the exchange-correlation functional.

In order to find a reasonable configuration for our calculation, we take test calculation to optimize the bulk structure of pyrite with GGA and LDA exchange-correlation functional. In the calculation, the plane wave cutoff energy set is 280 eV and the key point set is $4\times4\times4$, the convergence tolerances set is 10^{-5} eV/atom. The optimized cell parameter of the two methods is 0.5415 nm and 0.5425 nm respectively, which is in good agreement with the experiment data (0.5417 nm) reported. It indicates that this configuration is sufficient to satisfy the request of accuracy.

The most important FeS_2 surface is the (100) surface, which is the most common growth surface and is also the perfect cleavage surface. Research from Nesbitt et al. (1998) suggest that the (100) surface of pyrite exhibits good stability and only minimal relaxation from the truncated solid. Therefore, our adsorption calculation is based on FeS_2 (100) surface and the relaxation of surface is ignored.

The FeS_2 (100) surfaces are modeled using the supercell approximation. Surfaces are cleaved from a GGA optimized crystal structure of pyrite. A vacuum spacing of 1.5 nm is inserted in the *z*-direction to form a slab and mimic a 2D surface. This has been shown to be sufficient to eliminate the interactions between the mirror images in the *z*-direction due to the periodic boundary conditions.

Surface models used a $[1\times1\times1]$ crystal unit cell as a surface slab in the supercell. We have also investigated the effect of slab thickness on the calculation result. It shows that three Fe—S layer model can get almost the same result as four Fe—S layers model.

Two models of oxygen adsorption are considered, vertical form and parallel form, which are illustrated in Fig. 9.1 and Fig. 9.2. For all the cases, the adsorbate/substrate system is optimized by GGA. In optimization, all atoms on the pyrite substrate are fixed; only the O atoms are allowed to move. The initial O—O double bond length and the distance between Fe atom and O atom are 0.121 nm and 0.196 nm, respectively. To simplify the calculation, the adsorption coverage will not be considered.

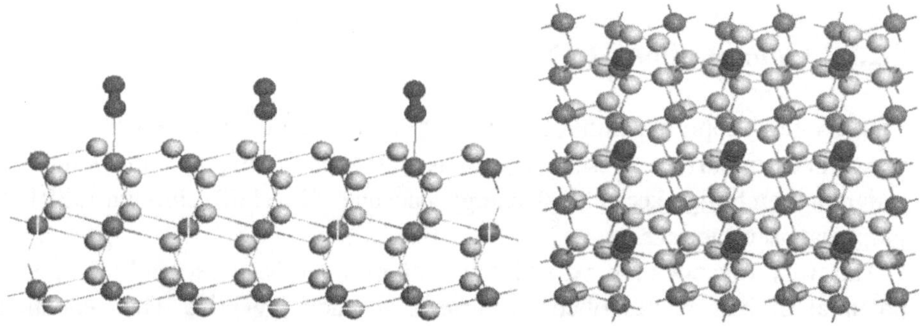

(a) (b)

Figure 9.1 Vertical form of O adsorption on FeS_2 (100) surface
(a) Side view; (b) Top view

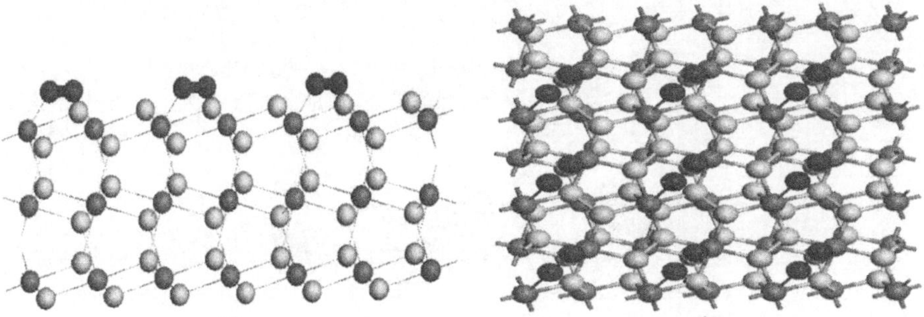

Figure 9.2 Parallel form of O adsorption on FeS_2 (100) surface
(a) Side view; (b) Top view

The adsorption energies are defined positive for a stable adsorbate/substrate system:

$$E_{ads} = E_{O_2} + E_{FeS_2} - E_{FeS_2-O_2} \tag{9-3}$$

where $E_{FeS_2-O_2}$ is the energy of the adsorbate/substrate system, E_{O_2} and E_{FeS_2} are the energy of the adsorbate and the substrate alone.

9.2.2 Bulk FeS_2 Properties

GGA property calculation on optimized bulk pyrite has been performed. The calculated band structure and frontier orbital distribution are presented in Fig. 9.3 and Fig. 9.4. The calculated band gap of FeS_2 is 0.97 eV, and it is in good line

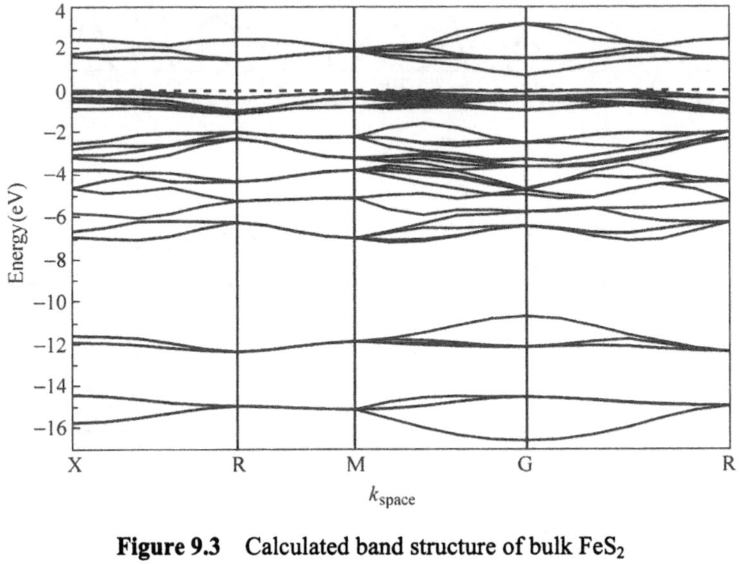

Figure 9.3 Calculated band structure of bulk FeS_2

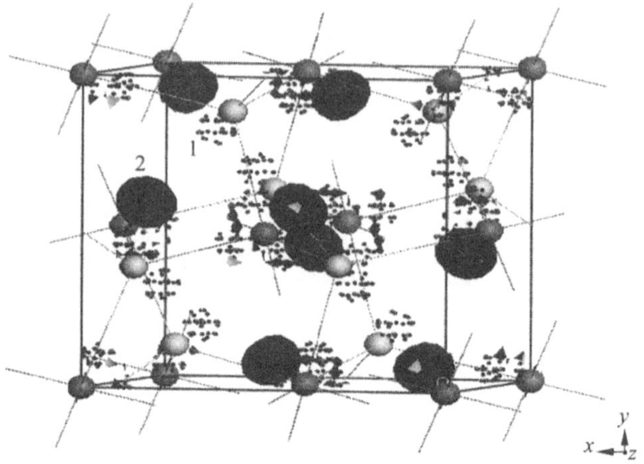

Figure 9.4 The frontier orbital distribution in bulk FeS_2
1–lowest unoccupied state; 2–highest occupied state

with the experimental data (0.96 eV). The density of states (DOS) diagram for bulk pyrite is shown in Fig. 9.5. These figures show that the Fe (3d) non-bonding orbitals are located at the top of the valence band and extend to just below the Fermi level. Another part of Fe (3d) orbital and antibonding sp^3 orbitals of sulfur contributes to the conduction band (CB). These findings are consistent with the crystal field theory (CFT) description of a metal cation surrounded by ligands in an octahedral array, with unoccupied Fe (3d) e_g states immediately above the

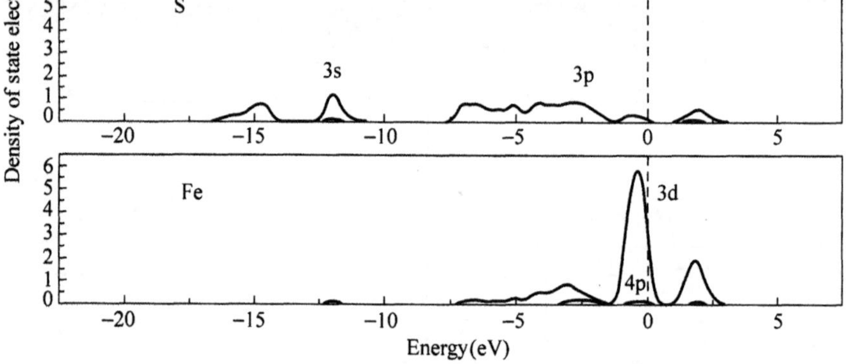

Figure 9.5 Total and partial density of state of bulk FeS_2

fully filled t_{2g} states. The same conclusion has been reported by Andrew Hung et al. (2002). The peaks from −11 eV to −17 eV come from S (3s) state. Below the valence band, there is a mixture of Fe (3d), S (3p) and S (3s) orbital; it is the area of Fe—S bond and S—S bond.

9.2.3 Property of FeS_2 (100) Surface

In bulk FeS_2, each Fe atom is coordinated with six S atoms and each S atom is coordinated with three Fe atoms and one S atom. Fe atom has two free electrons and S has 6 free electrons. For S atom, one electron is used to form S—S dimmer and the other 5 electrons are used to form Fe—S bond. For Fe atom, all the free electrons are shared by Fe—S bond. That is to say the electron contribution of Fe atom to each Fe—S bond is 2/6 and that of S atom is 10/6. When the crystal is cleaved from (100) directions, only Fe—S bonds is broken (see Fig. 9.6). Then the 6-fold Fe atom becomes 5-fold Fe atom and 4-fold S atom becomes 3-fold S atom. Thus, it will produce dangling bonds and surface states.

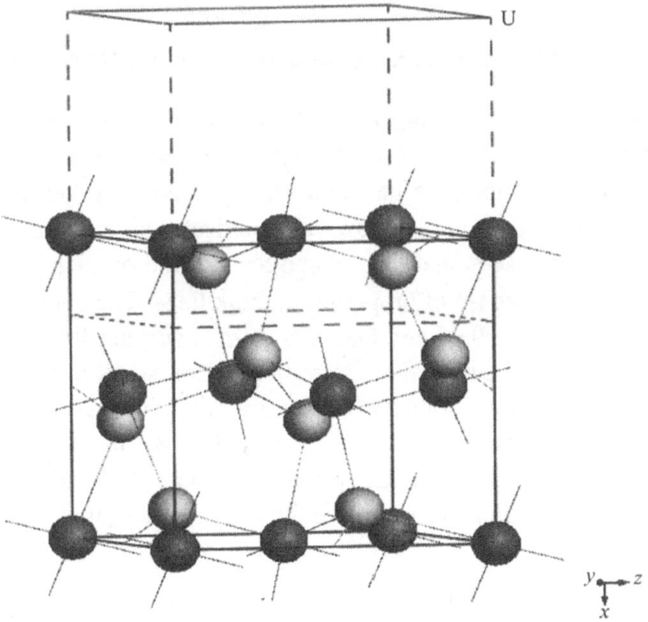

Figure 9.6 Schematic cleavage of FeS_2 from (100) directions

Figure 9.7 presents the DOS distribution of FeS_2 (100) surface. By comparing the PDOS (partial density of state) on the surface with the PDOS from the bulk, a slight narrowing and an upward shift of the Fe (3d) band can be seen because the surface Fe atom loses one neighbor S atom. However, the most important effects of these coordinative unsaturations appear with the S (3p) band, which shows a few peaks around the Fermi level and its bottom now starts only at −6 eV. These unsaturated S (3p) bands play a crucial role in frontier orbitals.

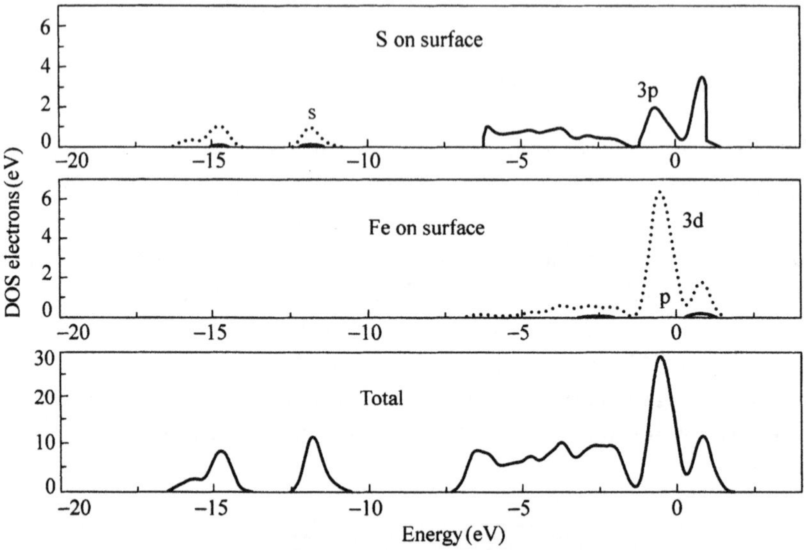

Figure 9.7 Total and partial density of state of FeS_2 (100) surface

The atomic and bond overlap population analysis of pyrite (100) surface is listed in Table 9.1 and Table 9.2. The overlap population may be used to assess the covalent or ionic nature of a bond. A high value of the bond population indicates a covalent bond, while a low value indicates an ionic interaction (Segall et al., 2002). It can be seen that Fe—S and S—S bond overlap population on the surface is smaller than those of properties on bulk FeS_2. The net charge of Fe atom is varying from −0.12 to 0.05 and that of S atom is varying from 0.09 to −0.12. It indicates that when a new surface is formed, there exists a process of electron transfer from Fe dangling bond to S dangling bond. This conclusion is in good correspondence with the Nesbitt's DFT works (1998). In this situation, surface Fe and S atom have more ionic properties. Both Fe^{2+} and S^{2-} have high electrochemistry reduction activity, as is the base for oxygen adsorption.

Table 9.1 Atomic population of bulk FeS_2 and FeS_2(100) surface

	Atom	s	p	d	Total	Net charge
Bulk	S	1.81	4.10	0.00	5.91	0.09
	Fe	0.31	0.59	7.29	8.19	−0.19
Surface	S^1	1.86	4.26	0.00	6.12	−0.12
	Fe^1	0.32	0.39	7.22	7.93	0.07

1–atom on the surface; 2–atom below the surface.

Table 9.2 Bond overlap population of FeS_2 and $FeS_2(100)$ surface

	Bond	Population overlap
Bulk	S—S	0.26
	Fe—S	0.34
Surface	S^1—S^2	0.26
	Fe^1—S^1	0.28

1–atom on the surface; 2–atom below the surface.

9.2.4 Oxygen Adsorption

The adsorption energy of oxygen on the $FeS_2(100)$ surface is presented in Table 9.3. It can be seen that the adsorption energy of the parallel form O is found to be 0.34 eV higher than that of vertical form O. It indicates that the dominant adsorption model may be the parallel form as shown in Fig. 9.1. It is a reasonable assumption model considering that the magnitude of adsorption energy in Table 9.3 is in the same order with that of O_2 adsorption on metal oxide surface reported by other authors (Zhang, 1996; Bechtold and Schennach, 1996; Sasaki et al., 2002). The bond length variation before and after optimization is also shown in Table 9.3.

Table 9.3 Summary of results for the FeS_2 (100)/O_2 adsorption system

	Adsorption geometry	Adsorption energy (eV)	O—O bond length (nm)	Fe^1—O bond length (nm)	Fe^2—O bond length (nm)
Initial configuration			0.121	0.196	0.196
			0.121	0.196	
Optimized	Vertical form	0.20	0.135	0.199	0.197
	Parallel form	0.34	0.137	0.187	

It follows that in the adsorption state, the O—O bond length varies from 0.121 to 0.135 and 0.137 nm for vertical and parallel adsorption forms, respectively. This bond length is close to the O—O bond length in peroxides (0.149 nm). It means that after adsorbed by the FeS_2 surface, the oxygen is more like peroxides rather than O_2. It is clearly in line with the works reported by Ahlberg and Broo (1996a,b,c).

The calculated DOS of the parallel O_2 adsorption structure is given in Fig. 9.8. It can be seen that the shoulder peaks at −10 eV is the bonding interaction between the O (2p) and the Fe (4s), (4p) and (3d) orbital. The antibonding counterpart starts at 0 eV above the Fermi level and is essentially composed of the Fe (3d) orbital and O (2p) orbital. The area from −7.5 eV to 0 eV (and above) concerns the interactions of the O (2p) with the Fe (4s) and (3d) for the peak. According to the above calculation and discussion, some conclusions can be summarized as follows.

The calculated bulk properties of FeS_2 are in agreement with the recent experimental value. The frontier orbital is predominantly occupied by Fe (3d) orbital. It is also an exemplification for the feasibility of using DFT in this system. When the crystal is cleaved from (100) directions, an only the Fe—S bond was broken. Thus, it will produce dangling bonds and surface states. In the formation of the surface, there exists a process of electron transfer from Fe dangling bond to S dangling bond. In this situation, surface Fe and S have more ionic properties than atom properties. Both Fe^{2+} and S^{2-} have high electrochemistry reduction activity, as is the base for oxygen adsorption. From the viewpoint of adsorption energy, the parallel form O adsorption is in preference. The result also shows that after adsorbed by the FeS_2 surface, the O_2 is more like peroxides.

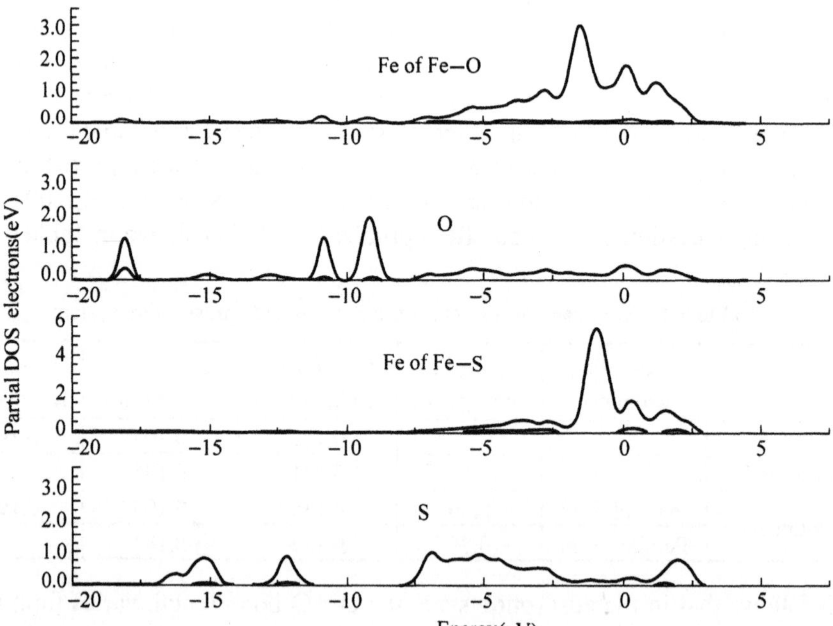

Figure 9.8 Total and partial density of state of O_2/FeS_2 (100) system

9.3 Density Functional Theory Research on Activation of Sphalerite

Sphalerite, which is also known as Blende, is an important mineral of zinc. Most natural sphalerite contains iron more or less in lattice with the amount depending on the chemical environment and temperature (Lusk et al., 1993). High iron sphalerite is called marmatite. The studies of the electronic structure and surface properties of ZnS and transition metal doped ZnS are of interest from both a fundamental and practical perspective. As discussed in Chapter 6, sphalerite has a poor flotation response to collectors due to the solubility of Zn-xanthate species, but it can be activated by other transition metal ions such as Cu(II), Pb(II) and Fe(II) (Finkelstein,1997a,b; 1999) resulting in an excellent flotation response. It is generally accepted that Cu^{2+} ion can replaces the Zn^{2+} at the mineral surface and the xanthate collector will react with the surface Cu ions. Although it is still not clear about the effect of Cu^{2+} on surface properties of ZnS, we believe that there exist some relation between the doped ions and surface properties of ZnS.

We employ density functional theory to calculate the electron structures of bulk sphalerite, and the pure and doped ZnS (110) surface in order to explain the effect of Cu and Fe ions on the properties of ZnS and the mechanism of activation flotation of sphalerite.

9.3.1 Computational Methods

The electric-structure-calculation presented here is performed using the CASTEP computer code, which is based on density functional theory, aided by the CERIUS2 graphical front-end. The wave functions are expended in a plane wave basis set, and the effective potential of ions is described by ultrasoft pseudo potential.

The structure of sphalerite is analogous to the diamond structure as shown in Fig. 9.9. The structure is a unit cube with zinc atoms at the corners and face centers. The space group of sphalerite is F-43 m. The initial ZnS lattice constant used in calculation is $a = 0.543$ nm. The distance between sulfur and its neighbor zinc is 0.234 nm and the distance between two neighbors S or Zn atom is 0.383 nm. The quality of pseudo potential is tested to optimize the structure of bulk ZnS and to compare with the experiment data. In all calculations, the generalized gradient approximation (GGA)-PW91 is used for exchange-correlation functional.

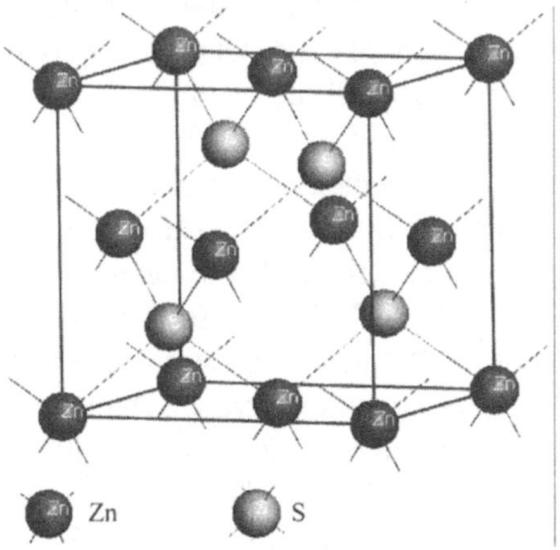

Figure 9.9 Crystal structure of sphalerite and marmatite

The most important ZnS surface is the (110), which is the most common growth surface and is also the perfect cleavage surface. Therefore, the calculation is based on the ZnS (110) surface. The surfaces are cleaved from the bulk ZnS with the optimum unit cell volume determined using the GGA with CASTEP. The Cu and Fe doped surfaces are built by the substitution of Cu or Fe for Zn atom on the cleaved surface. A vacuum spacing of 1.5 nm is inserted in the z-direction to form a slab and mimic a 2D surface. In order to eliminate the interactions between mirror images in the z-direction due to the periodic boundary conditions, in test calculations, we have done some total energy calculation to find a proper thickness of slab. The result shows that 1.5 nm is a desirable thickness.

In the calculation, we have examined the effect of some parameter on total energy and structure of surface and crystal. These parameters include plane wave cutoff energy, k-point and SCF convergence criteria, and the geometry optimization convergence criteria for the geometry optimization task.

The calculation results indicate that a plane wave cutoff energy of 280 eV and Monkhorst-Pack k-point sampling density of $4\times4\times4$ are sufficient for the lattice constant and total energy to converge to within 0.0005 Å and 10^{-5} eV respectively. For surface relaxation, a plane wave cutoff of 280 eV and a $4\times4\times1$ k-point mesh are sufficient to converge the surface geometry to within 0.001 Å and relaxed surface energies to within 0.001 J/m^2.

9.3.2 Bulk ZnS Properties

The energy bands of the frontier electrons in semiconductors consist of a valence band fully occupied by electrons at low levels, a vacant conduction band at high energy levels; and a forbidden band called the band gap separates these two bands. Figure 9.10 shows the calculated band structure of sphalerite along the selected high-symmetry lines within the first brillouin zone of the FCC lattice. The corresponding total density of state (DOS) and partial DOS of every element are shown in Fig. 9.11. In Fig. 9.10 the band gap is calculated to be 2.7 eV. It is smaller than the experiment band gap, which is measured to be 3.6 eV (David, 2000). The result is in good agreement with other DFT works (Edelbro et al., 2003; Vanghan et al., 1997). The S (3s) orbital is located in the interval from −14.5 eV to −12.5 eV in the valence band. This is also in agreement with the works reported by Jellinek et al., (1974). In bulk ZnS, each Zn atom is coordinated with 4 S atoms and each S atom is coordinated with 4 Zn atoms. Zn and S atoms have two free electrons and six free electrons respectively, which are shared by four Zn—S bonds. These bonds are composed of Zn (sp^3) hybridization orbital and S (sp^3) hybridization orbital. Figure 9.11 also shows some hybridization of Zn(4s, 4p) and S (3s, 3p) orbital in the interval from 0 to −5 eV, this is the area of Zn—S bond. The bottom of the conduction band has almost equal amounts of Zn(4s) and S (3p) orbital character and is Zn—S antibonding area. Zn (3d) orbitals are located at the narrow interval from −5 eV to −7.5 eV, it is more positive compared with the XPS study of sphalerite.

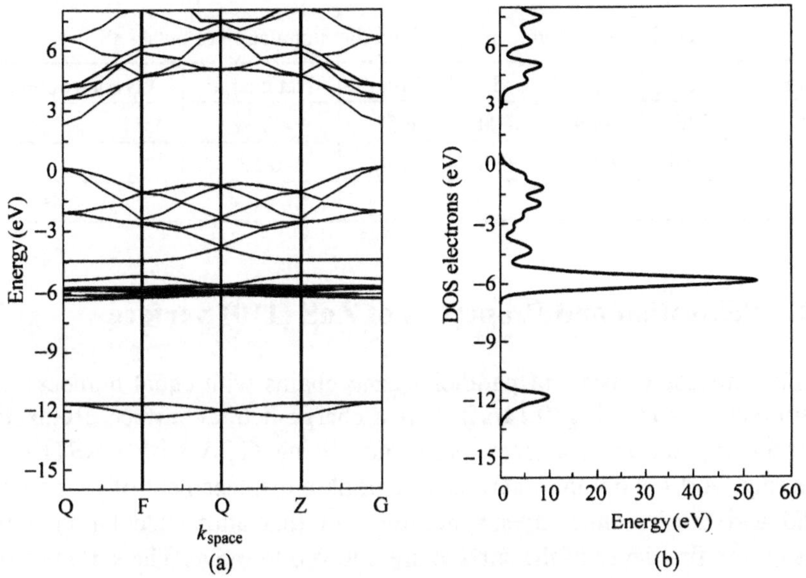

Figure 9.10 The calculated band structure of sphalerite

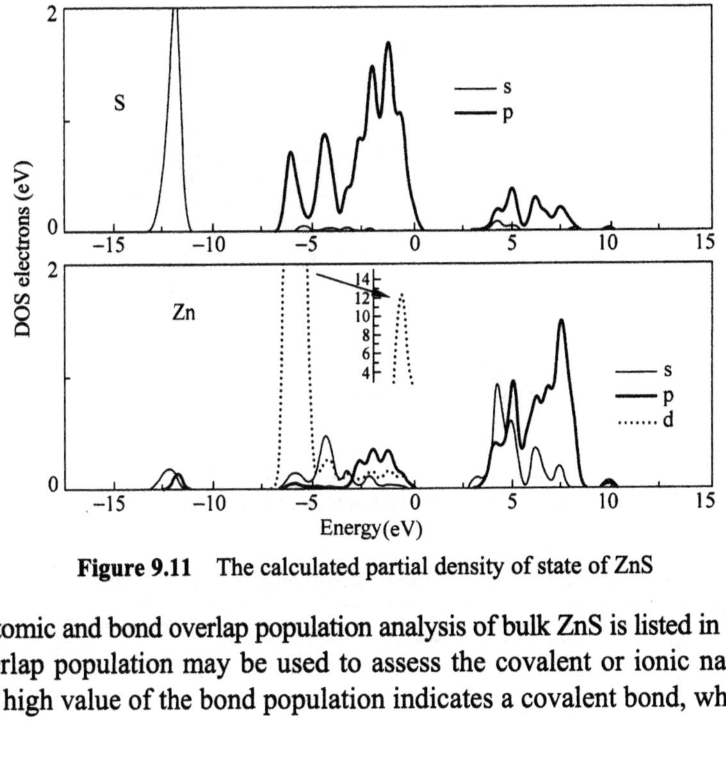

Figure 9.11 The calculated partial density of state of ZnS

The atomic and bond overlap population analysis of bulk ZnS is listed in Table 9.4. The overlap population may be used to assess the covalent or ionic nature of a bond. A high value of the bond population indicates a covalent bond, while a low value indicates an ionic interaction. Table 9.4 indicates that the Zn—S bond has both covalent and ionic nature. And the value of net charge suggests that in Zn—S covalent bond, S atom shares more electrons.

Table 9.4 Atomic and bond overlap population of bulk ZnS

Atom	s	p	d	Total	Net charge	Overlap population
S	1.83	4.69	0.00	6.52	−0.52	
Zn	0.86	0.65	9.98	11.48	0.52	
Zn—S						0.46

9.3.3 Relaxation and Properties of ZnS (110) Surface

ZnS(110) surface consists of parallel zigzag chains with equal numbers of zinc and sulphur ions (see Fig. 9.12(a)). It is a charge neutral surface. Relaxation of the ZnS (110) surface has been performed using GGA with CASTEP. Some pioneering works show that there is a negligible displacement of ions below the second and third atomic layer. Therefore, in relaxation calculation, only the atoms on the first layer of the surface are allowed to move. The surface structure and ionic displacement vectors for the (110) surface are shown in Fig. 9.12(b). Ionic displacements due to surface relaxation are presented in Table 9.5.

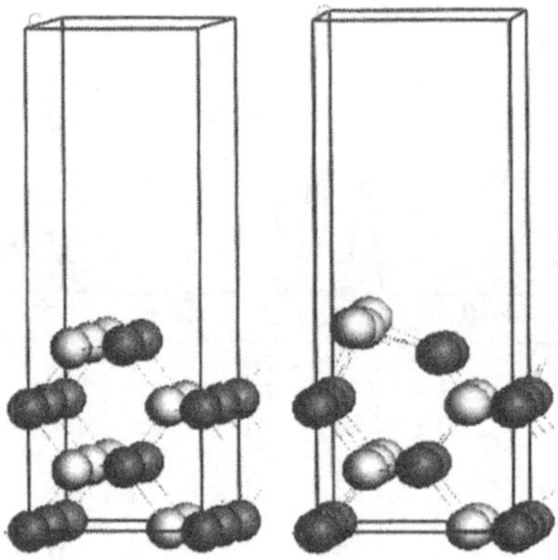

(a) (b)

Figure 9.12 The model of ZnS (110) surface and schematic relaxation of surface (a) ZnS (110) surface; (b) Ionic displacement vectors in surface relaxation

Table 9.5 The ZnS (110) surface ionic displacements due to surface relaxation

Atom	Coordination	Δx(nm)	Δy(nm)	Δz(nm)
Zn	3	0.001	0.001	−0.02
S	3	−0.001	0.005	0.009

These results suggest that the most significant relaxation of the ZnS (110) surface is a downward displacement of the surface Zn atoms by approximately 0.02 nm. The surface S atoms relax out the surface by about 0.01 nm. The band structure and partial density of state (PDOS) of relaxed ZnS (110) surface are illustrated in Fig. 9.13. The atomic and bond overlap population analysis of ZnS (110) surface is listed in Table 9.6. It shows that the band gap of ZnS (110) surface is 1.5 eV, and it is smaller than that of bulk ZnS. The reason for band gap

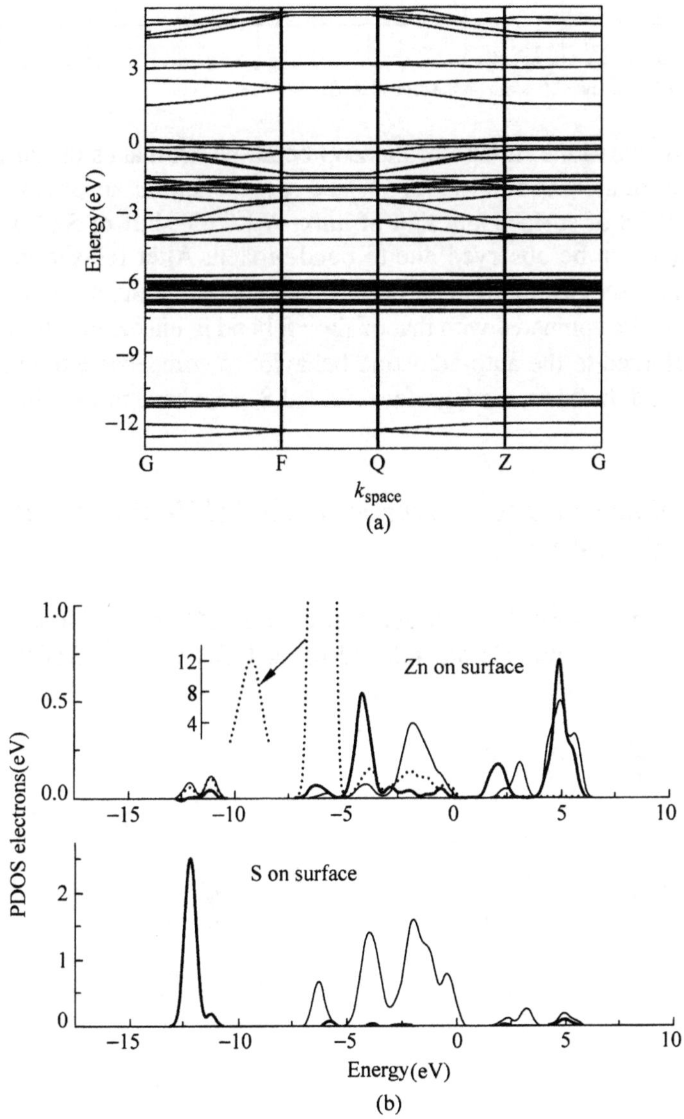

Figure 9.13 (a) Calculated band structure of ZnS (110); (b) Calculated PDOS of ZnS (110)

Table 9.6 Atomic and bond overlap population of ZnS(110) surface

Atom	s	p	d	Total	Net charge	Overlap population
S^1	1.85	4.73	0.00	6.58	−0.58	
S^2	1.84	4.70	0.00	6.54	−0.54	
Zn^1	0.86	0.65	9.98	11.48	0.52	
Zn^2	0.61	0.79	9.98	11.37	0.63	
Zn^1—S^1						0.40
Zn^1—S^2						0.44
Zn^2—S^2						0.58

1 – atom on the first layer; 2 – atom on the second layer.

reduction is due to the rupture of the ZnS bond, which makes the surface Zn and S atom unsaturated and produces surface state. There is an obvious difference between PDOS of surface and that of bulk. A 1.2 eV shift of S (4s) towards the valence band can be observed due to bond-broken. After relaxation, the surface Zn—S bond becomes more ionic because the overlap population of Zn—S bond becomes smaller compared with that of Zn—S bond in bulk ZnS. This phenomenon can be attributed to the auto-adjusting behavior to compensate the production of dangling bond. In this regard, surface Zn and S atom have more ionic properties.

9.3.4 Relaxation and Properties of ZnS (110) Surface Doped with Cu^{2+} and Fe^{2+}

The models of the ZnS (110) surface doped with Cu^{2+} and Fe^{2+} ions are presented in Fig. 9.14. These models are relaxed using GGA with CASTEP. The rule of

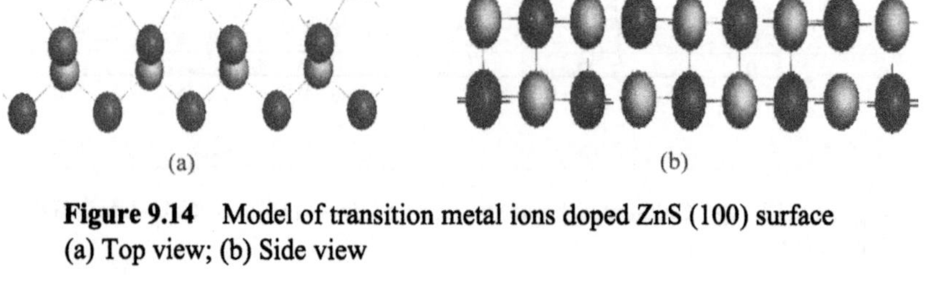

(a) (b)

Figure 9.14 Model of transition metal ions doped ZnS (100) surface
(a) Top view; (b) Side view

relaxation is the same as that of the ZnS (110) surface. The Ionic displacements due to surface relaxation are presented in Table 9.7. It can be seen that the most remarkable displacement is the shift of the Zn atom along *z*-direction; this is similar to the relaxation of the ZnS (110) surface. When the Zn atom is substituted by Cu or Fe ions, the extent of surface relaxation becomes smaller.

Table 9.7 Ionic displacements of the doped ZnS (110) surface due to surface relaxation

Surface	Atom	Coordination	Δx(nm)	Δy(nm)	Δz(nm)
Cu doped	Zn	3	0.001	0.001	−0.010
	S	3	−0.0011	0.0041	0.0090
	Cu	3	0.0047	0.0001	−0.0001
Fe doped	Zn	3	0.0012	0.0011	−0.015
	S	3	−0.0009	0.0035	0.0070
	Fe	3	0.0032	0.0004	0.0000

The PDOS of the relaxed surface doped with Cu^{2+} and Fe^{2+} is presented in Fig. 9.15 and Fig. 9.16 respectively. It can be seen that the valence band of the surface doped with Cu^{2+} is predominantly composed of Cu (3d) orbital which extends from 0 to –5 eV. This is in excellent agreement with Andrea's XPS studies (1999). There is no obvious gap between the occupied and unoccupied levels. The S (3p) and Zn (4p) orbits are located at the bottom of the conduction band.

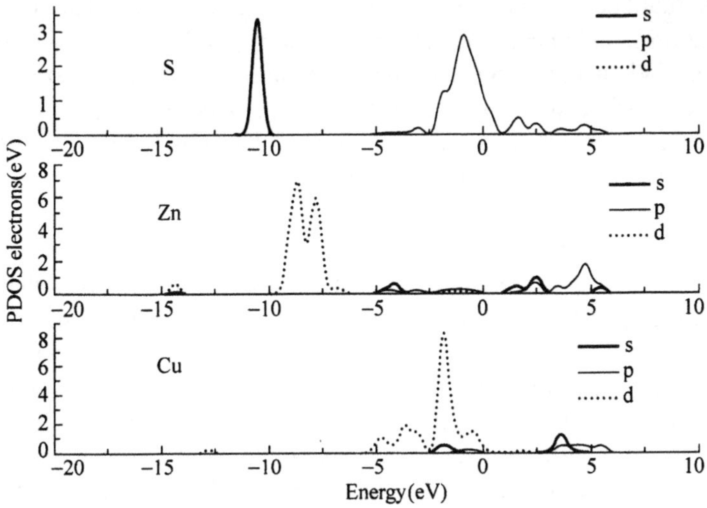

Figure 9.15 The PDOS of the Cu^{2+} doped ZnS (110) surface

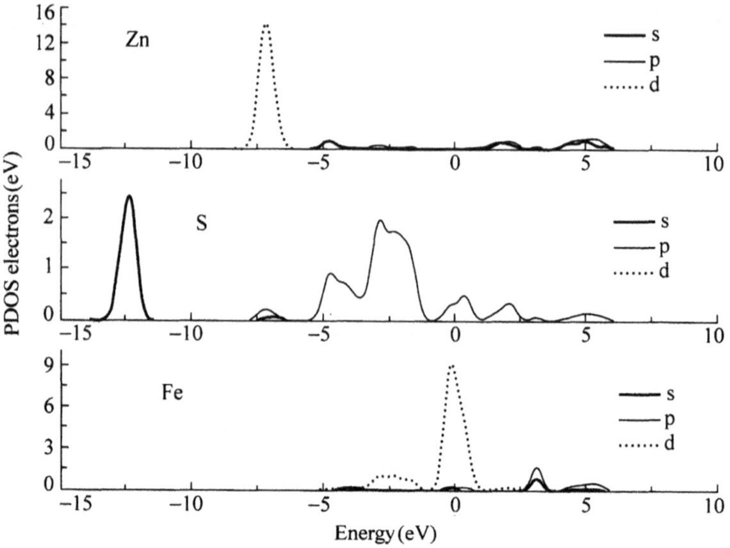

Figure 9.16 The PDOS of the Fe^{2+} doped ZnS (110) surface

Similar PDOS distribution can be seen on the ZnS surface doped with Fe ions. The dominant state in valence band is Fe (3d) orbital, and the conduction band is composed of S (3p) and Zn (4p) orbital. This result indicates that doping Cu or Fe ions on the ZnS surface reduces the band gap of the ZnS. This kind of reduction will produce lot of surface state in bulk ZnS forbidden band.

It is reported that the band structure of ZnS doped with transition metal ions is remarkably different from that of pure ZnS crystal. Due to the effect of the doped ions, the quantum yield for the photoluminescence of samples can be increased. The fact is that because more and more electron-holes are excited and irradiative recombination is enhanced. Our calculation is in good correspondence with this explanation. When the ZnS (110) surface is doped with metal ions, these ions will produce surface state to occupy the valence band and the conduction band. These surface states can also accept or donate electrons from bulk ZnS. Thus, it will lead to the improvements of the photoluminescence property and surface reactivity of ZnS.

In flotation, when sphalerite is activated by Cu^{2+} or Fe^{2+}, the ZnS surface will exhibit good reactivity to organic collector. Our calculation shows that when the surface is doped by transition metal ions, the surface ions will be rendered more ionic property, which benefits the interaction between the mineral surface and the collector anions. It gives more profound explanation for Cu^{2+} activated behavior to ZnS.

Because the oxidation of sulphide has a pronounced effect on sulphide mineral flotation, oxidation will produce metal ions on the mineral surface and these ions will react with collectors to render the surface hydrophobicity. From the DOS shown in Fig. 9.13, Fig. 9.15 and Fig. 9.16, marmatite and (Zn, Cu) S are the intrinsic semiconductors and sphalerite is a broad band semiconductor. The top of the valence band of above three materials are dominantly occupied by Fe (3d),

S (3p), and Cu (3d) orbit, respectively. According to the frontier orbital theory, the electrons in the highest occupied state are most easily bound and have an unexpectedly great significance for the chemical reactivity of materials. It indicates that the different reduction or oxidation would happen on the three mineral surfaces in the pulp during flotation system.

Based on Pattrick et al.(1999), marmatite is easily to be oxidized than sphalerite and Fe at the (Zn, Fe) S surface is predominantly bonded to O. There is also some literatures to show that the ZnS surface is resistant to oxidation but oxidation of the ZnS surface after Cu activation is apparent. All these studies imply that ZnS has low electrochemistry activity. This is because of the following reasons:

ZnS is a broad band semiconductor, the electron in the fully occupied valence band is difficult to be excited up to the conduction band. There is less free electron in ZnS, and then it is not easy to accelerate the electrochemistry oxidation of surface. That is to say, the dissolution of ZnS from mineral surface is difficult and has little chance to react with thio-collectors. When substituted by Cu or Fe ions, the forbidden gap of doped ZnS will decrease and the quantity of free electrons in ZnS will increase. This variation will enhance the electrochemistry activity of ZnS.

9.3.5 Effects of Doped Ions on Mixed Potential

Marmatite has a narrower band gap than (Zn, Cu)S, thus it should be more easily oxidized than (Zn, Cu)S. But in fact, the sphalerite after Cu activation has the most excellent flotation response using xanthate. These phenomena can be explained by the mixed potential theory.

According to the mixed potential theory, an anodic reaction can occur only if there is a cathodic reaction proceeding at finite rate at that potential (Rand and Woods, 1984). For the flotation systems, the cathodic reaction is usually given by the reduction of oxygen. The corresponding anodic reaction involves interaction of xanthate on the sulphide minerals in various ways, including the reaction of xanthate with the sulphide mineral (MS) to form metal xanthate and the oxidation of xanthate to dixanthogen (X_2) at the mineral surface.

The mixed potential of the sulphide mineral in the flotation pulp will determine the oxidation product on its surface. If the mixed potential of the mineral in the presence of oxygen, xanthate and other reagents is above the mixed potential for the X^-/X_2 redox couple, then the reaction will produce dixanthogen on the surface. If the mixed potential is lower than the X^-/X_2 redox couple, metal xanthate reaction will take place rendering the surface hydrophobic.

A lot of work show that the existence of Fe ions on solid surface can catalyze the reduction of oxygen. It indicates that the rate of oxygen reduction on marmatite is larger than rate of reduction on (Zn, Cu)S, this will cause the mixed potential of marmatite to be higher than that of (Zn, Cu)S and ZnS. This process can be illustrated in Fig. 9.17.

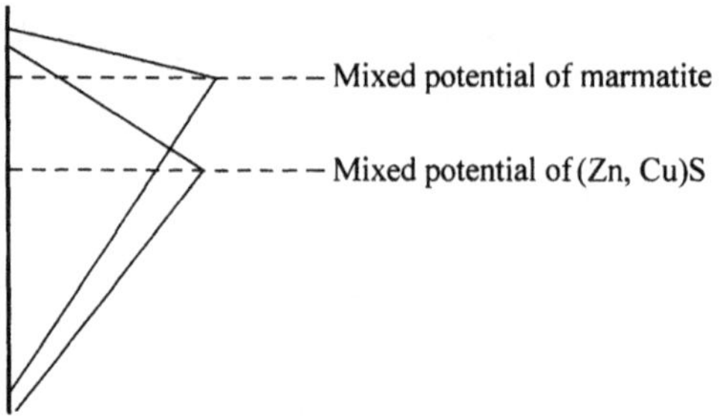

Figure 9.17 Schematic of mixed potential marmatite and (Zn, Cu) S

9.4 The Molecular Orbital and Energy Band Discussion of Electrochemical Flotation Mechanism of Sulphide Minerals

As the discussions in the previous chapters, some sulphide minerals possess stronger collectorless floatability and collector floatability at certain potential range. Normally, according to the mixed potential mode, an anodic reaction can occur only if there is a cathodic reaction occurring at that potential and at the same time. For the flotation system, the cathodic reaction is usually given by the reduction of hydration oxygen. The corresponding anodic reactions involve that in different ways. The self-oxidation of sulphide mineral results in the formation of elemental sulphur showing self-induced collectorless floatability or produced sulfur-oxy and hydroxyl species showing no flotation. The reaction of the collector such as xanthate on sulphide mineral may form metal salt or dixanthogen to exhibit collector flotation behavior. In the following section, we will discuss the anode oxidation and cathode reduction occurring at the sulphide/solution interface from molecular orbital and energy band theory.

9.4.1 Frontier Orbital of Collector and Oxygen

Dimethyl-dithio-amine-carbonate (DDTC) and butyl xanthate (BX) are often used in the flotation of sulphide minerals. Their structures are displayed in Fig. 9.18. Shown in the same figure. is the structure of hydration oxygen. We take ab initio theoretical calculations on the orbital energy of BX and DDTC and the hydration oxygen atom. All the calculations are taken by Gaussian98 software. The geometries of the compounds and the frequencies are evaluated using the DFT level of the three parameter compound functional of Becke (B3LYP). The 6-31G(d) basis set is used for all atoms. The geometry structures of neutral molecules and ion molecules are optimized under no constraint. The resulting HOMO and LUMO energy and orbital symmetry of BX and DDTC are listed in Table 9.8 and Fig. 9.19.

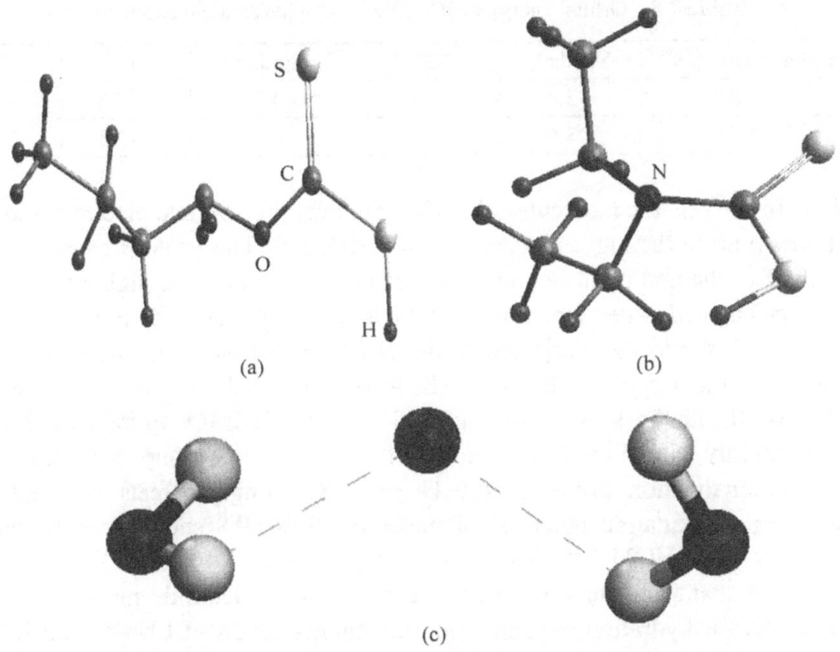

Figure 9.18 The structures of butyl xanthate, dithiocarbamate and hydration oxygen molecules
(a) Butyl xanthate; (b) Dithiocarbamate; (c) Hydration oxygen

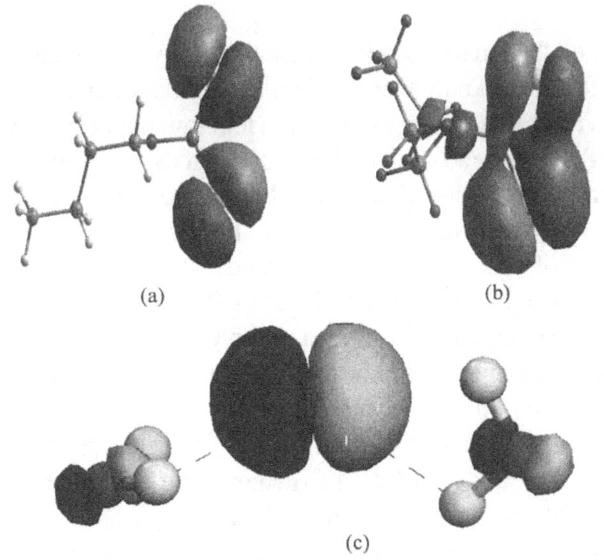

Figure 9.19 The highest occupied orbit of BX, DDTC and lowest unoccupied orbit of hydration oxygen atom
(a) Butyl xanthate; (b) Dithiocarbamate; (c) Hydration oxygen

Table 9.8 Orbital energy of BX, DDTC and hydration oxygen atom

Orbital energy (eV)	Xanthate	DDTC	Hydration O	OH^-	X_2
HOMO	−3.4	−3.6	−13.9	−7.23	−9.23
LUMO	3.6	3.8	−3.9	−1.23	−2.23

For simple organic molecules, the MOs are occupied by two electrons (spin up and down) up to the highest occupied MO (HOMO). The HOMO energy level is on behalf of the ability of the molecule donating electrons. The higher the HOMO is, the easier it donates electrons. All MOs above the HOMO are unoccupied, starting with the lowest unoccupied MO (LUMO). LUMO represents the ability of the molecule accepting electrons. The lower the LUMO is, the easier it accepts electrons. It can be seen from Fig. 9.19 that the HOMO orbitals of BX and DDTC mainly spread over three atoms of the C—S—S group. It indicates that the functionalization group of the C—S—S strongly affects the reduction potential and oxidation potential of molecule. Table 9.8 shows that the energy level of xanthate HOMO is –3.4 eV, which is higher than that of DDTC, –3.67 eV. It suggests that xanthate can donate electrons to an electrode more easily than DDTC. As for hydration oxygen atom, the energy levels of LUMO and HOMO are, respectively, −3.94 eV and −13.9 eV.

9.4.2 The Molecular Orbit and Energy Band Discussion of Collectorless Flotation of Galena and Pyrite

The structures of the energy bands of galena and pyrite are shown in Fig. 9.20. From Section 9.2.2, we know that the Fe (3d) non-bonding orbitals are located at the top of the valence band and extend to just below the Fermi level, and that the part of Fe (3d) orbital and antibonding sp^3 orbitals of sulfur contribute to the conduction band. Similar calculations can be applied to galena. The valence band of galena is mainly made up of S (3p) orbit. Pb (6s) band and (6p) comprise the conduction band of PbS. It should be pointed out here that we obtained all the energy levels of band using a Fermi level as 0 reference in the calculations in Section 9.2.2. The energy level is only a relative value. If we discuss the electron transfer between mineral electrode and pulp electrolyte, we should know the energy level of band using vacuum level as 0 reference. According to the research of Chernyshova (2003), the Fermi levels of pyrite and galena are, respectively, −4.4 eV and −3.86 eV in vacuum. The energy levels of galena and pyrite in Fig. 9.20 are the value using vacuum level as 0 reference. Like the HOMO in a molecule, the Fermi level in sulphide mineral is on behalf the ability to donate or accept electrons, The more negative the Fermi level, the more up-shifted the electro potential, and so the greater the tendency to accept electrons from the adsorbate to the electrode.

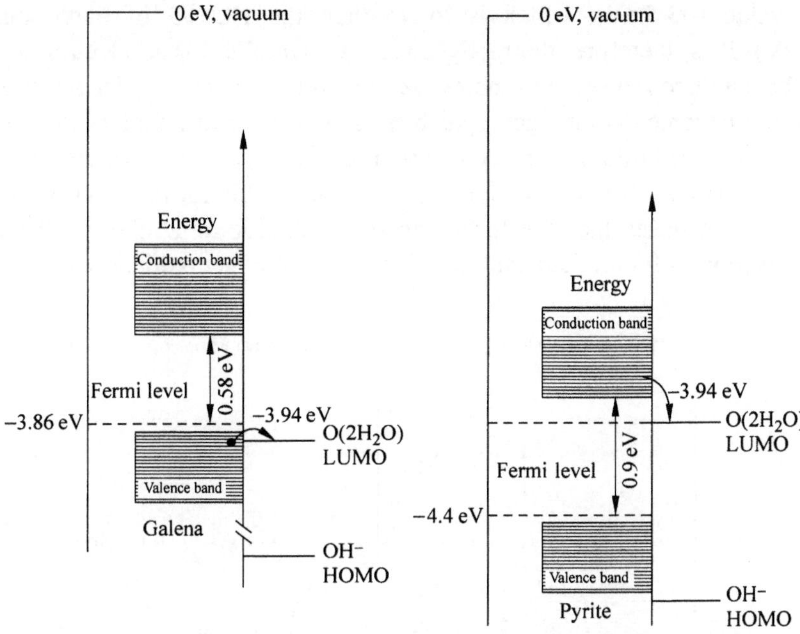

Figure 9.20 Illustration of electron transfer between sulphide surface and hydration oxygen atom showing the mechanism of collectorless flotation of galena and pyrite

It can be seen from Fig. 9.20 that the top valence band of galena is −3.86 eV, which is higher than that of the LUMO of hydration O. It is feasible in thermodynamics to transfer electrons from galena valence band to LUMO of hydration O. The valence band of galena is mainly made up of S (3p) orbit. Galena is oxidized to form elemental sulphur, which shows surface hydrophobic and good collectorless floatability. For pyrite, it is easy to see that the LUMO of hydration O atom is lower than the bottom of pyrite conduction band (−3.5 eV) but higher than the top valence band (−4.4 eV). Therefore, the electron transfer process occurring at pyrite surface is different from the process at galena surface. The conduction band of pyrite consists of the part of Fe (3d) orbital and antibonding sp^3 orbitals of sulfur. The oxidation of pyrite easily produces iron hydroxyl species or sulphur-oxy species to exhibit surface hydrophilic and poor self-induced collectorless floatability.

9.4.3 The Molecular Orbit and Energy Band Discussion of Collector Flotation of Galena and Pyrite

Figure 9.21 shows the conduction band and valence band within pyrite along with the orbital energies (HOMO and LUMO) of BX and DDTC in solution. On the left side, the bottom of the conduction band (unoccupied band) of pyrite has a lower value (−3.5 eV), which is lower than the HOMO of butyl xanthate (−3.4 eV). It is, therefore, thermodynamically favorable for an electron to jump from the xanthate molecule to the pyrite electrode. Xanthate molecule loses one electron to become dixanthogen, which renders pyrite hydrophobic. However on the right side, the bottom of the conduction band of pyrite is above the HOMO of DDTC (−3.61 eV). It is thermodynamically unfavorable for the electron transfer to occur. It suggests that the formation of disulphide of DDTC on the pyrite surface is more difficult than xanthate. It requires stronger oxidation atmosphere.

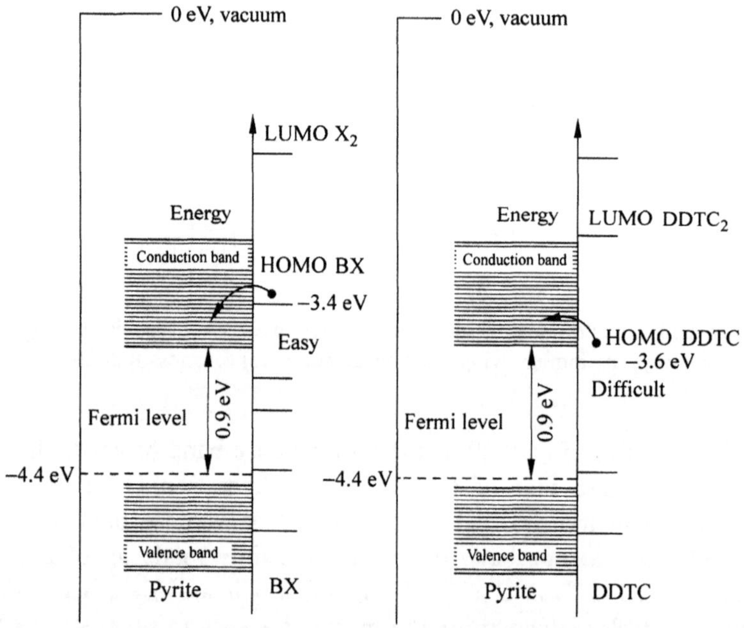

Figure 9.21 Illustration of electron transfer between pyrite surface and collector showing the mechanism of collector flotation of pyrite

Figure 9.22 shows the band energy level of galena and frontier MO of BX and DDTC ion. It can be seen that the highest occupied orbital of BX and DDTC molecule is higher than the bottom conduction band of galena (−3.28 eV). It is difficult to transfer electrons from molecule to mineral surface, i.e., the disulphide of collector such as xanthate and dithiocarbamate may not be easily formed on the galena surface. However, on the other hand, the top valence band of galena is −3.86 eV, which is higher than that of the LUMO of hydration oxygen. The oxidation of the galena surface produces lead ion and sulphur or sulphur-oxy species. The collector metal salts can be formed on the galena surface through the reactions between lead ion and collector. Takahashi (1991) also reported that in the adsorption of EX^- on the ionic solids electron transferred from the solids to EX^-, while they transfer from EX^- to the solids in the adsorption of EX^- on the covalent solids.

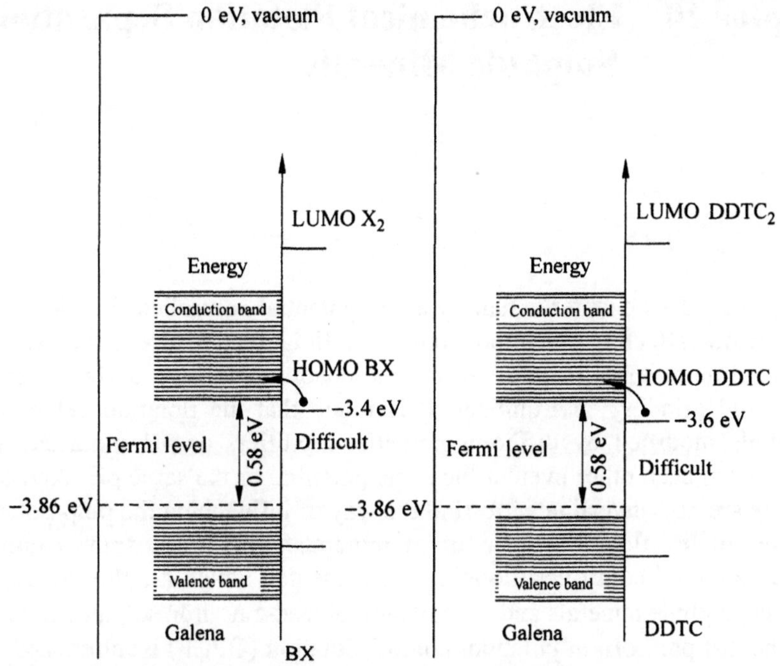

Figure 9.22 Illustration of electron transfer between galena surface and collector showing the mechanism of collector flotation of galena

Chapter 10 Electrochemical Flotation Separation of Sulphide Minerals

Abstract This chapter is mainly about potential controlled flotation. First, the factors affecting potential controlled flotation are discussed, such as potential modifiers, pH modifier, frother, conditioning time, surface pretreatment and grinding environment. It is found that the flotation behavior of minerals modified by different potential modifiers or pH modifiers may differ from each other even at the same potential or the same pH. And other factors are also investigated to find the way of influencing the pulp potential to control the flotation behavior of minerals. Thereafter, some examples about potential controlled flotation are presented which are the flotation of copper sulphide minerals and the flotation of lead-zinc-iron sulphide minerals. As the last part, origin potential control flotation (OPCF) is introduced. The factors coupled with OPCF including grinding, lime dosage, aeratation and reagent schemes are all dicussed, which are different from the conventional flotation. In the end listed are some applications of OPCF technology.

Keywords potential controlled flotation; potential modifier; original potential; original potential control flotation (OPCF)

10.1 Technological Factors Affecting Potential Controlled Flotation Separation of Sulphide Ores

The behaviors and mechanisms of electrochemical flotation of sulphide minerals have been extensively investigated in laboratories. In order to improve the selectivity of separation of a mixture of sulphide minerals and to optimize sulphide ore flotation by electrochemistry control, it is required to understand the effects of technological parameters such as potential modifier, pH modifier, frother, galvanic interaction between sulphide minerals, conditioning, pretreatment and grinding media etc. (Trahar, 1984; Nakazawa and Iwasaki, 1986; Yelloji and Natarajan, 1990; Wang et al., 1991a,b,c,d; Wang et al., 1992).

10.1.1 Potential Modifiers

Potential modifiers are one of the most important factors affecting the electrochemical controlled flotation of sulphide minerals by the addition of which a genuine redox potential is established which has shown its significance in the flotation of sulphide minerals. It should be noted here that the influence of potential modifiers on the electrochemical controlled flotation of sulphide minerals is different for various modifiers even if under the same potential.

Figure 10.1 reflects the flotation behaviors of chalcopyrite in the presence and absence of butyl xanthate as a function of potential. It is evident that the flotation behavior of chalcopyrite is different at the same potential modified by ammonium persulphate and hydrogen peroxide. The initial potential of collectorless and butyl xanthate flotation of chalcopyrite are around 0 V. However, the upper limit potential of collectorless and butyl xanthate flotation of chalcopyrite is different using different potential modifiers. Chalcopyrite appears almost complete flotation at 0.3 V extended to 0.6 V when the potential is modified by ammonium persulphate, but it is completely depressed above 0.3 V when the potential is modified by hydrogen peroxide. Perhaps hydrogen peroxide is a stronger oxidizing agent than ammonium persulphate and eliminates the over potential of oxidation of sulphur to oxy-sulphur species which confers the chalcopyrite surface hydrophilic easier in the presence of hydrogen peroxide than in ammonium persulphate. Heyes and Trahar (1977) reported that the recovery of chalcopyrite is much different at the same potential modified respectively by potential raising agent.

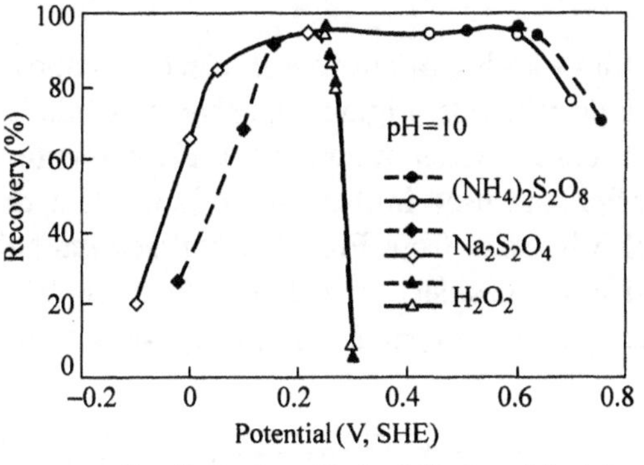

Figure 10.1 Recovery of collectorless (dashed line) and butyl xanthate induced flotation (solid line) of chalcopyrite as a function of pulp potential. Solid line-butyl xanthate: 2.9×10^{-5} mol/L, dashed line-collectorless flotation: Butyl ether alcohol 7.5 mg/L (Wang, 1992)

Figure 10.2 demonstrates the flotation recovery of pyrite vs. potential with butyl xanthate as a collector. It is follows that pyrite exhibits good floatability over a wider potential range extending from −0.1 V to 0.3 V when the potential is modified by sodium sulphide and ammonium persulphate. However, the potential range of pyrite flotation is diminished from 0.2 V to 0.3 V when the potential is modified by sodium dithionite and ammonium persulphate. It suggests that the reduction environment induced by reducing agents may be different for the collector action even if the potential value is the same. Pyrite exhibits better floatability by using sodium sulphide as a potential modifier than by using sodium dithionite because the former exhibits in fact sulphur-induced flotation behavior in addition to collector action.

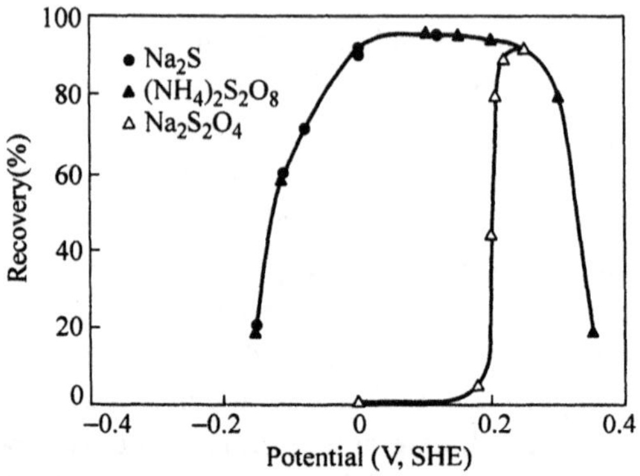

Figure 10.2 Recovery of flotation of pyrite as a function of pulp potential (Butyl xanthate: 2.9×10^{-5} mol/L; Butyl ether alcohol: 7.5 mg/L; from Wang, 1992)

10.1.2 pH Modifier

pH modifier also plays an important role in electrochemical controlled flotation separation of sulphide minerals. Figure 10.3 shows that the upper limiting pH for the depression of collectorless flotation of chalcopyrite is decreased when different pH modifiers are used in the order of sodium hydroxide, lime, sodium sulphide. It can also be seen from Fig. 10.3 that sodium sulphide decreases the pulp potential evidently besides for pH modification. When pulp potential is decreased to reducing environment at high pH by Na_2S, the collectorless flotation

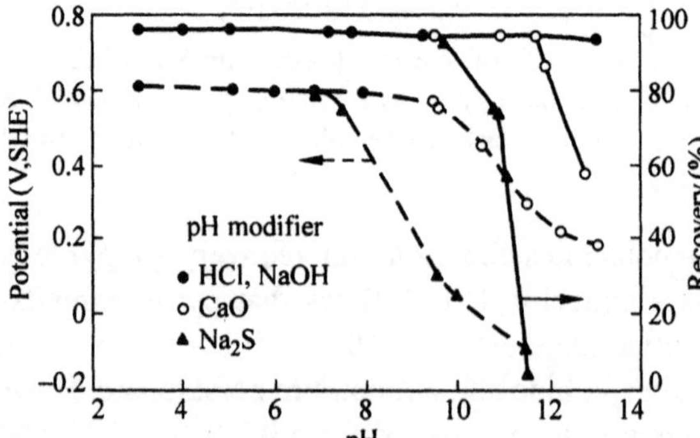

Figure 10.3 Recovery of collectorless flotation of chalcopyrite and pulp potential as a function of pH (Wang et al., 1991a,b,c,d)

of chalcopyrite is suppressed. The decrease of pulp potential with the increase of pH is related to the kinds of pH modifiers in the descending order of sodium hydroxide, lime and sodium sulphide which is in agreement with the order of upper limiting pH for depression.

Figure 10.4 presents the flotation recovery of galena, sphalerite and pyrite vs. butyl xanthate concentration with different pH modifiers at pH = 12. At the same pH condition, lime shows stronger depressing on these three minerals than sodium hydroxide with butyl xanthate as a collector. Pyrite can be completely depressed at pH = 12 modified by NaOH or lime. Sphalerite can be effectively depressed only by using lime modifying pH to 12. At pH = 12 modified by NaOH or lime, galena remains with good flotation response. Therefore, lime may be the effective pH modifier and depressant for the flotation separation of lead-zinc-iron sulphide ores. Fuerstenau et al. (1968) reported that complete depression of ethyl xanthate flotation of pyrite may occur whether KOH, NaOH, K_2CO_3 or CaO are used for pH control, but the upper limiting pH is variable with different pH modifier.

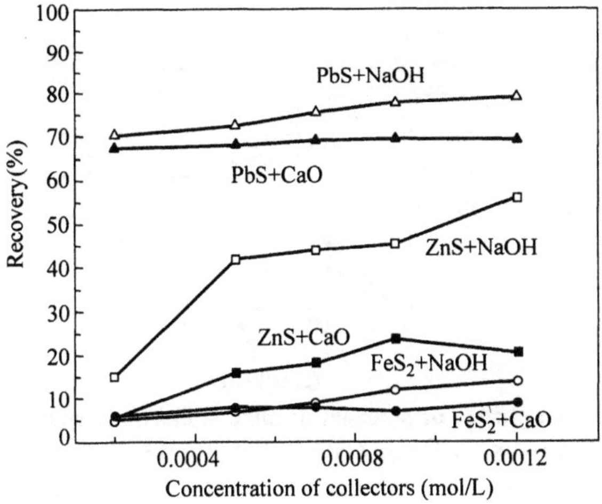

Figure 10.4 Flotation recovery of galena, sphalerite and pyrite as a function of butyl xanthate concentration with different pH modifiers at pH = 12

At the same pH made by different pH modifiers, the different flotation response of one sulphide mineral may arise from its effect on the potential of this mineral electrode. The change of potential of mineral electrodes with time at pH = 12 modified by NaOH and $Ca(OH)_2$ is measured and demonstrated in Fig. 10.5. It follows that a mineral electrode potential increases faster with the time at pH = 12 modified by sodium hydroxide and changes a little at the same pH modified by calcium hydroxide. When pH is adjusted by NaOH, the electrode potential of galena, sphalerite and pyrite increase rapidly, respectively, from −30, −12, and 70 mV at the initial stage to −10, 10, and 110 mV after 50 min. When pH is modified by lime, the electrode potential of galena, sphalerite and pyrite slightly increase respectively, from −40, −12, 60 mV at the initial stage to −36, −8, and 75 mV after 50 min. It is the different influence of pH modifiers on the potential that they show different influence on the flotation of sulphide minerals.

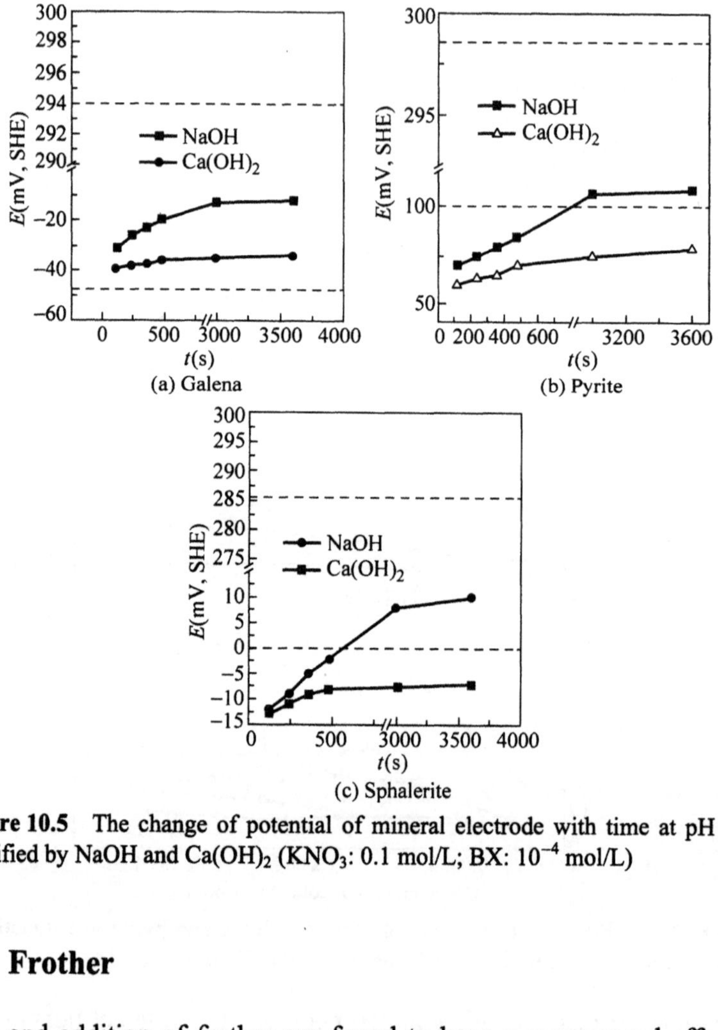

Figure 10.5 The change of potential of mineral electrode with time at pH = 12 modified by NaOH and $Ca(OH)_2$ (KNO_3: 0.1 mol/L; BX: 10^{-4} mol/L)

10.1.3 Frother

The type and addition of frother are found to have a pronounced effect on the collectorless floatability of chalcopyrite (Heyes and Trahar, 1977). The recovery of collectorless flotation of chalcopyrite is much higher using PPG40 than amyl alcohol. The effects of several frothers on the collectorless flotation of some minerals have been tested and the results are presented in Table 10.1. It further provides the evidence that the type of frother produces a markable influence on collectorless flotation of sulphide minerals. The frothers with lower surface tension are more effective in enhancing the recovery of collectorless flotation of sulphide minerals.

Table 10.1 Influence of frother on the collectorless flotation of the sulphide minerals (Wang et al., 1991a,b,c,d)

Frother (100 mg/L)		Butyl ether alcohol	Yarmour	MIBC	Mixed alcohol	Pine oil	Alcohol
Surface tension (10^{-3} N/m)		61.78	67.43	68.30	68.74	68.74	70.91
Separation efficiency of chalcopyrite/pyrite		84.85	63.66	63.65	56.06	52.59	
Recovery(%)	Stibnite	94.0	87.0	85.0	61.0	74	32.3
	Arsenopyrite	96.0		94.0		74	32.0

10.1.4 Conditioning Time

Because the electrochemical controlled flotation of sulphide mineral is possible under adequate oxidizing environment, longer time for conditioning may be useful for the flotation of some sulphide minerals that require oxidation atmosphere. Since iron balls are usually being used as the grinding media in mineral processing operations, the environment created in an iron mill is highly reducing. Figure 10.6 demonstrates the pulp potential as a function of flotation time. It can be seen that if the chalcopyrite ore is ground in a iron mill which set the pulp potential at −0.3 V and the flotation is conducted immediately after grinding, the initial pulp potential is in strong reducing environment −0.3 V and increases slowly to −0.2 V after flotation for 2 min, and continues to increase to +0.2 V after flotation for 8 min. The maximum recovery of chalcopyrite under these conditions can reach only to 61.32%. If the chalcopyrite ore is ground in a iron mill but the pulp is strongly agitated in a tank or in flotation cell for 10 min before flotation, it will

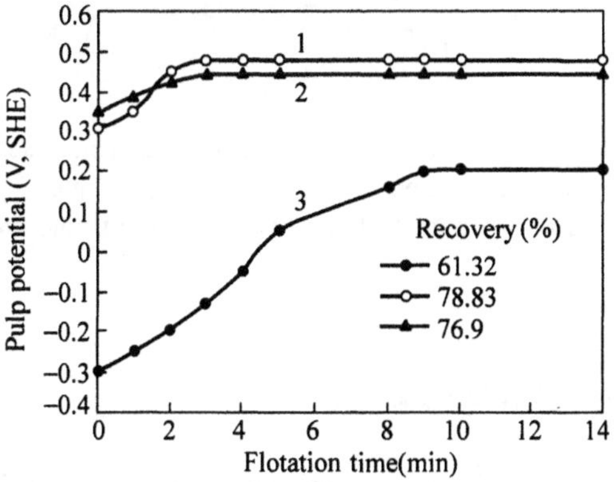

Figure 10.6 Pulp potential as a function of flotation time
1–strong agitating 10 min under air; 2–aeration conditioned 10 min; 3–flotation immediately after grinding

set the initial potential in oxidizing environment at 0.36 V or 0.3 V respectively, the pulp potential is soon increased to 0.48 V and 0.45 V after flotation for 2 min. The maximum recovery of chalcopyrite can reach to 78.83% and 76.9% respectively. It suggests that the potential controlled flotation of sulphide ores is affected by agitating condition due to its influence on redox atmosphere.

10.1.5 Surface Pretreatment

The oxidized or aged sulphide minerals usually appear to have poor floatability which may be improved by surface pretreatment. The influence of surface pretreatment on the collectorless flotation of chalcopyrite and galena is given in Fig. 10.7. It follows that oxidized galena and aged chalcopyrite exhibit poor flotation, which is greatly improved by pretreating with ultrasonic wave, the action of which is to remove the hydrophilic oxidized film on the oxidized mineral surface. Wang and Forssberg (1989) reported that no flotation was observed for oxidized galena, pyrite and arsenopyrite in the absence of EDTA, but flotation was recovered by the addition of EDTA, which was attributed to the removal of the hydrophilic metal hydroxide layers by EDTA from these sulphide surfaces and thereby exposing a sulphur-rich surface.

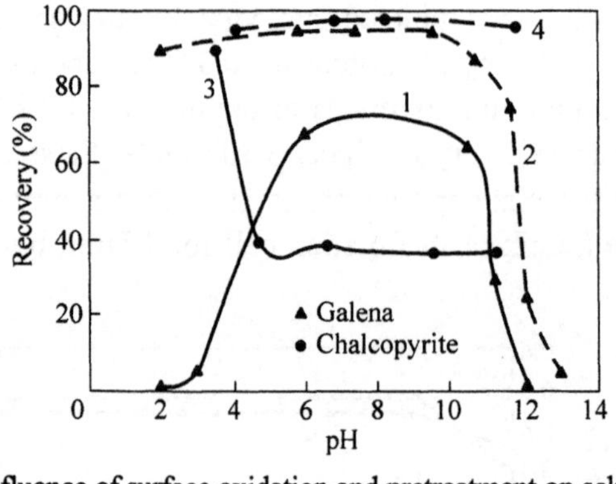

Figure 10.7 Influence of surface oxidation and pretreatment on self-induced flotation of galena and chalcopyrite. Galena: 1–untreated; 2–pretreated by ultrasonic wave. chalcopyrite: 3–aged two month; 4–pretreated by ultrasonic wave (Wang et al., 1991a,b,c,d)

10.1.6 Grinding Environment

When a complex sulphide ore is subjected to wet grinding, electrochemical interactions between the grinding media and various sulphide minerals and chemical or electrochemical interactions involving dissolved species and mineral surface could be expected. Subsequent flotation of such sulphide minerals would be influenced by the type of grinding media used and dissolved metal ions (Nakazawa and Iwasaki, 1986;Yelloji Rao and Natarajan, 1989a,b,c, 1990).

Different grinding media will result in different redox environments of pulp and hence affect potential controlled flotation behavior of sulphide minerals. Heyes and Ralston (1988) reported that chalcopyrite ground in a reducing environment provided by 5×10^{-3} mol/L sodium dithionite solutions at $pH=9.2$ exhibited weak collectorless floatability and was made strongly floatable by increasing the potential above $0-0.1$ V. Galena in a reducing environment provided by 5×10^{-3} mol/L sodium dithionite solutions at $pH=9.2$ is not floatable but can be readily floated by increasing the E_h to 0.45 V. In practice, sulphide minerals are frequently ground in an iron mill under strongly reducing conditions owing to the operation of one of the following reactions:

$$Fe^{2+} + 2e^- = Fe \tag{10-1}$$

$$E^0 = -0.44(V)$$

$$Fe(OH)_2 + 2H^+ + 2e^- = Fe + 2H_2O \tag{10-2}$$

$$E^0 = -0.047(V)$$

$$Fe_3O_4 + 8H^+ + 8e^- = 3Fe + 4H_2O \tag{10-3}$$

$$E^0 = -0.084(V)$$

which are comparable to the dithionite environment. They show that chalcopyrite and galena are not floatable when ground in an iron mill in which the potential is about -0.25 V but are made strongly floatable by increasing the potential to above 0.1 V for chalcopyrite and 0.2 V for galena. Generally speaking, chalcopyrite and galena are less floatable when ground in an iron mill than in a ceramic mill.

Grinding media also have an influence on the collector flotation of sulphide minerals. Grano et al. (1990) found that the floatability of butyl xanthate of chalcopyrite and iron sulphides is lower when the copper ore was ground in an iron mill than in a ceramic mill. They concluded that the introduction of highly electrochemically active grinding media in cast iron would enhance the generation of ferric hydroxides as a result of the anodic dissolution of the media which may completely coat the sulphide mineral surfaces leading to depression for the mineral. It would be expected, therefore, that the removal of some of these hydroxides by grinding in an inert environment would allow greater flotation recoveries to be achieved. Cases et al. (1990) reported that only dixanthogen was observed on pyrite after using stainless steel rods, while ferric iron xanthate and physically absorbed xanthate ions were found to be formed when iron rods were used.

In lead-zinc-iron sulphide ores, iron ball is a typical grinding medium. Besides, any one of the three minerals may be regarded as the grinding medium of the other two minerals. The change of potential of mineral electrode with time in different grinding media at pH = 12 modified by $Ca(OH)_2$ was measured and given in Fig. 10.8. It is obvious that by grinding in the iron ball medium the electrode potential of galena, sphalerite and pyrite tends to decrease with the time, respectively descending from −35, 10, and 130 mV at the initial stage to −60, −10, and −30 mV after 7 min. The electrode potential of galena and sphalerite increases slowly with the time when grinding in pyrite medium. The electrode potential of sphalerite and pyrite decreases slowly with the time when grinding in galena medium. Whereas when grinding in sphalerite medium, the electrode potential of galena tends to increase and that of pyrite tends to decrease. Therefore, even if in iron ball grinding medium, different mineral compositions may have variable potential environment accounting for one reason of difficulty in flotation separation of poly-metal sulphide ores.

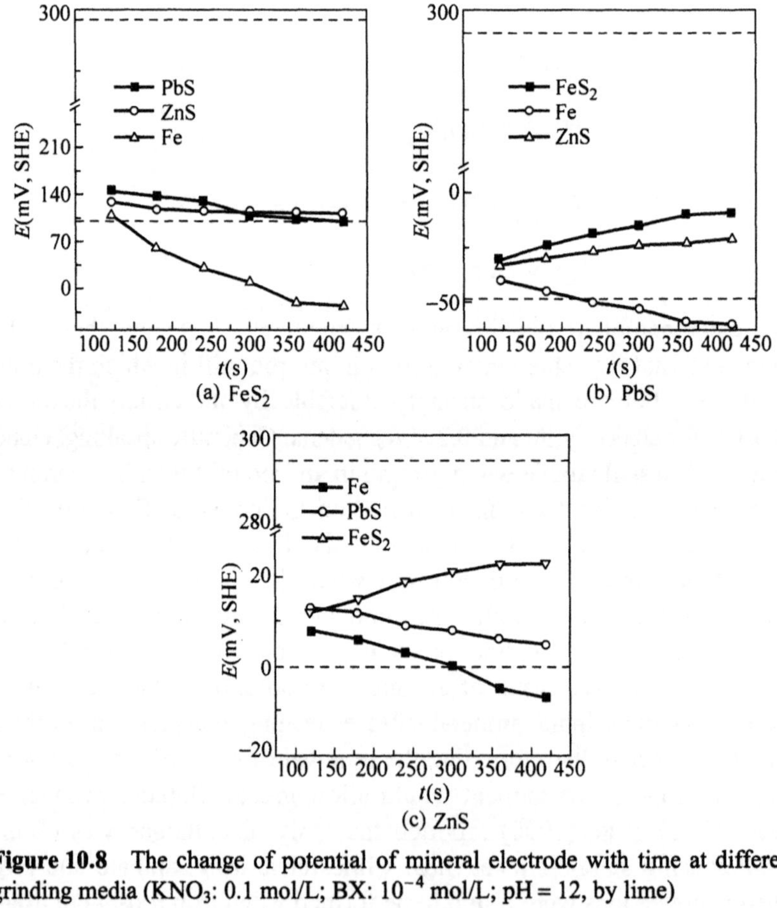

Figure 10.8 The change of potential of mineral electrode with time at different grinding media (KNO_3: 0.1 mol/L; BX: 10^{-4} mol/L; pH = 12, by lime)

10.2 Flotation Separation of Sulphide Minerals and Ores

10.2.1 Copper Sulphide Minerals and Ores

1. Chalcopyrite /Galena

Flotation separation of chalcopyrite from galena, especially in the copper-lead mixed concentration, is very difficult in plant practice. The potential controlled flotation provides a possible way to separate them effectively. Figure 10.9 presents a flowsheet of collectorless flotation separation of chalcopyrite and galena. Both chalcopyrite and galena are floated in an adequate oxidizing environment at natural pH. The separation of mixed concentrate is conducted at higher oxidizing potential, which is usually modified by H_2O_2 or other strong oxidizing agents. Table 10.2 shows the results of flotation separation for the mixture of chalcopyrite-galena and quartz according to the flowsheet in Fig. 10.9. The obtained copper concentration assayed 25.78% and recovery is 81%, and the obtained lead concentration assayed 65.7% and recovery is 90% after separation.

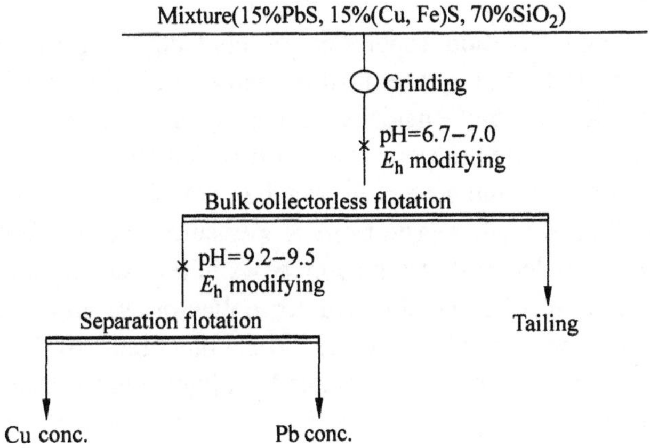

Figure 10.9 Flowsheet of collectorless flotation separation of a mixture of chalcopyrite and galena

Table 10.2 Separation results of collectorless flotation of a mixture of chalcopyrite, galena and quartz (Feng et al., 1991)

Products	Cu		Pb		Bulk flotation		Separation	
	Grade (%)	Recovery (%)	Grade (%)	Recovery (%)	E_h (mV)	pH	E_h (mV)	pH
Ceramic mill	25.78	81.00	65.70	90.00	300	6.7	480	9.2
Iron mill	20.28	80.88	56.70	73.48	265	7.0	410	9.5

Tests have been done further on the separation of a Cu-Pb mixed concentration of ethyl xanthate flotation of copper-lead-iron sulphide ore by E_h control modifying with H_2O_2. Test results are presented in Table 10.3. It indicates the possibility of selective flotation separations of copper-lead flotation concentration by E_h control. The feed of copper-lead mixed concentrated assayed Cu 6.53% and Pb 62.38%. Using hydrogen peroxide as a potential modifier, a copper concentration with 24.19% Cu and recovery with 89% can be obtained after separation.

Table 10.3 E_h control separation of a mixed concentration of collector flotation of chalcopyrite-galena ore (Wang, 1992a,b)

Products	Test 1# (H_2O_2: 476 g/t)				Test 2# (H_2O_2: 408 g/t)			
	Grade (%)		Recovery (%)		Grade (%)		Recovery (%)	
	Cu	Pb	Cu	Pb	Cu	Pb	Cu	Pb
Cu conc.	24.19	16.88	89.00	6.43	23.83	18.19	85.98	7.84
Pb conc.	0.95	76.84	11.00	93.57	1.36	74.98	14.02	92.66
Feed	6.53	62.38	100	100	7.2	60.23	100	100

2. Chalcopyrite / Pyrite

Potential controlled flotation separation of chalcopyrite-pyrite ores has been extensively tested to be one of the most probable to obtain industrial application on the basis of the fact that chalcopyrite has strong collectorless and collector floatability whereas pyrite has poor collectorless floatability.

Figure 10.10 presents the schematic rougher flowsheet of potential controlled flotation separation of chalcopyrite from copper-sulphur ores. Table 10.4 lists the results of flotation chalcopyrite from many types of copper-sulphur ores. It follows that almost similar results are obtained for collectorless and xanthate flotation chalcopyrite from copper-sulphur ores. When the grade of feed ore is about 0.5% – 1.5% Cu, the grade of the obtained rougher concentration can reach to 7% – 17% Cu and recovery is above 90%. When the grade of feed ore is about 3% – 5% Cu, the grade of the obtained rougher concentration can reach to above 17% Cu and recovery is above 95%.

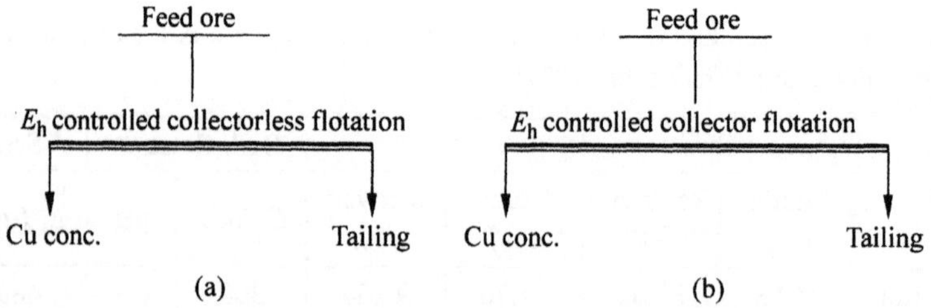

Figure 10.10 Principal flowsheets of potential controlled flotation of copper-sulphur ores

Table 10.4 Batch tests of induced flotation separation results of copper-sulphur ore (Wang et al., 1991a,b,c,d)

Ore type	Flowsheet in Fig. 10.10	Cu grade (%)		Cu recovery (%)
		Feed	Rougher conc.	
Porphyry copper ore	I	0.53	7.23	90.75
	II	0.53	6.66	90.76
	I	1.55	15.05	93.21
	II	1.63	17.25	91.68
Disseminated copper ore	I	5.33	19.41	97.23
	II	5.22	17.64	97.82
	I	3.45	22.95	95.3
	II	3.45	22.14	95.2

Figure 10.11 presents the schematic flowsheets of potential controlled flotation separation to recover chalcopyrite and pyrite from a copper-sulphur ore. Flowsheet I is collectorless flotation of chalcopyrite and then collector floatation of pyrite. Flowsheet II is collectorless flotation of chalcopyrite and then sodium sulphide-induced flotation of pyrite. Batch flotation results are illustrated in Table 10.5. It is evident that both flowsheets are suitable for flotation separation of copper-sulphur ore. The feed ore assayed 0.38% Cu and about 6% S, the copper concentrate obtained assayed 18%–19% Cu with a recovery of 89%. For sulphur concentrate, the grade is 37%–43% S with a recovery of 82%–85%. Interestingly, the grade of sulphur concentrate is higher using sodium sulphide induced flotation than collector flotation.

It is further substantiated by the pilot test results for a copper sulphide ore (see Table 10.6). To maintain the pulp potential in the range of 142–189 mV, the flotation separation of copper-sulphur ore can obtain good results in collectorless or xanthate flotation system. The grade of feed ore is about 3.5% Cu, the grade of the obtained concentrate can reach to 27%–28% Cu and recovery is above 95%.

Table 10.5 Induced-flotation separation results of copper/sulphur ore

	Yield (%)		Grade (%)				Recovery (%)			
			Cu		S		Cu		S	
Flowsheet (%)	I	II	I	II	I	II	I	II	I	II
Cu conc.	1.75	1.91	19.51	18.31	25.54	25.67	88.95	89.24	7.48	7.62
S conc.	13.57	12.06	0.15	0.20	37.74	43.91	5.32	6.13	85.71	82.33
Tailing	84.68	86.03	0.026	0.021	0.48	0.76	5.73	4.63	6.81	10.15
Feed	100	100	0.38	0.38	5.98	6.44	100	100	100	100

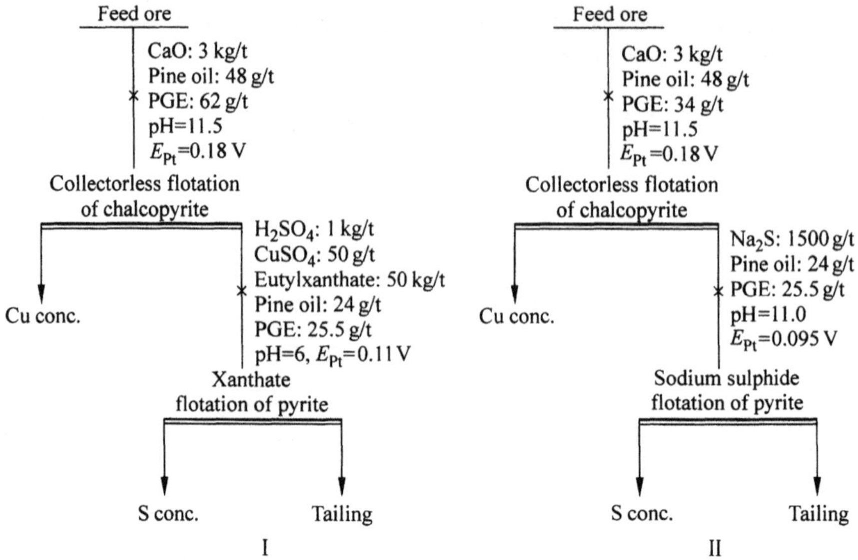

Figure 10.11 Principal flowsheets of potential controlled flotation separation of copper/sulphur ore

Table 10.6 Pilot plant test results of potential controlled-flotation of copper sulphide ore

Flowsheet	Cu grade (%)		Cu recovery (%)	Reagents (g/t)			
	Feed	Conc.		BX	CaO	Pine oil	E_{Pt} (mV)
Collectorless flotation	3.85	29.33	95.88	0	2.6	163	142 – 189
Xanthate flotation	3.53	27.85	95.47	44.9	1.8	118	142 – 189

3. Chalcopyrite/Arsenopyrite

Table 10.7 shows the flotation separation results of a mixture of chalcopyrite and arsenopyrite. It may be seen that at high pH, the chalcopyrite can be separated effectively from arsenopyrite by controlling the potential at about 280 mV. The floated product assays 29.4% Cu and recovery is 93.48%.

Table 10.7 Separation results of the mixture of chalcopyrite and arsenopyrite (1 : 1) by butyl xanthate (1.5×10^{-5} mol/L) induced flotation

Products	Grade (%)		Recovery (%)		CaO pH	E_h (mV)
	Cu	As	Cu	As		
Floated	29.4	3.80	93.48	8.52	11	280
Unfloated	2.06	42.42	6.52	92.48		

10.2.2 Lead-Zinc-Iron-Sulphide Minerals and Ores

Under adequate reducing environment dixanthogen will not be formed on the pyrite surface but lead xanthate is still formed on the galena surface, which underlies the basis of potential controlled flotation separation of galena from pyrite by xanthate flotation.

The results in Table 10.8 show that the selective separation of galena from pyrite may be accomplished in alkaline medium and reducing potential. For the 1:1 mixture of galena and pyrite, the floated product assayed 76.31% Pb with a recovery of 90% at pH = 10.5 and potential −210 mV. Similar results can be achieved for the flotation separation of a mixture of galena and arsenopyrite as shown in Table 10.9. The floated product assayed 74.36% Pb with a recovery of 92% at pH = 9.5 and potential −300 mV.

Table 10.8 Separation results of the mixture of galena and pyrite (1 : 1) by collector flotation (BX: 1.5×10^{-5} mol/L)

Products	Grade (%)		Recovery (%)		CaO pH	$Na_2S_2O_4$ E_h (mV)
	Pb	S	Pb	S		
Floated	76.31	5.26	90.01	11.98	10.5	−210
Unfloated	7.69	39.40	8.99	88.42		

Table 10.9 Separation results of a mixture of galena and arsenopyrite (1 : 1) by collector-induced flotation

Products	Grade (%)		Recovery (%)		CaO pH	$Na_2S_2O_4$ E_h (mV)
	Pb	As	Pb	As		
Floated	74.36	6.43	92.21	14.86	9.5	−300
Unfloated	6.95	40.76	7.79	85.14		

Table 10.10 Separation results of the mixtures of two minerals (1 : 1) among galena, sphalerite and pyrite by E_h controlled flotation (pH modified by lime, grinding time for 5 min, in Fe medium)

Mixtures	Products	Grade (%)			Recovery (%)			Collector (mol/L)	pH	E_h (mV)
		Pb	Zn	S	Pb	Zn	S			
PbS-ZnS	Pb	74.4	3.2		88.14	4.7		DDTC: 10^{-4}	12	90
	Zn	8.87	57.51		11.86	95.3			12	
$ZnS-FeS_2$	Zn		58.1	31.68		92.1	40.17	BX: 10^{-4}	12	150
	S		5.13	48.56		7.9	59.83		12	

For the flotation separation of the mixtures of galena-sphalerite, using DDTC as a collector and lime as a pH modifier, the mixture is ground in Fe medium and the potential is made at 90 mV. The lead concentrate assayed of 74.4% Pb and zinc concentrate assayed of 57.51% Zn can be obtained. The recovery of lead and zinc are, respectively, 88.14% and 95.3%. The flotation separation of a mixture of sphalerite and pyrite is conducted at pH = 12 modified by lime and using butyl xanthate as a collector. The mixture is ground in Fe medium and the potential is 150 mV. The zinc concentrate assayed of 58.1% Zn and sulphur concentrate assayed of 48.56% S can be obtained. The recovery of zinc and sulphur are, respectively, 92.1% and 59.83%.

10.3 Applications of Potential Control Flotation in Industrial Practice

The investigations on electrochemistry related to sulphide mineral flotation have been widely reported for a long time. However, pilot tests and industrial plant operations of potential control flotation (PCF) of sulphide ores have made little progress due to the absence of an applicable control potential method. The addition of oxidants and reductants will result in a big consumption of the chemicals. In the meantime, adjusting the pulp potential by external field is also of low efficiency.

Through many year's efforts, a new method has been developed in our group and applied successfully in several concentrators of galena-sphalerite-pyrite multi-sulphide ore. The key technological points are discussed as below.

10.3.1 Original Potential in Grinding Process

Pulp potential in grinding process, is controlled by many factors, such as oxygen dissolved in pulp, oxidation of iron fine powders from the grinding media and particles of sulphide minerals, whose electrochemical interactions are very complex in the grinding-flotation system of sulphide minerals ore (Nakazawa and Iwasaki, 1985).

Taking galena, sphalerite and pyrite system as an example, anodic reactions include (Gu et al., 2000):

$$PbS + 2H_2O = HPbO_2^- + S + 3H^+ + 2e^- \qquad (10\text{-}4)$$

$$E^0 = 1.182(V)$$

$$ZnS + 6H_2O = ZnO_2^- + SO_4^{2-} + 12H^+ + 8e^- \qquad (10\text{-}5)$$

$$E^0 = 0.635(V)$$

$$FeS_2 + 11H_2O \!=\!=\! Fe(OH)_3 + 2SO_4^{2-} + 19H^+ + 15e^- \tag{10-6}$$

$$E^0 = 0.402(\mathrm{V})$$

Cathodic reaction is usually the reduction of O_2 in pulp:

$$\frac{1}{2}O_2 + H_2O + 2e^- \!=\!=\! 2OH \tag{10-7}$$

$$E^0 = 0.401(\mathrm{V})$$

Moreover, due to the rise in the electrode potential difference galvanic interaction between galena, sphalerite and pyrite, it will further accelerate the oxidation of the sulphides.

In addition, the following anodic oxidation reactions of iron from steel lining and balls can occur:

$$Fe \!=\!=\! Fe^{2+} + 2e^- \tag{10-8}$$

$$E^0 = -0.44(\mathrm{V})$$

$$Fe^{2+} + 3H_2O \!=\!=\! Fe(OH)_3 + 3H^+ + e^- \tag{10-9}$$

$$E^0 = 1.042(\mathrm{V})$$

These reactions will consume oxygen (O_2), resulting in lower pulp potential value. The mixed potential of the above reactions is originated from intrinsic electrochemical behavior of sulphide minerals in grinding-flotation systems, so it is called "original potential". The magnitude of "original potential" is lower when measured in overflow of grinding due to grinding usually in iron medium and results in an reduction atmosphere at the initial stage of flotation, which is suitable for the flotation of galena. However, the potential will increase gradually with the time due to the action of oxygen in the open system and make the potential unstable and flotation of galena worse. Therefore, the stable and control of the potential is very important for flotation separation of poly-sulphide minerals. The process using the original potential to improve the separation of sulphide minerals is named "original potential control flotation" (OPCF).

10.3.2 Effect of Lime Dosage on "Original Potential"

The pulp potential is related to pH of the flotation pulp. In grinding and flotation process of PbS-ZnS-FeS_2 ore, the pulp potential decreases with the increase of pH. For Nanjing concentrator of China, when the pulp pH reaches to 12.5 – 12.8, the original pulp potential range is about 0.10 – 0.20 V. But for the concentrator located in high altitude area, since saturated oxygen is low, the original potential will decrease to low range from –0.06 – –0.10 V. Figure 10.12 shows the pulp pH and pulp potential in Xitieshan concentrator in Qinghai Province of China where the altitude was more than 3000 m. It can be seen that, the original pulp potential is less than −0.095 V when pH is modified to 12.8 by slime, which is the better potential region for flotation of galena. Figure 10.13 further shows that the pulp pH and pulp potential in galena flotation process can maintain stable under a higher dosage of lime, which provides the basis for original potential control flotation.

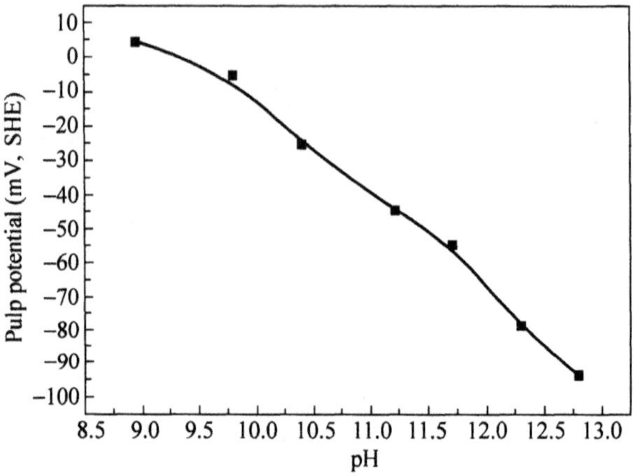

Figure 10.12 Pulp potential as a function of pH in Xitieshan flotation concentrator

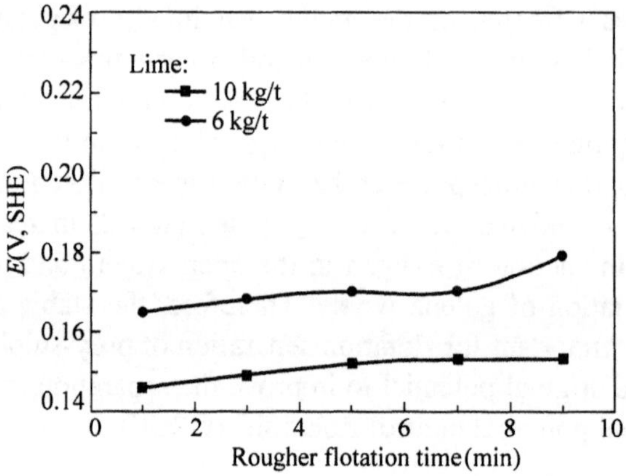

Figure 10.13 Effect of lime dosage on the pulp potential in galena rougher flotation process of Nanjing concentrator

10.3.3 Coupling with Other Flotation Process Factors

In the open system, the pulp potential after grinding will increase with the process of aeration of agitation and flotation (see Fig. 10.14). Because the flotation of galena needs relatively low potential in which lead xanthate is formed easily on the galena surface. Therefore, the flotation operational time must couple with the change of the pulp potential so as to selectively separate galena from other minerals requiring relatively high potential for flotation like pyrite. Table 10.11 lists oxygen dissolved and pulp potential in the grinding-flotation system tested in Beishan flotation concentrator. The results indicated that the dissolved oxygen and potential in mill charge and classifier overflow are lower, and increased with the flotation time. It is advisable for the flotation time of lead circuit to be less than 18 min.

Table 10.11 Oxygen dissolved and the pulp potential in Beishan concentrator of PbS-ZnS ore

Test sites	Time (min)	O_2 conc. (%)	*E* (V, SHE)
Mill discharge	—	3	0.147
Classifier overflow	—	9	0.156
Stirrer	4	30	0.166
PbS flotation Rougher	8	35	0.174
Scavenger I	4	60	0.181
Scavenger II	3	74	0.190
Scavenger III	3	92	0.198

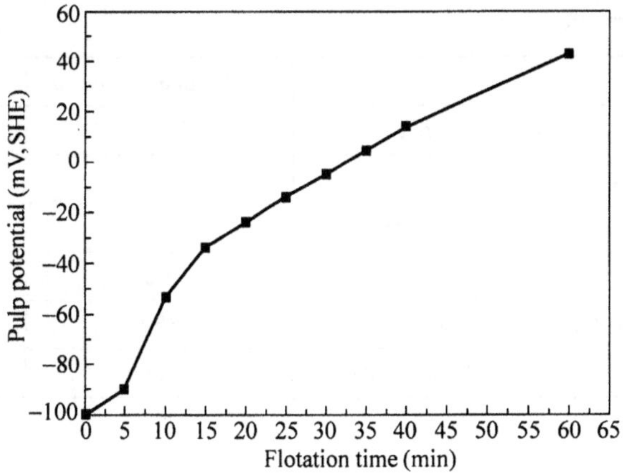

Figure 10.14 Change of the pulp potential with the operation time in Xitieshan flotation concentrator

10.3.4 Coupling with Reagent Schemes

Lower original potential in grinding is beneficial to flotation selectivity of galena. Thus, collectors should be added into the mill directly. Table 10.12 is the results of flotation separation of Beishan concentrator of PbS-ZnS ore for collector addition in different places. It shows that the recovery and grade of galena concentrate are improved obviously when collectors are added in mill compared to in agitator.

Table 10.12 The flotation results in Beishan concentrator of PbS-ZnS ore

	PbS conc.	
Collector adding site	Agitator	Mill
Grade (%)	61.13	64.23
Recovery (%)	74.91	76.07

10.3.5 Coupling with Flotation Circuit

Under suitable pulp potential in the OPCF process, the flotation speed of galena is very fast and most galena has been floated just in rougher flotation stage. So the flotation cell of galena scavenger flotation operation can be diminished and the flotation circuit is further optimized. The comparison of operation parameters between traditional flowsheet and OPCF flotation is presented in Table 10.13. As can be seen, the flotation potential and time are lower in OPCF than in traditional flowsheet.

Table 10.13 Comparison of the operation factors between traditional flotation and OPCF flotation

		Traditional flotation	OPCF flotation
PbS rougher flotation	pH	8 – 9	12.5 – 12.8
	E_h (mV)	350 – 280	180 – 150
	Time (min)	20	8
PbS cleaner flotation	pH	8.0	12.0
	E_h (mV)	350	195
	Time (min)	20	10

10.3.6 Applications of OPCF Technology in Several Flotation Concentrators

1. Nanjing Lead-Zinc Concentrator

Figure 10.15 is the OPCF technology used in the flotation concentrator of Nanjing lead-zinc mine and the operation results of the OPCF technology is shown in Table 10.14. The OPCF separation of galena from sphalerite and pyrite is conducted in the strongly alkaline pulps of pH = 12.4 – 12.5 modified by lime, with sodium diethyl dithiocarbamate (DDTC) as a collector and without the addition of $ZnSO_4$ and Na_2SO_3 as the depressants of ZnS and FeS_2.

It can be seen from Table 10.14 that in common flow sheet a depressants is used and pH is neutral, the flotation separation of galena, sphalerite and pyrite is not so good. The OPCF can much improve the results of selective flotation of PbS-ZnS. Compared with common flowsheet, the grade of concentration increases respectively from 52.1% to 60.0% for Pb, from 52.6% to 53.0% for Zn and from 37.4% to 46.5% for S and the recovery of concentration increases respectively from 85.9% to 88.9% for Pb, 87.0% to 91.9% for Zn and 68.8% to 71.9% for S.

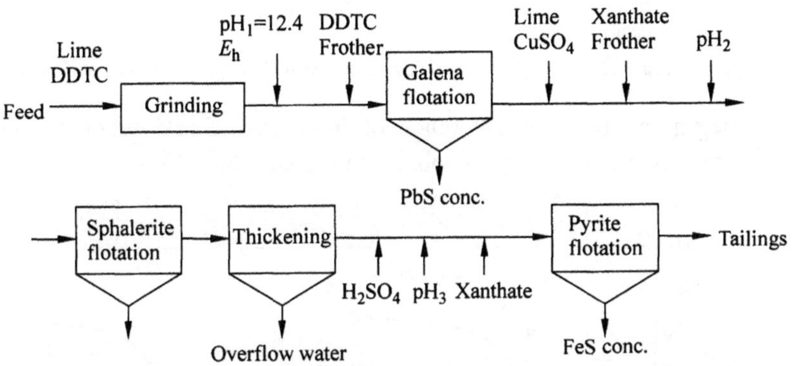

Figure 10.15 OPCF process in concentrator of Nanjing PbS-ZnS-FeS$_2$ ore

Table 10.14 Comparison of the flotation results in concentrator of Nanjing PbS-ZnS-FeS$_2$ ore

Flowsheet	Products	Grade (%)			Recovery (%)			Conditions for PbS flotation
		Pb	Zn	S	Pb	Zn	S	
Traditional technology	Pb conc.	52.1	5.59	20.8	85.9	4.27	5.61	pH = 7.0 DDTP: 40 g/t Frother: 20 g/t ZnSO$_4$: 1.1 kg/t Na$_2$SO$_3$: 0.7 kg/t
	Zn conc.	1.23	52.6	32.4	4.41	87.0	18.95	
	S conc.	0.61	1.38	37.4	6.91	7.20	68.80	
OPCF technology	Pb conc.	60.0	4.68	17.4	88.9	3.47	4.22	pH = 12.4 E_h = 170 mV Lime: 5 kg/t DDTC: 60 g/t Frother: 20 g/t
	Zn conc.	1.03	53.0	30.6	3.58	91.9	17.42	
	S conc.	0.55	0.60	46.5	5.16	2.80	71.9	

2. Xitieshan Lead-Zinc Concentrator

The industrial test of OPCF technology is employed in Xitieshan lead-zinc mine, which is located in Qinghai Province of China with high altitude above sea level. Figure 10.16 shows the flowsheets of OPCF. The effect of pulp potential on flotation separation of galena and sphalerite in industrial test can be seen in Fig. 10.17. It follows that the suitable pulp potential range of galena flotation is located in less than −5 mV, at where the floatability of sphalerite is very poor. The sphalerite

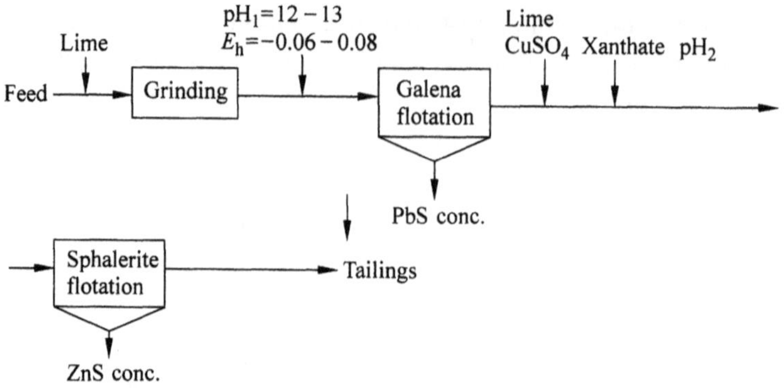

Figure 10.16 OPCF flowsheet of concentrator in Xitieshan PbS-ZnS-FeS_2 ore

flotation begins at the potential range of more than 5 mV, when the potential reaches to 60 mV, the recovery of sphalerite is more than 90%.

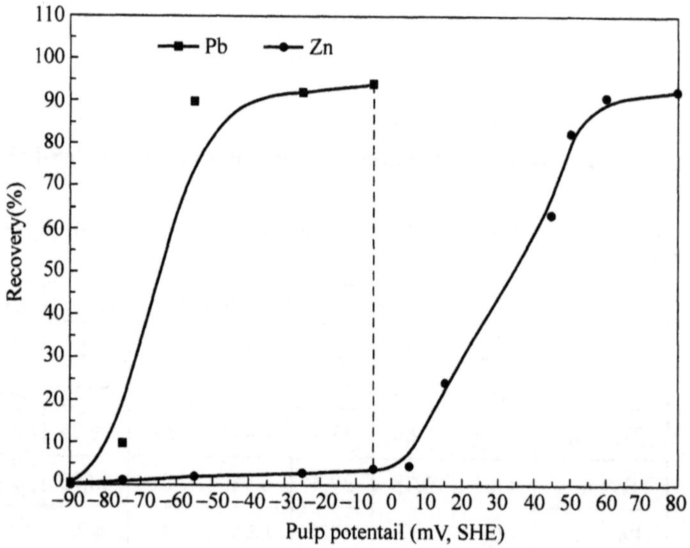

Figure 10.17 Effects of pulp potential on the flotation separation of PbS-ZnS in Xitieshan concentrator

Figure 10.18 further shows the flotation separation efficiency of galena and sphalerite from the date of the quantity containing Pb or Zn each other in the corresponding concentration. It may be seen that the recovery of ZnS in PbS concentration and PbS recovery in ZnS concentration are very low, indicating that the flotation separation of PbS and ZnS is selective in this new flowsheet.

The comparison of flotation results using OPCF and traditional flowsheet is shown in Table 10.15. It can be seen that either PbS, ZnS flotation recovery or the grade of the concentration, is improved obviously in OPCF. The grade and recovery of lead concentration, respectively, increase from 73.77% to 79.19% and from 91.86% to 93.69%. The grade and recovery of zinc concentration are, respectively, enhanced from 49.63% to 49.85% and 86.44% to 91.93%.

In addition, the flotation rate of galena in the OPCF process is much faster than that in the traditional through the measurement of galena recovery of each cell as shown in Fig. 10.19. The Pb recovery is more than 90% only through 4 flotation cells operation and Zn recovery reaches to 87% only through 12 flotation cells operation in OPCF. But for the traditional process, 8 flotation cells are needed to reach 90% Pb recovery and 20 flotation cells are needed to reach 87% Zn recovery.

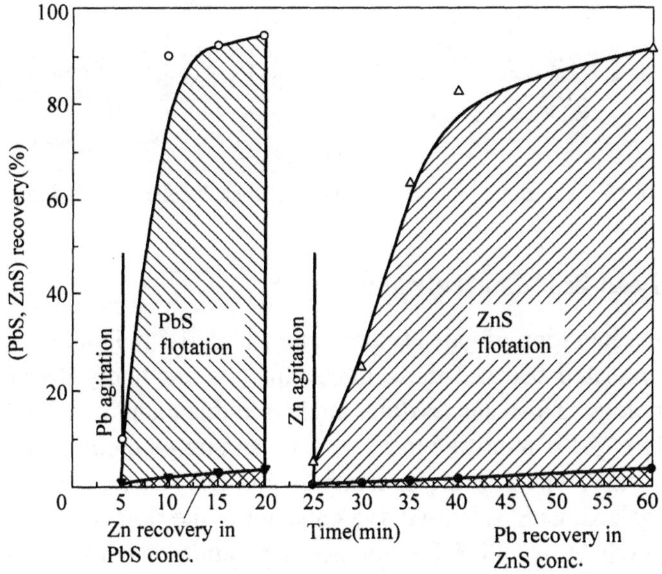

Figure 10.18 The separation of PbS-ZnS and flotation time by OPCF in Xitieshan concentrator

Table 10.15 Comparisons of flotation results of OPCF and traditional flotation in concentrator of Xitieshan lead-zinc mine

Flowsheet	Products	Grade (%)		Recovery (%)		Conditions for PbS flotation
		Pb	Zn	Pb	Zn	
Traditional technology	Pb conc.	73.77	2.54	91.86	4.27	pH = 7.0 – 8.0 Frother: 130 g/t Collector:130 g/t
	Zn conc.	0.81	49.63	4.41	86.44	
OPCF technology	Pb conc.	79.19	2.10	93.69	3.47	pH = 12.0 – 13.0 E_h = −0.08 – −0.06 V Lime: 5 – 7 kg/t DDTC: 28 – 35 g/t
	Zn conc.	0.52	49.85	3.58	91.93	

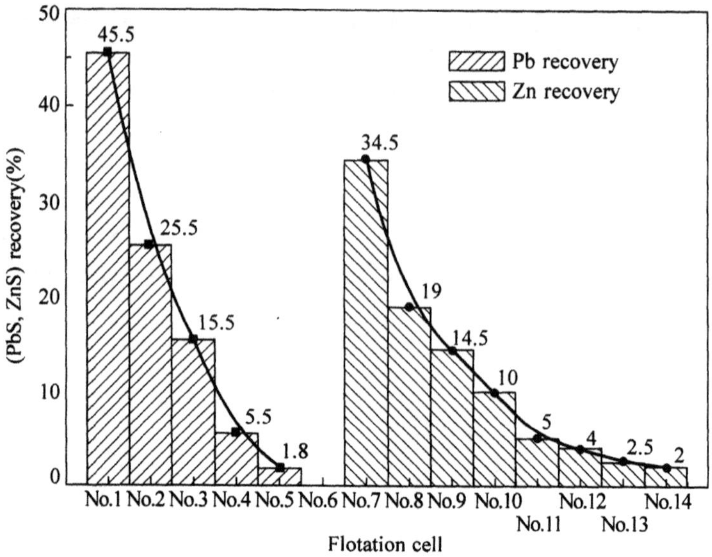

Figure 10.19 The flotation rate (the recovery of each cell) for OPCF in Xitieshan concentrator

3. Fankou Lead-Zinc Concentrator

The flotation separation of galena, sphalerite and pyrite in Fankou lead-zinc mine is very complicated because these three minerals are finely disseminated. The OPCF technology is also successfully applied to this plant to separate these three minerals. Here, pH is modified to 12 by lime and pulp potential is maintained as less than 170 mV. The mixture of xanthate and DDTC is used as a collector in flotation of galena. $CuSO_4$ is used as a collector in the flotation of sphalerite. The principal flowsheet of OPCF for flotation separation of Fankou lead-zinc ore is given in Fig. 10.20. The comparison of results of plant production for OPCF and old flowsheet is listed in Table 10.16. It can be seen that the OPCF technique

Table 10.16 The comparison of results of plant production for OPCF and old flowsheet

Products	Grade (%)		Recovery (%)		Flowsheet
	Pb	Zn	Pb	Zn	
Pb conc.	58.02	3.97	82.4	2.54	Old flowsheet
Zn conc.	1.29	53.03	4.98	92.27	
Tailings	0.66	0.59	16.62	5.19	
Feed	4.25	9.43	100	100	
Pb conc.	61.44	2.56	87.66	1.61	OPCF technology
Zn conc.	0.83	56.34	3.16	94.62	
Tailings	0.61	0.57	9.18	3.77	
Feed	4.25	9.43	100	100	

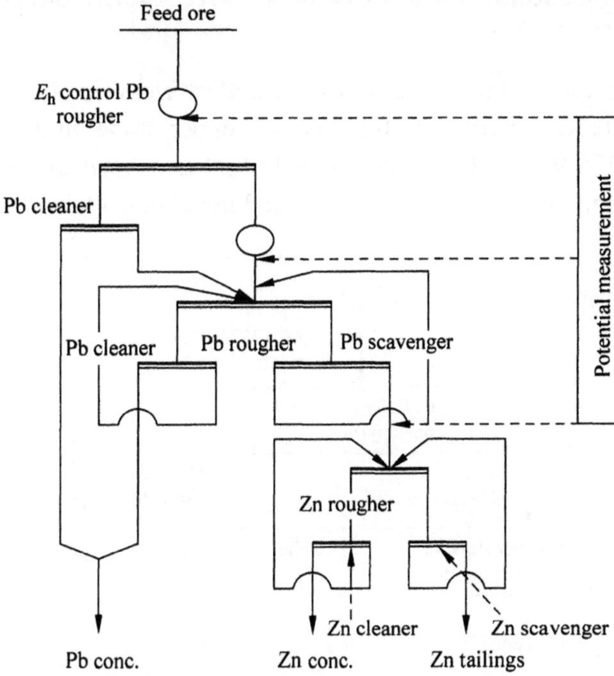

Figure 10.20 The principal flowsheet of OPCF for flotation separation of Fankou lead-zinc ore

enhances greatly the efficiency of flotation separation of lead-zinc in Fankou mine. The grade and recovery of lead concentration are, respectively, increased from 58.02% to 61.44% and from 82.4% to 87.66%. The grade and recovery of zinc concentration are, respectively, enhanced from 53.03% to 56.34% and 92.27% to 94.62%. Moreover, the reagent consumption is diminished in OPCF relatively to old flowsheet as given in Table 10.17. It not only decreases the dosage of collector and CaO, but also cancels the $ZnSO_4$. The total volume of flotation cell required for flotation of galena descends due to the rapid flotation rate of galena in OPCF and hence the number of cell is abbreviated.

Table 10.17 The comparison of reagent consumption plant production for OPCF and old flowsheet

Flowsheet	$CuSO_4$	$ZnSO_4$	DS	EX	DDTC	CaO	$2^{\#}$oil
Old	0.61	0.35	0.079	0.102	0.062	8.8	0.128
OPCF	0.65	0	0.02	0.08	0.04	8	0.128

4. Tonglushan Concentrator of $CuFeS_2$-FeS_2 Ore Application of Collectorless Flotation

Figure 10.21 presents the flow sheet for the collectorless flotation and separation of chalcopyrite and pyrite in Tonglushan copper mine in China. The pulp potential is 0.175 V and pH is 11.5. The industrial test shows that the collectorless flotation can obtain better results than the traditional collector flotation.

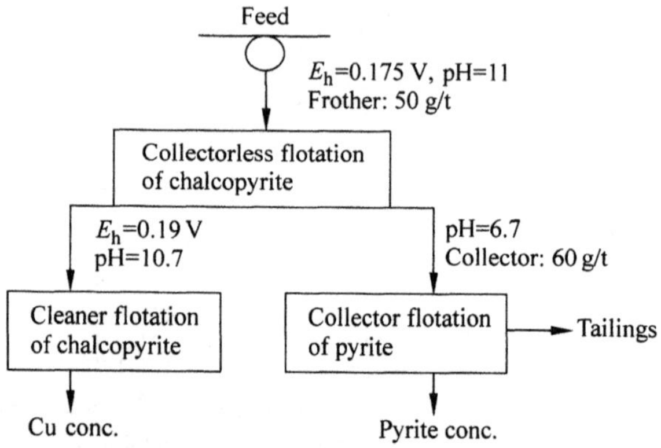

Figure 10.21 Flow sheet of collectorless flotation for concentrators of Tonglushan $CuFeS_2$-FeS_2 ore

References

Abramov, A. A., Leonov, S. B., Avdohin, V. A. and Kurscakova, G. M., 1977. Electrochemistry and thermodynamics of sulphide minerals and sulphydry with hydro-sulphide and cation collector. 12th IMPC, 22 – 26

Adam, K., Natarajan, K. A. and Iwasaki, I., 1984. Grinding media wear and its effect on the flotation of sulphide minerals. International Journal of Mineral Processing, 12(1 – 3): 39 – 54

Ahlberg. E. and Broo. A. E., 1988. Proc. Inter. Symp. Electrochem. In: Mineral and Metal Process. II (Eds. P.E. Richardson and R. Woods). ECS: Pennington, N. J., 36 – 48

Ahlberg, E. and Broo, A. E., 1996a. Oxygen reduction at sulphide minerals. 1. A rotating ring disc electrode (RRDE) study at galena and pyrite. Inter. J. Miner. Process, 46(1 – 2): 73 – 89

Ahlberg, E. and Broo, A. E., 1996b. Oxygen reduction at sulphide minerals. 2. A rotating ring disc electrode (RRDE) study at galena and pyrite in the presence of xanthate. Inter. J. Miner. Process, 47(1 – 2): 33 – 47

Ahlberg, E. and Broo, A. E., 1996c. Oxygen reduction at sulphide minerals. 3. The effect of surface pre-treatment on the oxygen reduction at pyrite. Inter. J. Miner. Process, 47(1 – 2): 49 – 60

Ahlberg, E. And Broo, A. E. 1997. Electrochemical reaction mechanisms at pyrite in acidic perchlorate solutions. J. Electrochem. Soc., 144(4): 1281 – 1286

Ahmed, S. M., 1978. Electrochemical studies of sulphides, II. Measurement of the galvanic currents in galena and the general mechanism of oxygen reduction and xanthate adsorption on sulphides in relation to flotation. Inter. J. Miner. Process, 5: 175 – 182

Ahn, J. H. and Gebhardt, J. E., 1991. Effect of grinding media-chalcopyrite interaction on the self-induced flotation of chalcopyrite. Inter. J. Miner. Process, 33: 243 – 262

Allison, S. A. and Finkelstein, N. P., 1971. Study of the products of reaction between galena and aqueous xanthate solutions. Trans. IMM, Sec. C, 80: 235 – 239

Allison, S. A., Goold, L. A., Nicd, M. J. and Granville, A., 1972. A determination of the products of reaction between various sulphide minerals and aqueous xanthate solution, and a correlation of the products with electrode rest potential. Metallurgical Transactions, 3: 2613 – 2618

Aloson, F. N. and Trevino, T. P., 2002. Pulp potential control in flotation. The Metallurgical Quarterly, 41(4): 391 – 398

Andrea, P. G., Angela, G. L., Athryn, E. P. K. and Roger, C. S., 1999. The mechanism of copper activation of sphalerite. Applied Surface Science, 137: 207 – 223

Andrew, H., Joseph, M., Irene, Y., Russo, S. P., 2002a. Density-functional theory studies of pyrite FeS_2(100) and (110) surfaces. Surface Science, 513(3): 511 – 524

Andrew, H., Joseph, M., Irene, Y., Russo, S. P., 2002b. Density-functional theory studies of pyrite FeS_2 (111) and (210). Surface Science, 520 (1 – 2): 111 – 119

Arce, E. M. and Gonzalez, I., 2002. A comparative study of electrochemical behavior of chalcopyrite, chalcocite and bornite in sulfuric acid solution. Inter. J. Miner. Process, 67: 17 – 28

Baldauf, H. and Schubert, H., 1980. Fine Particles Processing, 1(39): 767 – 786

Ball, B. and Richard, R. S., 1976. The chemistry of pyrite flotation and depression. In: Flotation, A. M. Gaudin Memorial volume, M. C. Fuerstanau(eds.), AIME, Inc., 1: 458 – 484

Basiollio, C., Pritzker, M. D., Yoon, R. H., 1985. Thermodynamics, electrochemistry and flotation of the chalcocite-potassium ethyl xanthate system. SME-AIME Annual Meeting, New York, Preprint No. 85 – 86

Beattie, M. J. V. and Poling, G. W., 1987. A study of the surface oxidation of arsenopyrite using cyclic voltammetry. Inter. J. Miner. Process, 20: 87 – 108

Beattie, M. J. V. and Poling, G. W., 1988. Flotation depression of arsenopyrite through use of oxidizing agents. Trans. IMM, Sec., C, 97: 15 – 20

Bechtold, E. and Schennach, R., 1996. Adsorption of oxygen on chlorine-modified Pt (100) surfaces. Surface Science, 369(1 – 3): 277 – 288

Biegler, T., Rand, D. A., Woods, R., 1975. Oxygen reduction on sulphide minerals. Part 1: Kinetic and mechanism at rotated pyrite electrodes. Electrochemistry and Interfacial Electrochemistry, 60: 151 – 163

Biegler, T. and Horne, M. D., 1985. The electrochemistry of surface oxidation of chalcopyrite. J. Electrchem. Soc., 132: 1363 – 1369

Buckley, A. N. and Woods, R., 1984. An X-ray photoelectron spectroscopic investigation of the surface oxidation of sulphide minerals. In: P. E. Richardson, S. Srinivasan and R. Woods(eds.). Electrochemistry in Mineral and Metal Processing, The Electrochemical-Society, Inc., 84 – 10: 286 – 302

Buckley, A. N., Hamilton, I. C., Woods, R., 1985. Investigation of the surface oxidation of sulphide minerals by linear potential sweep and X-ray photoelectron. In: K. S. E. Forssberg(ed.), Flotation of Sulphide Minerals, Elsevier. Amsterdam, 6: 41 – 60

Buckley, A. N. and Woods, R., 1990. X-ray photoelectron spectroscopic and electrochemical studies of the interaction of xanthate with galena in relation to the mechanism. Int. J. Miner. Process, 28: 301 – 311

Buckley, A. N. and Riley, K. W., 1991. Self-induced floatability of sulphide minerals: examination of recent evidence for elemental sulfur as the hydrophobic entity. Surf. Interface Anal., 17: 655 – 659

Buckley, A. N., 1994. A survey of the application of X-ray photoelectron spectroscopy to flotation research. Colloids Surf., 93: 159 – 172

Buckley, A. N. and Woods, R., 1995. Identifying chemisorption in the interaction of thiol collectors with sulphide minerals by XPS: adsorption of xanthate on silver and silver sulphide. Colloids and Surfaces A: Physicochemical and Engineering Aspects, 104, 2 – 3

Buckley, A. N. and Woods, R., 1996. Relaxation of the lead-deficient sulphide surface layer on oxidized galena. Journal of Applied Electrochemistry, 26(9): 899 – 907

Buckley, A. N. and Woods, R., 1997. Chemisorption—the thermodynamically favored process in the interaction of thiol collectors with sulphide minerals. Inert. J. Miner. Process, 51: 15 – 26

Bulut, G. and Atak, S., 2002. Role of dixanthogen on pyrite flotation: solubility, adsorption studies and E_h, FTIR measurements. Minerals & Metallurgical Processing, 19(2): 81 – 86

Cases, J. M., Kongolo, M., de Donato, P., Michot, L. and Erre, R., 1990. Interaction between firely ground galena and pyrite with potassium amylxanthate in relation to flotation, 2. Influence of grinding media at natural pH. Inter. J. Miner. Process, 30: 35 – 67

Cases, J. M. and de Donato, P., 1991. FTIR analysis of sulphide mineral surfaces before and after collection galena. in: K. S. E. Forssberg(ed.), Flotation of Sulphide Minerals. Inter. J. Mineral Process, 33: 49 – 65

Cata, M., Ciccu, R., Delfa, G., 1973. Improvement in electric separation and flotation by modification of energy level in surface layers. 10th International Mineral Processing Congress, London, 1: 1 – 22

Chander, S. and Fuerstenau, D. W., 1975. Sulphide minerals with thiol collectors: the chalcocite diethy dithiophosphate system. 11th International Mineral Processing Congress, 1: 583 – 603

Chander, S., Wie, J. M., Fuerstenau, D. W., 1975. On the native floatability and surface properties of naturally hydrophobic solids. In: P. Somasunfaran and R. G. Grieves(eds.), Advances in Interfacial Phenomena of Particulate/Solution/Gas Systems. AIME Symp., Ser., 150 (71): 183 – 188

Chander, S., Kar, G., Fuerstenau, D. W., 1979. The corrosion and wetting behaviour of copper in the presence of some thiol reagents. Corrosion Science, 19: 405 – 416

Chander, S., Pang, J., Briceno, A., 1988. Proc. Inter. Symp. Electrochem. In: Mineral and Metal Process. II (Eds. P. E. Richardson and R. Woods). ECS: Pennington (N. J.). 247 – 263

Chander, S., Briceno, A., Pang, J., 1993. Mechanism of sulfur oxidation in pyrite. Miner. Metall. Process, (3): 113 – 118

Chander, S., 1999. Advances in Flotation Technology (Eds. B. K. Parekh and J. D. Miller). SME: Littleton (CO). 129 – 145

Chang Chung-Liang, Lee Tai-Cheng, Huang Ta-Jen, 1998. Oxygen reduction mechanism and performance of $Y_1Ba_2Cu_3O_{7-d}$ as a cathode material in a high-temperature solid-oxide fuel cell. Journal of Solid State Electrochemistry, 2(5): 291 – 298

Chen Jin, Feng Jimin, Li Shikun, 1986. Study on the behaviour of collectorless flotation of galena. Nonferrous Metals(quarterly), 38(2): 33 – 39

Cheng Jianhua, Feng Qiming, Lu Yiping, 1998. Structure of organic depressant and its depressing on sulphide minerals. Nonferrous Metals, 3(50): 60 – 64

Cheng Jianhua, 1999. Energy band theory of pulp potential controlled flotation and its application in the study of organic depressants. Ph.D thesis. Central South University of Technology, Changsha

Cheng Jianhua, Feng Qiming, Lu Yiping, 2002. Molecular structure design and interaction principle sulphide minerals organic depressant. Journal of Guangxi University, 4(27): 276 – 280

Cheng, W. and Liu, Q., 1994. Research on the small molecular organic depressor of coal pyrite. Coal Science and Technology, 9(220): 33 – 36

Cheng, X. and Iwasaki, I., 1992. Pulp potential and its implications to sulphide flotation. Mineral Processing and Extractive Metallurgy Review, 11: 187 – 210

Cheng, X., Iwasaki, I., Smith, K. A., 1994. An electrochemical study on cathodic decomposition behaviour of pyrrhotite in deoxygenated solutions. Mineral & Metellurgical Processing, 8(1): 160 – 167

Cook, M. A. and Nixon, J. C., 1950. Theory of water-repellent films on solids formed by adsorption from aqueous solutions of heteropolar compounds. J. Phys. Colloid Chem., 54: 445 – 459

Cook, M. A. and Wadsworth, M. E., 1957. Hydrolytic and ion pair adsorption processes in flotation, ion exchange and corrosion. In: Proceedings of the 2nd Inter. Congr. Surf. Activity. Butterworth & Co. Publisher, London, 3: 228 – 242

Costa, M. C., Botelho, do Rego A. M., Abrants, L. M., 2002. Characterization of a natural and an electro-oxidized arsenopyrite: a study on electro-chemical and X-ray photoelectron spectroscopy. Inter. J. Miner. Process, 65: 83 – 108

Dai Jingping, Sun Wei, Cao Limei, Hu Yuehua, 2000. Influence of mechanical excitation on adsorption of sodium diethyl dithioformate on galena and sphalerite. Trans. Nonferrous Met. Soc., China, 10: 101 – 105

David, R. Lide (Eds.), 2000. Handbook of Chemistry and Physics. 81rd Edition. CRC Press, 31 – 35

Ding Dunghuang, Long Xiangyun, Wang Dianzuo, 1993. Mechanism of pyrite oxidation and flotation. Nonferrous Metals, 45(4): 4 – 30 (in Chinese)

Dong Qing-hay, Sun Wei, Hu Yuehua, Wang Dianzuo, 2006. Research on mechanical and electrochemical behavior during the flotation of pyrite. Mining and Metallurgical Engineering, 26(1): 32 – 36 (in Chinese)

Du Tietz, C., 1957. Progress in Mineral Dressing. Stockholm, 417 – 439

Du Tietz, C., 1975. 11th International Mineral Processing Congress. 375 – 403

Duun, J. G., Fernandez, P. G., Hughes, H. C., Ling, H. G., 1989. Aqueous oxidation of arsenopyrite in acid. The Aus. IMM Annual Conference, Perth-kalgoorlie, W. A., 217 – 220

Eadington, P. and Prosser, A. P., 1969. Oxidation of lead sulphides in aqueous suspensions. Trans. IMM, Sect. C: Min. Pro. Ext. Metall., 78: 74 – 82

Edelbro, R., Sandström, Å., Paul, J., 2003. Full potential calculations on the electron bandstructures of sphalerite, pyrite and chalcopyrite. Applied Surface Science, 206(1 – 4): 300 – 313

Ekmekci, Z. and Demirel, H., 1997. Effects of galvanic interaction on collectorless flotation behaviour of chalcopyrite and pyrite. Int. J. Miner. Process, 52: 31 – 48

Eligillani, D. A. and Fuerstenau, M. C., 1968. Mechanisms involved in cyanide depression of pyrite. AIME, Soc. Min. Eng., 241: 437 – 445

Eliseev, N. I., Kirbitova, N. V., Sharapova, N. D., Glazyrina, L. N., Shramm, E. O., 1982. The effect of sodium sulphide on the flotation behaviour of sulphide minerals. Tsvetn. Met., 9: 99 – 102 (in Russian)

Esposito, M. C., Chander, S., Aplan, F. F., 1987. Characterization of pyrite from coal source. In: Process Metallurgy, Ⅶ, Transaction of Metallurgical Society/American Institute of Mining Engineers, 475 – 493

Fahlstrom, P. H., 1974. Autogenous grinding of base metal ores at Boliden aktiebolag. Can. Min. Metall. Bull., 67(743): 128 – 141

Fahlstrom, P. H., Fagremo, O., Gjerdrum, A. S., 1975. Autogenous grinding at Boliden's Aitik plant, Parts Ⅰ and Ⅱ, Mining World, 42

Feng Qiming, 1989. Pulp electrochemistry of flotation of sulphide minerals: theory and technology (Ph.D. thesis). Central South University of Technology

Feng Qiming, Chen Jin, Xu Shi, 1991. Relationship between thermodynamics, property of reactive products on sulphide and collectorless flotation separation of sulphide minerals. J. CSIMM (suppl.), 22(3): 28 – 35

Fereshteh, R., Caroline, S., James, A. F., 2002. Sphalerite activation and surface Pb ion concentration. Inter. J. Miner. Process, 67: 43 – 58

Fierro, R. E., Tryk, D., Scherson, D., Yeager, E., 1988. Perovskite-type oxides: oxygen electrocatalysis and bulk structure. Journal of Power Sources, 22 (3 – 4): 387 – 398

Finkelstein, N. P. and Goold, L. A., 1972. The reaction of sulphide minerals with thiol compounds. National Institute of Metallurgy, South Africa, Reported No. 1439

Finkelstein, N. P., Allsion, S. A., Lovell, V. M., Stewart, B. V., 1975. Advances in Interf Pheno. of Particulate/Solution/Gas Systems (Eds. P. Somasundaran and R. B. Grieves). AIChE: New York. 71(150): 165 – 175

Finkelstein, N. P. and Poling, G. W., 1977. The role of ditholates in the flotation of sulphide minerals. Miner. Sci. Eng., 9: 177 – 197

Finkelstein, N. P., 1997a. The activation of sulphide minerals for flotation: a review. Inter. J. Miner. Process, 52: 81 – 120

Finkelstein, N. P., 1997b. The activation of sulphide minerals for flotation: a review. Inter. J. Miner. Process, 52: 120 – 181

Finkelstein, N. P., 1999. Addendum to: The activation of sulphide minerals for flotation: a review. Inter. J. Miner. Process, 55(4): 283 – 286

Fornasiero, D., Montalti, M., Ralston, J., 1995. Kinetics of adsorption of ethyl xanthate on pyrrhotite: in situ UV and infrared spectroscopic studies. Langmuir, 11: 467 – 478

Forssberg, K. S. E., Antti, B. M., Palsson, B., 1984. Computer-assisted calculations of thermodynamic equilibria in the chalcopyrite-ethyl xanthate system. In: M. J. Jones and R. Oblatt (eds.), Reagents in the Minerals Industry. IMM, Rome, Italy, 251 – 264

Fuerstenau, M. C., Kuhn, M. C., Elgillani, D. A., 1968. The role of dixanthogen in xaomthate flotation of pyrite. Trans. AIME, 241: 437

Fuerstenau, M. C. and Sabacky, B. J., 1981. Inter. J. Miner. Process, 8: 79 – 84

Fuerstenau, M. C., Misra, M., Palmer, B. R., Xanthate adsorption on selected sulphides in the presence of oxygen. Inter. J. Miner. Process

Gardner, J. R. and Woods, R., 1971. An electrochemical investigation of contact angle and floatation in the presence of alkyxanthates, II. galena and pyrite surfaces. Aust. J. Chem., 30: 981 – 991

Gardner, J. R. and Woods, R., 1979. An Electrochemical investigation of the natural floatability of chalcopyrite. Inter. J. Miner. Process, 6: 1 – 16

Garrels, R. M. and Christ, C. L., 1965. Solutions, Minerals and Equilibria. Harper & Row, New York, 403 – 429

Gaudin, A. M., 1932. Flotation. McGraw-Hill Book Co., New York, 552

Gaudin, A. M. and Schuhmann, R. Jr., 1936. The action of potassium *n*-amyl xanthate on chalcocite. J. Phys. Chem., 40: 257 – 275

Gaudin, A. M., 1957. Flotation (2nd ed.). McGraw-Hill Book Co., New York, 573

Gaudin, A. M., Miaw, H. L., Spedden, H. R., 1957. Native floatability and crystal structure. In: J. H. Schuman(Ed), Proc. 2nd Int. Congr. Surf. Activity. III, Electrical phenomena and Solid/Liquid Interface. Butterworths, London, 202 – 219

Gebhardt, J. E. and Shedd, K. B., 1988. Effect of solution composition on redox potential of platinum and sulphide mineral electrodes. In: Richardson and R.Woods (Eds), Electrochemistry in Mineral and Metal Processing II. Electro. Soc. Inc., 88 – 21, 84 – 100

Goold, L. A. and Finkelstein, N. P., 1972. Product of reaction between galena and aqueous xthante solutions. South Africa, IMM, Report No.1439

Grano, S., Ralston, J., Smart, R. st. C., 1990. Influence of electrochemical environment on the flotation behaviour of Mt. Isa. copper and lead-zinc ore. Inter. J. Miner. Process, 30: 69 – 97

Gu Guohua, 1998. Redox reaction in grinding and flotation system of sulphide minerals and original potential flotation (Ph.D Thesis). Central South University of Technology (in Chinese)

Gu Guohua, Hu Yuehua, Qiu Guanzhou, Liu Ruyi, 2000. Electrochemistry of sphalerite activated by Cu^{2+} ion. Transaction of Nonferrous Metals Society of China. 10(special issue): 64 – 67

Gu Guohua, Hu Yuehua, Qiu Guanzhou, Wang Hui, Wang Dianzuo, 2002a. Potential control flotation of galena in strong alkaline media. Journal of Central South University of Technology, 9(1): 16 – 20

Gu Guohua, Hu Yuehua, Qiu Guanzhou, Wang Dianzuo, 2002b. Electrochemistry of galena in high alkaline flotation. Mining and Metallurgical Engineering, 22(1): 52 – 55 (in Chinese)

Gu Guohua, Hu Yuehua, Wang Hui, Qiu Guanzhou, Wang Dianzuo, 2002c. Original potential flotation of galena and its industrial application. Journal of Central South University of Technology, 9(2): 91 – 94

Gu Guohua, Wang Hui, Qiu Guanzhou, Hu Yuehua, 2004. Collector matching in origin potential flotation. Trans. Nonferrous Met. Soc. China, 14(3): 576 – 581

Gupta, S., Tryk, D., Bae, I., Aldred, W., Yeager, E., 1989. Heat-treated polyacrylonitrile-based catalysts for oxygen electroreduction. Journal of Appl. Electrochem., 19(1): 19 – 27

Guy, P. J. and Trahar, W. J., 1984. The influence of grinding and flotation environments on the laboratory batch flotation of galena. Inter. J. Miner. Process, 12: 15 – 38

Guy, P. J. and Trahar, W. J., 1985. The effects of oxidation and mineral interaction on sulphide flotation. In: Flotation of Sulphide Minerals, K. S. E. Forssberg(ed.), Elsevier, 6: 91 – 109

Hamilton, I. C. and Woods, R., 1981. An investigation of surface oxidation of pyrite and pyrrhotite by linear potential sweep voltammetry. J. Electroanal. Chem. Interfacial Electrochem., 118: 327 – 343

Hamilton, I. C. and Woods, R., 1984. A voltammetric study of the surface oxidation of sulphide minerals. In: P. E. Richardson, S. Srinivasan and R. Woods(eds.), Electrochemistry in Mineral and Metal Processing, The electrochem. Soc. Inc., 84 – 10: 259 – 285

Harris, P. J. and Finkelstein, N. P., 1977. The interaction of chalcocite, oxygen and xanthate. National Institute for Metallurgy, Randburg, South Africa, Report No.1896, 33 – 37

Harris. P. J., 1988. Reagents in Mineral Technology (Eds. P. Somasundaran and B. M. Moudgil). Marcel Dekker: New York. Ch11: 371 – 384

Hayes, R. A., Price, M. D., Ralston, J., 1987. The collectorless flotation of sulphide minerals. Miner. Process Extra. Metall. Rev., 2: 203 – 234

Hayes, R. A. and Ralston, J., 1988. The collectorless flotation and separation of sulphide minerals by E_h control. Inter. J. Miner. Process, 23: 55 – 84

Hepel, T. and Pomianowski, A., 1977. Diagrams of electrochemical equilibria of the system copper-potassium ethyl xanthate-water at 25℃. Int. J. Miner. Process, 4: 345 – 361

Heyes, G. W. and Trahar, W. J., 1977. The natural floatability of chalcopyrite. Int. J. Miner. Process, 4: 317 – 344

Heyes, G. W. and Trahar, W. J., 1979. Oxidation-reduction effects in the flotation of chalcocite. Inter. J. Miner. Process, 6: 229 – 252

Heyes, G. W. and Trahar, W. J., 1984. The flotation of pyrite and pyrrhotite in the absence of conventional collectors. In: P. E. Richardson et al. eds., Electrochemistry in Mineral and Metal Processing. Electro. Chem. Soc., 84 – 10: 219 – 232

Hodgson, M. and Agar, G. E., 1984. An electrochemical investigation into the natural floatability of pyrrhotite. In: P. E. Richardson, S. Srinivasan, R. Woods (eds.), Electrochemistry in Mineral and Metal Processing. The Electrochemistry Society Inc., 84 – 10: 185 – 201

Hoey, G. R., Dingley, W., Freeman, C., 1977. Corrosion behaviour of various steels in ore grinding. CIM Bull., 2: 105

Hoyack, M. E. and Raghavan, S., 1987. Interaction of sphalerite. Trans. IMM Sec. C, 96: C173 – C178

Hu Yuehua, Zhang Shunli, Qiu Guanzhou, Wang Dianzuo, 1995. The mechanism of activation flotation of pyrite depressed by lime. J. Cent. South Univ. Technol., 26(2): 176 – 180

Hu Yuehua, Qiu Guanzhou, Sun Shuiyu, Wang Dianzuo, 2000. Recent development in researches of electrochemistry of sulphide flotation at Central South University of Technology. Trans. Nonferrous Met. Soc. China, 10: 1 – 7

Hu Yuehua, Sun Wei, Qin Wenqing, 2002. Mechanics-electrochemistry action in PbS flotation. The Chinese Journal of Nonferrous Metals, 12(5): 1060 – 1064(in Chinese)

Hu Yuehua, Dai Jingping, Zhang Qin, 2004. Electrochemical flotation of diethyldithiocarbamate-pyrrhotite system. J. Cent. South Univ. Technol., 11(3): 270 – 274

Janetski, N. D., Woodburn, S. I., Woods, R., 1977. An electrochemical investigation of pyrite flotation and depression. Inter. J. Miner. Process, 4: 227 – 239

Jellinek, F., Pollak, R. A., Shafer, M. W., 1974. X-Ray photoelectron spectra and electronic structure of zirconium trisulphide and triselenide. Materials Research Bulletin, 9(6): 845 – 856

Jiang Hao, Hu Yuehua, Xu Jing, 2000. Electrochemical characteristics of couple electrode of galena-pyrite in different solutions. Trans. Nonferrous Met. Soc. China, 10: 87 – 88

Johnson, N. W. and Munro, P. D., 1988. E_h-pH measurements for problem solving in a zinc reverse flotation process. Austral Inst. Min. Metall., 239(3): 53 – 58

Karkovsky, I. A., 1957. Physicochemical properties of some flotation reagents and their salts with ions of heavy iron-ferrous metals. Proc. 2nd Int. Congr. Surface Activity, London, 4: 225 – 237

Kartio, I. J., Basilio, C. I., Yoon, R. H., 1996. An XPS study of sphalerite activation by copper. In: R. Woods, F. Doyle, P. E. Richardson (eds.), Electrochemistry in Mineral and Metal Processing IV. The Electro-Chemical Society, 25 – 34

Kelebek, S., 1987. Wetting behaviour, polar characteristics and flotation of inherently hydrophobic minerals. Trans. IMM, Sec. C, 96: 103 – 107

Kelebek, S. and Smith, G. W., 1989. Collectorless flotation of galena and chalcopyrite: correlation between flotation rate and the amount of extracted sulfur. Miner. Metall. Process, 6(3): 123 – 129

Kelsall, G. H. and Page, P. W., 1985. Aspects of chalcopyrite ($CuFeS_2$) electrochemistry. In: P. E. Richardson, S. Srinivasan, R. Woods (eds.), Electrochem. in Miner. and Metal Processing. The Electrochemical Soc. Inc., 84 – 10: 303 – 320

Kelsall, G. H., Yin, Q., Vaughan, D. J., England, K. E. R., 1996. Electrochemical oxidation of pyrite in acidic aqueous electrolytes. In: Proceeding 4th International Symposium on Electrochemistry in Mineral and Metal Processing. Electrochemical Society, 196(6): 131 – 142

Kelsall, G. H., Yin, Q. et al., 1999. Electrochemical oxidation of pyrite in aqueous electrolytes. J. of Electro-analytical Chemistry, 471: 116 – 125

King, F., Quinn, M. J., Litke, C. D., 1995. Oxygen reduction on copper in neutral NaCl solution. Journal of Electro-analytical Chemistry, 385(1): 45 – 55

Kneer, E. A., 1997. Electrochemical measurements during the CMP of Tungsten thin films. J. Electrochem. Soc., 144: 3041 – 3049

Kostina, G. M. and Chernyak, A. S., 1979. Investigation of the mechanism of electrochemical oxidation of arsenopyrite and pyrite in caustic soda solution. Zhurnal Prikladnoi Khimii, 52: 1532 – 1535

Kowal, A. and Domianowski, A., 1973. Cyclic voltammetry of ethyl xanthate on a natural copper sulphide electrode. Electrocnal Chem. Interf. Electrochem., 46: 411 – 420

Laajalehto, K., Nowak P., Pomianowski, A., Suonien, E., 1991. Xanthate adsorption at the PbS/aqueous interface: comparison of XPS, infrared and electrochemical results. Colloids Surf., 57: 319 – 333

Laajalehto, K., Nowak, P., Suoninen, E., 1993. On the XPS and IR identification of the products of xanthate sorption at the surface of galena. Int. J. Miner. Process, 37: 123 – 147

Labonte, G. and Finch, J. A., 1988. Measurement of electrochemical potential in flotation systems. CIM Bulletin, 81(920): 78 – 83

Lamache, M., Lam, O., Bauer, D., 1984. Electrochemical oxidation of galena and ethylxanthate. In: R. E. Richardson, S. Srinivasan, R. Woods (eds.), Electrochemistry in Mineral and Metal Processing. The Electrochemistry Society Inc., 84 – 10: 54 – 65

Learmont, M. E. and Iwasaki, I., 1984. The Effect of grinding media on flotation of galena. Minerals & Metallurgical Processing, 1(3): 136 – 143

Lepetic, V. M., 1974. CIM Bulletin, 67: 71

Leppinen, J. O., Basilio, C. I., Yoon, R. H., 1989. In-situ FTIR study of ethyl xanthate adsorption on sulphide minerals under conditions of controlled potential. Int. J. Miner. Process, 26: 259 – 274

Leppinen, J. O., 1990. FTIR and flotation investigation of the adsorption of ethyl xanthate on activated and non-activated sulphide minerals. Inter. J. Miner. Process, 30: 245 – 263

Leppinen, J. O., Basilio, C. I., Yoon, R. H., 1998. FTIR study thiocarbamate adsorption on sulphide minerals. Colloids and Surfaces A, 32: 113 – 125

Li Haipu, Jiang Yuren, Cao Xuefeng, Hu Yuehua, 2001. Synthesization of modified starch and its performance. Mining and Metallurgical Engineering, 4(21): 29 – 32 (in Chinese)

Lin, Q., Jin, H., Wang, D., 1991. Preparation of some new micromolecular organic depressant and their reactivity. Journal of Central-South Institute Mining and Metallurgy, 3(22): 256 – 262 (in Chinese)

Liu Runqing, Sun Wei, Hu Yuehua, Xiong Daoling, 2006. Depression mechanism of small molecular mercapto-organic depressants on flotation behaviour of complex sulphides. Chinese Journal of Nonferrous Metals, 16(4): 746 – 751 (in Chinese)

Lowson, R. T., 1982. Aqueous oxidation of pyrite by molecular oxygen. Chemistry Review, 82(5): 461

Lusk, J., Scott, S. D., Ford, C. E., 1993. Phase relations in the Fe-Zn-S system to 5 kbars and temperatures between 325 and 150℃. Econ. Geol., 88: 1880 – 1903

Luttrell, G. H. and Yoon, R. H., 1984a. Surface studies of the collectorless flotation of chalcopyrite. Colloids and Surfaces, 12: 239 – 254

Luttrell, G. H. and Yoon, R. H., 1984b. The collectorless flotation of chalcopyrite ores using sodium sulphide. Inter. J. Miner. Process, 13: 271 – 283

Ma Guoyin, Sun Wei, Hu Yuehua, 2006. Investigation of collecting nature of aryl thiolate collectors on jamesonite and marmatite. Mining and Metallurgical Engineering, 26(8): 156 – 157 (in Chinese)

Majima, H. and Takeda, M., 1968. Electrochemistry studies of the xanthate-dixanthogen system on pyrite. Trans. AIME, 241: 431 – 436

Majima, H., 1969. How oxidation affects selective flotation of complex sulphide ores. Can. Met. (Quarterly) , 8(3): 269 – 273

McCarron, J. J., Walker, G. W., Buckley, A. N., 1990. An X-ray photoeletron spectroscopic investigation of chalcopyrite and pyrite surfaces after conditioning in sodium sulphide solutions. Inter. J. Miner. Process, 30: 1 – 76

Mielezarski, J. A. and Yoon, R. H., 1989. Spectroscopic studies of the structure of the adsorption layer of thionocarbamate. J. Coll. Inter. Sci., 131(2): 423 – 432

Mielezarski, J. A., 1997. Reply to comment on In situ and ex situ infrared studies of nature and structure of thiol layers adsorbed on cuprous sulphide at controlled potential, simulation and experimental results. Langmuir, 13: 878 – 880

Mielezarski, J. A., Mielezarski, E., Cases, J. M., 1998. Influence of chain length on adsorption of xanthates on chalcopyrite. Inert. J. Miner. Process, 52: 215 – 274

Mishra, K., 1992. Journal of the Electrochemical Society, 139(6): 749 – 752

Mory, M. S., Grano, S. R., Ralston, J., Prestidge, C. A., Verity, B., 2001. The electrochemistry of Pb^{2+} activated sphalerite in relation to flotation. Minerals Engineering, 14(9): 1009 – 1017

Muscat, J., Hung, A., Russo, S., Yarovsky, I., 2002. First-principles studies of the structural and electronic properties of pyrite FeS_2. Phys. Rev., B65, 054107

Nagaraj, D. R., Wang, S. S., Avotins, P. V., Dowling, E., 1986. Structure activity relationships for copper depressments. Trans. IMM, Sec. C, 95: 17 – 26

Nagaraj, D. R. and Brinen, J. S., 2001. SIMS study adsorption of collectors on pyrite. Inter. J. Miner. Process, 63: 45 – 47

Nakazawa, H. and Iwasaki, I., 1986. Galvanic contact between nickel arsenide and pyrrlotite and its effects on flotation. Inter. J. Miner. Process, 18: 203 – 215

Nakazawa, H. and Iwasaki, I., 1985. Effect of pyrite-pryyhotite contact on their floatabilities. Minerals & Metallurgical Processing, 2(4): 206 – 211

Nagaraj, D. R., et al., 1982. Copper depresants: correlation between structure and activity. 112th SME-AIME Annual Meeting, 89 – 93

Natarajan, K. A., Riemer, S. C., Iwasaki, I., 1984. Influence of pyrrhotite on the corrosive wear of grinding balls in magnetite ore grinding. Inter. J. Miner. Process, 13(1): 73 – 81

Nesbitt, H. W., Bancroft, G. M., Pratt, A. R., Scaini, M. J., 1998. Sulfur and iron surface states on fractured pyrite surfaces. American Mineralogist, 83: 1067 – 1076

Neeraj, K. M., 2000. Kinetic studies of sulphide mineral oxidition and xanthate adsorption. Doctor thesis of Virginia Polytechnic Institute and State University. A Bell & Howell Company UMI dissertation Services

Nixon, J. C., 1957. Discussion in Proceedings of the 2nd Inter. Congr. Surf. Activity. Butterworth & Co. Publisher, London, 3: 369

O'Dea, A. R., Prince, K. E., Smart, R. S. C., Gerson, A. R., 2001. Secondary ion mass spectrometry investigation of the interaction of xanthate with galena. Inter. J. Miner. Process, 61: 121 – 143

O'Dell, C. S., Dooley, R. K., Walker, G. W., Richardson, P. E., 1984. Chemical and electrochemical reactions in the chalcocite-xanthate system. In: P. E. Richardson, S. Srinivasan, R. Woods (eds.). Electrochemistry in Mineral and Metal Processing. The Electrochemistry Society, Inc. 26 – 53

O'Dell, C. S., Walker, G. W., Richardson, P. E., 1986. Electrochemistry of the chalcocite-xanthate system. J. Appl. Electrochem., 16: 544 – 554

Opahle, I., Koepernik, K., Eschrig, H., 2000. Full potential band structure calculation of iron pyrite. Computational Materials Science, 17(2 – 4): 206 – 210

Page, P. W. and Hazell, L. B., 1989. X-ray photoelectron spectroscopy (XPS) studies of potassium amyl xanthate (KAX) adsorption on precipitated PbS related to galena flotation. Inter. J. Miner. Process, 25: 87 – 100

Pang, J. and Chander, S., 1990. Oxidation and wetting behavior of chalcopyrite in the absence and presence of xanthate. Miner. Metall. Process, 7(3): 149 – 155

Pattabi, M., Castellanos, R. H., Castillo, R., Ocampo, A. L., Moreira, J., Sebastian, P. J., McClure, J. C., 2001. Electrochemical characterization of tungsten carbonyl compound for oxygen reduction reaction. Inter. J. Hydrogen Energy, 26(2): 171 – 174

Pattrick, R. A. D., England, K. E. R., Charnock, J. M., Mosselmans, J. F. W., 1999. Copper activation of sphalerite and its reaction with xanthate in relation to flotation: an X-ray absorption spectroscopy (reflection extended X-ray absorption fine structure) investigation. Inter. J. Miner. Process, 55(4): 247 – 265

Persson, I., Persson, P., Valli, M., Fozo, S., Malmensten, B., 1991. Reaction on sulphide mineral surfaces in connection with xanthate flotation studied by diffuse reflectance FTIR spectroscopy, atomic absorption spectrophotometry and calorimetry. In: K. S. E. Forssberg (ed.). Flotation of Sulphide Mineral. Inter. J. Miner. Process, 33: 67 – 81

Peters, E., 1977. The electrochemistry of sulphide minerals. In: J. O'M. Bockris, D. A. J. Rand, B. J. Weich (eds.). Trends in Electrochemistry. New York: Plenum Press, 267 – 290

Peters, E., 1986. Leaching of sulphides. In: P. Sommasundaran (ed.). Advances in Mineral Processing. Proc. Sym. Honoring N. Arbiter on His 75th Birthday, SME, Inc. Colorado, 445 – 462

Pillai, K. C., Young, V. Y., O'M Bockris, J., 1983. XPS studies of xanthate adsorption on galena surfaces. Appl. Surf. Sci., 16: 322 – 344

Plaksin, I. N. and Bessonov, S. V., 1957. Role of gases in flotation reactions. In: Proceedings of the 2nd Inter. Congr. Surf. Activity (J. H.Schulman ed.). Butterworth & Co. Publisher, London, 3: 361 – 367

Plaksin, I. N. and Shafeev, R. Sh., 1963. Bull IMM, 72, 715 – 722

Poling, G. W., 1976. Reactions between thiol reagents and sulphide minerals. In: Flotation. A.M. Gaudin Memorial Volume, M. C. Fuerstenau (ed.), AIME, 1: 334 – 363

Povarennyk, A. S., 1972. Crystal Chemical Classification of Minerals (two vols.). New York: Plenum Press

Pozzo, R. L. and Iwasaki, I., 1987. Effect of pyrite and pryyhotite on the corrosive wear of grinding media. Minerals & Metallurgical Processing, 4(2): 166 – 171

Pozzo, R. L., Malicsi, A. S., Iwasaki, I., 1988. Pyrite-pryyhotite-grinding media contact and its effect on flotation. Minerals & Metallurgical Processing, 5(1): 16 – 21

Pozzo, R. L., Malicsi, A. S., Iwasaki, I., 1990. Pyrite-pyrrhotite-grinding media contact and its effect on flotation. Minerals & Metallurgical Processing, 7(1): 16 – 21

Prestidge, C. A., Thiel, A. G., Ralston, J., Smart, R. S. C., 1994. The interaction of ethyl xanthate with copper (II)—activated zinc sulphide: kinetic effects. Colloids Surface, A. Physicochem. Eng. Aspects, 85: 51 – 68

Prestidge, C. A., Skinner, W. M., Ralston, J., Smart, R. S. C., 1997. Copper (II) activation and cyanide de-activation of zinc sulphide under mildly alkaline conditions. Appl. Surf. Sci., 108: 333 – 344

Pritzker, M. D. and Yoon, R. H., 1984a. Thermodynamic calculations on sulphide flotation systems: I. galena-ethyl xanthate system in the absence of metastable species. Inter. J. Miner. Process, 12: 95 – 125

Pritzker, M. D. and Yoon, R. H., 1984b. Thermodynamic calculations and electrochemical studies on the galena-ethyl xanthate system. In: P. E. Richardson et al. (eds.), Electrochemistry in Mineral and Metal Processing. Electrochem. Chem. Soc., 84-10: 26 – 53

Qin Wenqing, 1998. Electrochemical behavior of sulphide minerals and pulp potential controlled flotation. Ph. D. Thesis, Central South University of Technology, Changsha

Qin Wenqing, Qiu Guanzhou, Xu Jing, Hu Yuehua, 1998. Galvanic interaction of contacting sulphide particle and its effect on flotation. Transactions of Nonferrous Metals Society of China, 8(4): 661 – 665

Qin Wenqing, Qiu Guanzhou, Hu Yuehua, 2000. Electrochemical oxidation of sodium sulphide on rotating ring-disc electrode. Trans. Nonferrous Met. Soc. China, 10: 80 – 82

Qin Wenqing, Qiu Guanzhou, Hu Yuehua, Xu Jing, 2001. Kinetics of electrochemical process of galena electrode in diethyldithiocarbamate solution. Trans. Nonferrous Met. Soc. China, 11(4): 587 – 590

Qiu Guanzhou, Yu Runlan, Hu Yuehua, Qin Wenqing, 2004. Corrosive electrochemistry of jamesonite. Trans. Nonferrous Met. Soc. China, 14(6): 1169 – 1173

Qiu Guanzhou, Xiao Qi, Hu Yuehua, 2004. First-principles calculation of the electronic structure of the stoichiometric pyrite FeS_2 (100) surface. Computation Materials Science, 03 – 11: 2989 – 2994

Rand, D. A., 1975. Oxygen reduction on sulphide minerals, part III: comparisson of activity of various copper, iron, lead, nickle mineral electrodes. Electrochemistry and Interfacial Electrochemistry, 60: 265 – 275

Rand, D. A. and Woods, R., 1984. E_h measurements in sulphide mineral slurries. Inter. J. Miner. Process, 13: 29 – 42

Rao, S. R., Moon, K.S., Leja, J., 1976. Effect of grinding media on the surface reactions and flotation of heavy metal sulphides. In: Flotation. A. M. Gaudin Memorial Volume (M. C. Fuerstenau ed.), 509 – 522

Ravitz, S. F. and Porter, R. R., 1933. Oxygen-free flotation I: flotation of galena in absence of oxygen. Am. Inst. Min. Metall. Eng., Tech. Publ., No.513

Rey, M. and Formanek, V., 1960. Some factors affecting the selectivity in the differential flotation of lead-zinc ores in the presence of oxidized lead mineral. Proc. 5th Int. Min. Proc. Congr., Inst. Mining and Met., London, 343 – 352

Richardson, P. E. and Maust, E. E. Jr., 1976. Flotation – Gaudin Memorial Volume (Ed. M. C. Fuerstenau). New York : AIME, 1: 364

Richardson, P. E., Stout, J. V. III, Proctor, C. L., Walker, G. W., 1984. Electrochemical flotation of sulphides: chalcocite-ethylxanthate interactions. Inter. J. Miner. Process, 12: 73 – 93

Richardson, P. E. and Walker, G. W., 1985. The flotation of chalcocite, bornite, chalcopyrite and pyrite in an electrochemical-flotation cell. 15th Inter. Miner. Process Congr., Cannes, France, 2: 198 – 210

Richardson, P. E. and Yoon, R. H., Li, Y-Q, 1993. The photoelectrochemistry of pyrite and galena. 18th International Mineral Processing Congress, Sydney, 757 – 766

Richardson, P. E., Hu, Q., Finkelstein, N. P., Yoon, R. H., 1994. An electrochemical method for the study of the flotation chemistry of sphalerite. Inter. J. Miner. Process, 41: 71 – 76

Rubio, J. and Kitchener, J. A., 1964. Electrochemical study of galena-xanthate-oxgen flotation system. Trans. Ins. Min. Metall., 3: 313 – 322

Salamy, S. G. and Nixon, J. C., 1953. The application of electrochemical methods to flotation research. Recent Developments in Mineral Dressing, Institution of Mining and Metallurgy, London, 503 – 516

Salamy, S. G. and Nixon, J. C., 1954. Reaction between a mercury surface and some flotation reagents: an electrochemical study. Aust. J. Chem., 7: 146 – 156

Salvador, P. and Tafalla, D., 1991. Reaction Mechanism at n-FeS_2/I interface. Journal of the Electrochemical Society, 132(6): 1350 – 1356

Sanchez, V. M. and Hiskey, J. B., 1988. Electrochemical study of the surface oxidation of arsenopyrite in alkaline media. Metallurgical Transactions B, 19(6): 943 – 949

Sanchez, V. M. and Hiskey, J. B., 1991. Electrochemical behavior of arsenopyrite in alkaline media. Minerals & Metallurgical Processing, 8(1): 1 – 6

Sasaki Takehiko, Goto Yoshio, Tero Ryugo, Fukui Ken-ichi, Iwasawa Yasuhiro, 2002. Oxygen adsorption states on Mo(112) surface studied by HREELS. Surface Science, (502 – 503): 136 – 143

Segall, M. D., Lindan, P. L. D., Probert, M. J., Pickard, C. J., Hasnip, P. J., Clark, S. J., Payne, M. C., 2002. First-principles simulation: ideas, illustrations and the CASTEP code. J. Phys.: Cond. Matt., 14, 2717 – 2743

Shimoiizaka, T., Usui, S., Matsuoka, I., Sasaki, H., Sendai, A., 1976. Depression of galena flotation by sulphite or chromate ion. In: Flotation. A. M. Gaudin Memorial Volume, M. C. Fuerstenau (ed.), 1: 393 – 413

Skinner, W. M., Prestidge, C. A., Smart, R. S. C., 1996. Irradiation effects during XPS studies of copper(II) activation of zinc sulphide. Surf. Int. Sci., 24: 620 – 626

Sun Shuiyu, 1990. Sulphide flotation by electrochemical adjustment and collectorless flotation. Ph.D thesis. Central South University of Technology (in Chinese)

Sun Shuiyu, Li Baidan, Wang Dianzuo, 1990. Collectorless flotation of sulphide minerals. J. Cent. South Inst. Min. Metall., 21(5): 473 – 478 (in Chinese)

Sun Shuiyu, Wang Dianzuo, Li Baidan, 1991. An electro and quantum-chemical investigation on the collectorless flotation of pyrite. J. Cent. South. Inst. Min. Metall., 22(3) suppl: 43 – 48 (in Chinese)

Sun Shuiyu, Wang Dianzuo, Li Baidan, 1992. Principle of collectorless flotation of sulphides and design of separation schemes. Nonferrous Metals (Bimonthly), 6: 4 – 8 (in Chinese)

Sun Shuiyu, Wang Dianzuo, Li Baidan, 1993a. The collectorless flotation and separation and separation of chalcopyrite and pyrite by potential control. J. Cent. South Inst. Min. Metall., 24(4): 466 – 471

Sun Shuiyu, Wang Dianzuo, Li Baidan, 1993b. Studies on surface oxidation of sulphide minerals. Nonferrous Metals (Quarterly), 45(4): 42 – 49 (in Chinese)

Sun Shuiyu, Wang Dianzuo, Li Baidan, 1993c. Electrochemical and quantum-chemical investigations on self-induced flotation of galena. Nonferrous Metals (Quarterly), 45(2): 31 – 37 (in Chinese)

Sun Shuiyu, Li Baidan, Wang Dianzuo, 1993d. Considerations of semiconductor energy band theory on collectoless flotation of sulphide minerals. J. BGRIMM, 2(3): 38 – 43 (in Chinese)

Sun Shuiyu, Wang Dianzuo, Li Baidan, 1994a. Hydrophobicity-hydyphilicity balance relationships for collectorless flotation of sulphide minerals. J. Cent. South Univ. Technol., 1(1): 68 – 73

Sun Shuiyu, Wang Dianzuo, Long Xiangyun, 1994b. Frontier molecular orbital theory consideration for electron transfer process across sulphide mineral-solution interface. J. BGRIMM, 3(1): 34 – 39 (in Chinese)

Sun Wei, Hu Yuehua, Qiu Guanzhou, Xu Jing, 2002. Dynamic simulation of ion adsorption on ZnS(110). The Chinese Journal of Nonferrous Metals, 12(1): 187 – 190 (in Chinese)

Sun Wei, Hu Yuehua, Qiu Guanzhou, Qin Wenqing, 2004a. Oxygen adsorption pyrite (100) surface by density functional theory. J. Cent. South Univ. Technol., 11(4): 385 – 390

Sun Wei, Hu Yuehua, Qin Wenqing, 2004b. DFT research on activation of sphalerite. Trans. Nonferrous Met. Soc. China, 14(2): 376 – 382

Sun Wei, Liu Runqing, Hu Yuehua, 2005. Research on depression mechanism of jamesonite and pyrrhotite by organic depressant DMPS. Mining and Metallurgical Engineering (in Chinese)

Sun Wei, Liu Runqing, Cao Xuefeng, Hu Yuehua, 2006. Flotation separation of marmatite from pyrrhotite using DMPS as depressant. Transactions of Nonferrous Metal Society of China, 16(3): 671 – 675

Sutherland, K. L. and Wark, I. W., 1955. Principles of flotation. Austral-asian Inst. Min. Metall., Melboure, 489 – 499

Taggart, A.F. et al., 1930. Am. Inst. Min. Metall. Engrs. Tech. Publ., 312: 3 – 33

Taggart, A. F., 1945. Handbook of Mineral Dressing. New York: John Wiley & Sons Co.

Takahashi, K., 1991. A trend in the adsorption of ethyl xamthato on solids from the point of their energy levels by the molecular orbital method. 17th International Mineral Processing Congress, Dresden, Germany, Preprints II: 393 – 408

Thornton, E., 1973. The effect of grinding media on flotation selectivity. Proc. 5th Annual Meeting of Canadian Mineral Processors, Ottawa, 224 – 229

Tolun, R. and Kitchener, J. A., 1963-1964. Electrochemical study of the galena-xanthate-oxygen flotation system. Trans. Inst. Min. Metall., 73: 313 – 322

Toperi, D. and Tolun, R., 1969. Electrochemical study and thermodynamic equilibrium of the galena-oxygen-xanthate flotation system. Trans. IMM, Sec. C, 78: 191 – 197

Trahar, W. J., 1984. The influence of pulp potential in sulphide flotation. In: Principles of Mineral Flotation. The Wark symposium. M. H. Jones and J. T. Woodcock (Eds.), Australa. Inst. Min. Metall. Parkville, Victoria. Australia, 117 – 135

Trahar, W. J., Senior, G. D., Heyes, G. W., Creed, M. D., 1997. The activation of sphalerite by lead—a flotation perspective. Inter. J. Miner. Process, 49: 121 – 148

Usul, A. H. and Tolun, R., 1974. Electrochemical study of the pyrite-oxygen-xanthate system. Inter. J. Miner. Process, 1: 135 – 140

Vanghan, D. J. and Craig, J. R., 1978. Mineral Chemistry of Metal Sulphides. London: Cambridge University Press, 51 – 62, 376 – 412

Vanghan, D. J., England, K. E. R., Kelsall, G. H., Yin Q., 1995. Electrochemical oxidation of chalcopyrite and the related metal-enriched derivatives $Cu_4Fe_5S_8$, $Cu_9Fe_9S_{16}$ and $Cu_9Fe_8S_{16}$. Am. Miner., 80: 725 – 731

Vanghan, D. J., Becker, U., Wright, K., 1997. Sulphide mineral surfaces: theory and experiment. Inter. J. Miner. Process, 51: 1 – 4

Vathsala, K. A. Natarajan, 1989. Some electrochemical aspects of grinding media corrosion and sphalerite flotation. Inter. J. Miner. Process, 26(3 – 4): 193 – 203

Walker, G. W., Stout, J. V. Ⅲ, Richardson, P. E., 1984. Electrochemical flotation of sulphides: reaction of chalcopyrite in aqueous solution. Inter. J. Miner. Process, 12: 55 – 72

Wang Dianzuo, 1983. Structure and reactivty of organic depressant on flotation. Nonferrous Metals, (2): 47 – 51

Wang Dianzuo and Hu Yuehua, 1989. Solution Chemistry of Flotation. Changsha: Hunan Sci. and Techno. Press, 343 – 345

Wang Dianzuo and Sun Shuiyu, 1990. Quantum chemical mechanism on surface oxidation and flotation of sulphide minerals. Transaction of NFSOC, 1: 1 – 12

Wang Dianzuo, Hu Yuehua, Li Baidan, Sun Shuiyu, 1991a. Mechanism of collectorless flotation of sulphide minerals. Nonferrous Metals (Quarterly), 43(3): 34 – 39 (in Chinese)

Wang Dianzuo, Hu Yuehua, Li Baidan, Huang Kaiguo, Sun Shuiyu, 1991b. Study of technological factor of collectorless flotation of sulphide mineral. Nonferrous Metal (Quarterly), 43(4): 21 – 26 (in Chinese)

Wang Dianzuo, Long Xiangyun, Sun Shuiyu, 1991c. Quantum chemical mechanism on the surface oxidation and flotation of sulphide minerals. Trans. of NFSOC, 1(1): 20 – 27

Wang Dianzuo, Hu Yuehua, Li Baidan, Sun Shuiyu, 1991d. Study of mechanism of collectorless flotation of sulphide mineral. Nonferrous Metal (Quarterly), 43(4): 34 – 39 (in Chinese)

Wang Dianzuo, 1992. Development of Flotation Theory. Beijing: Science Press, 79 – 143

Wang Dianzuo, Hu Yuehua, Li Baidan, 1992. Collectorless flotation of sulphide minerals and challenge to classical flotation theory. Nonferrous Metals (Quarterly), 44(1): 22 – 27 (in Chinese)

Wang Dianzuo, Sun Shuiyu, Li Baidan, Huang Kaiguo, Bai Shibin, 1993a. Design and test of flotation and separation flowsheets for molybdenum, bismuth and iron sulphides. J. CSIMM, 24(3): 306 – 311 (in Chinese)

Wang Dianzuo, Sun Shuiyu, Huang Kaiguo, Bai Shibin, Li Baidan, 1993b. A study on the Na_2S-induced flotation and separation of a sulphide containing Mo, Bi and Fe. J. CSIMM, 24(3): 312 – 317

Wang Jianqi, 1992. Introduction to Electron Energy Spectra (XPS/XAES/UPS). Beijing: Defence Industrial Press

Wang, X., and Xie, Y., 1990. The effect of grinding media and environment on the surface properties and flotation behaviour of sulphide minerals. Miner. Process Extra Metall. Rev., 7: 49 – 79

Wang, X. H. and Forssberg, E., 1989. A study of the natural and induced hydrophbicity of some sulphide minerals by collectorless flotation. In: Processing of Complex Ores, G. S. Dobby and S. R. Rao (eds.), CIM, Halifax, 3 – 17

Wang, X. H. and Forssberg, K. S. E., 1991. Mechanism of pyrite flotation with xanthates. In: K. S. E. Forssberg (ed.), Flotation of Sulphide Minerals. Inter. J. Miner. Process, 33: 275 – 290

Wark. I. W. and Cox, A. B., 1933. The chemical basis of flotation. PAIMM, 90: 80 – 123

Wei, Q. and Asare, K. O., 1996. Semiconductor electrochemistry of particle pyrite: dissolution via hole and electron pathways. J. Electrochem. Soc., 143(10): 3193 – 3198

Witika, L. K. and Lombe, W. C., 1989. Electrochemical behaviour of carrollite ($CuCo_2S_4$) in aqueous buffer solutions. Trans. IMM, Sec. C. 98: C26 – C32

Woods, R., 1971. The oxidation of ethyl xanthate on platinum, gold, copper, and galena electrodes, relation to the mechanism of mineral flotation. J. Phys. Chem., 75: 354 – 362

Woods, R., 1984. Electrochemistry of sulphide flotation. In: M. H. Jones and J. T. Woodcock (Eds.), Principle of Mineral Flotation. The Wark Symposium, Australia's Inst. Min. Metal, Parkville, Victoria, Australia. 40: 91 – 115

Woods, R., 1991. Electrochemistry of flotation of sulphide minerals. Lectures in Central South University of Technology

Woods, R., 1996. Chemisorption of thiols on metal and metal sulphide. In: J. O'M Bockris, B. E. Conway, R. E. White (eds.). Modern Aspects of Electrochemistry. 29: 401 – 453

Woods, R., Young, C. A., Yoon, R. H., 1990. Ethyl xanthate chemisorption isotherms and E_h-pH diagrams for the copper/water/xanthate and chalcocite/water/xanthate systems. Inter. J. Miner. Process, 30: 17 – 33

Woods. R. and Yoon, R. H., 1994. Chemisorption of ethyl xanthate on copper electrodes. Inter. J. Miner. Process, 42(3 – 4): 215 – 223

Woods, R., Basilio, C. I., Kim, D. S., Yoon, R. H., 1994. Chemisorption of ethyl xanthate on silver-gold alloys. Colloids and Surfaces A: Physicochemical and Engineering Aspects, 83(1): 1 – 7

Woods, R. and Yoon, R. H., 1997. Comment on "In situ and ex situ infrared studies of nature and structure of thiol layers adsorbed on cuprous sulphide at controlled potential. Langmuir, 13: 876 – 877

Woods, R., Hope, G. A., Brown, G. M., 1998. Spectro-electro-chemical investigations of the interaction of ethyl xanthate with copper, silver and gold: III SERS of xanthate adsorbed on gold surfaces. Colloids and Surfaces A, 137: 329 – 337

Woods, R., Hope, G. A., Watling, K., 2000. Surface enhanced Raman scattering spectroscopic studies of the adsorption of flotation collectors. Minerals Engineering, 13(4): 345 – 356

Xiao Qi, Qiu Guanzhou, Hu Yuehua, 2001. Computational simulation to mechanical activation of pyrite (I)—relation of structural strain to chemistry reaction activity. The Chinese Journal of Nonferrous Metals, 11(5): 900 – 905 (in Chinese)

Xiao Qi, Qiu Guanzhou, Hu Yuehua, Wang Dianzuo, 2002a. Theoretical study on the geometry and the electronic structure of FeS_2 (100) surface. Acta Physica Sinica, 51(9): 2134 – 2137 (in Chinese)

Xiao Qi, Qiu Guanzhou, Hu Yuehua, Wang Dianzuo, 2002b. FeS_2 band gap modulation under pressure perturbation. Chinese Journal of High Pressure Physics, 16(3): 188 – 193 (in Chinese)

Xiong Daoling, Hu Yuehua, He Zhiguo, Zhan Xuehui, 2004. Advances in research on organic depressor's depressing of arsenopyrite in sulphide flotation. Mining and Metallurgical Engineering, 24 (2): 42 – 44 (in Chinese)

Xiong Daolin, Hu Yuehua, Qin Wenqing, Sun Wei, Liu Runqing, 2006. Selective flotation separation of marmatite from arsenopyrite by organic depressant PALA. Journal of Central South University, 37(4): 670 – 674 (in Chinese)

Xu, B. and Huang, K., 1995. Role of *a*-amino aryl phosphorated organic depressant in flotation of high nickel matte. Nonferrous Metals, 1(47): 21 – 23 (in Chinese)

Xu Jing, Sun Wei, Zhang Qin, Liu Hui, Hu Yuehua, 2003. Research on depression mechanism of pyrite and pyrrhotite by new organic depressant RC. Mining and Metallurgical Engineering, 23(6) (in Chinese)

Xuan Cheng and Iwasaki, I., 1992. Pulp potential and its implications to sulphide flotation. Miner. Process Extra. Metall. Rev., 11: 187 – 210

Yarar, B., Haydon, B. C., Kitchener, J. A., 1969. Electrochemistry of the galena-diethyldithio carbamate-oxygen flotation system. Trans. Instn. Min. Metall. Sec.C, 78: C181 – C184

Yekeler Meftuni, Sonmez Ibrahim, 1997. Effect of the hydrophobic fraction and particle size in the collectorless column flotation kinetics. Colloids and Surfaces A: Physicochemical and Engineering, 121(1): 9 – 13

Yelloji Rao M. K. and Natarajan, K. A., 1988. Influence of galvanic interaction between chalcopyrite and some metallic materials on flotation. Minerals Engineering, 1(4): 281 – 294

Yelloji Rao M. K. and Natarajan, K. A., 1989a. Effect of electrochemical interactions among sulphide minerals and grinding medium on chalcopyrite flotation. Minerals & Metallurgical Processing, 6(3): 146 – 151

Yelloji Rao M. K. and Natarajan, K. A., 1989b. Electrochemical effects of mineral-mineral interactions on the flotation of chalcopyrite and sphaterite. Inert. J. Miner. Process, 27: 279 – 293

Yelloji Rao M. K. and Natarajan, K. A., 1989c. Effect of galvanic interaction between grinding media and minerals on sphalerite flotation. Inter. J. Miner. Process, 27(1 – 2): 95 – 109

Yelloji Rao M. K. and Natarajan, K. A., 1990. Effect of electrochemical interactions among sulphide minerals and grinding medium on the flotation of sphalerite and galena. Inter. J. Miner. Process, 29: 175 – 194

Yoon, R. H., 1981. Collectorless flotation of chalcopyrite and sphalerite by using sodium sulphide. Inter. J. Miner. Process, 8: 31 – 48

Young, C. A., Woods, R., Yoon, R. H., 1998. A voltammetric study of chalcocite oxidization to metalstable copper sulphides. In: P. E. Richardson, R. Woods (ed.), Int. Symp. Electrochemistry in Mineral and Metal Processing II. Pennington: Electrochem. Soc., 3 – 17

Yu Runlan, Hu Yuehua, Qiu Guanzhou, Qin Wenqing, 2004a. An electrochemical study of DDTC adsorption jamesonite. Electrochemistry, 10 (2) (in Chinese)

Yu Runlan, Hu Yuehua, Qiu Guanzhou, Qin Wenqing, 2004b. A voltammetric study of corrosion and interaction of marmatite with collector. Mining and Metallurgical Engineering, 24(1) (in Chinese)

Yu Runlan, Hu Yuehua, Qiu Guanzhou, Qin Wenqing, 2004c. Interaction mechanism of jamesonite with flotation collectors by cyclic voltammetry. Journal of Central South University, 35(2): 202 – 206 (in Chinese)

Yu Runlan, Qiu Guanzhou, Hu Yuehua, Qin Wenqing, 2004d. Electrochemistry of copper activation of marmatite. Metal Mine, (2) (in Chinese)

Yu Runlan, Qiu Guanzhou, Hu Yuehua, Qin Wenqing, 2004e. Interface electrochemistry of interaction of collector with jamesonite. The Chinese Journal of Nonferrous Metals, 14(1): 127 – 131 (in Chinese)

Yu Runlan, Qiu Guanzhou, Hu Yuehua, Qin Wenqing, 2004f. Study of electrochemistry of jamesonite in DDTC and saturation $Ca(OH)_2$. The Chinese Journal of Nonferrous Metals, 14(10): 1763 – 1769 (in Chinese)

Yu Runlan, Qiu Guanzhou, Hu Yuehua, Qin Wenqing, 2004g. The corrosive electrochemical study of marmatite. Journal of Chinese Society For Corrosion and Protection, 24(4): 226 – 229

Yu Runlan, Qiu Guanzhou, Hu Yuehua, Qin Wenqing, 2005. Electrochemical adsorption behaviour and mechanism of diethyldithiocarbamate on surface of marmatite. The Chinese Journal of Nonferrous Metals, 15(9): 1452 – 1457 (in Chinese)

Yu Runlan, Qiu Guanzhou, Hu Yuehua, Qin Wenqing, 2006. Study on mechanism of diethyldithiocarbamate with jamesonite by spectroelectro-chemistry. Mining and Metallurgical Engineering, 26(1): 29 – 31 (in Chinese)

Zhang, D.H., 1996. Adsorption and photodesorption of oxygen on the surface and crystallite interfaces of sputtered ZnO films. Materials Chemistry and Physics, 45(3): 248 – 252

Zhang Qin, Hu Yuehua, Gu Guohua, Xu Jing, 2004a. The study on the Interaction between ethyl xanthate and pyrrhotite in electrochemical flotation by FTIR spectroscopy. Mining and Metallurgical Engineering, 24(5): 42 – 44 (in Chinese)

Zhang Qin, Hu Yuehua, Gu Guohua, Xu Jing, 2004b. Selective flotation separation of jamesonite from pyrrhotite by potassium cyanide. Journal of Central South University of Technology, 35(3): 372 – 375

Zhang Qin, Hu Yuehua, Gu Guohua, Nie Zhenyuan, 2004c. Study of the pyrrhotite self-induced flotation electrochemical. Nonferrous Metals, (2): 4 – 6 (in Chinese)

Zhang Qin, Hu Yuehua, Gu Guohua, Nie Zhenyuan, 2004d. Electrochemical flotation of ethyl xanthate-pyrrhotite system. Trans. Nonferrous Met. Soc. China, 14 (6): 1174 – 1179

Zhang Qin, Hu Yuehua, Gu Guohua, Xu Jing, 2004e. Selective flotation separation of jamesonite from pyrrhotite by lime. Mining and Metallurgical Engineering, 24(2): 30 – 32 (in Chinese)

Zhang Qin, Hu Yuehua, Gu Guohua, Xu Jing, 2004f. Mechanism of Cu^{2+} ion activation flotation of marmatite in absence and presence of ethyl xanthate. The Chinese Journal of Nonferrous Metals, 14(4): 676 – 679 (in Chinese)

Zhang Qin, Xu Jing, Hu Yuehua, Chen Tiejun, 2006. Study on separation of jamesonite from pyrrhotite by using RC as a new inhibitor. Mining and Metallurgical Engineering, 26(1): 27 – 28 (in Chinese)

Zhao Jing et al., 1988. Research on the mechanism of chalcopyrite depressed by sodium mercaptoacetic. Nonferrous Metals (part of mineral processing), (3): 42 – 45

Zhuo Chen and Yoon, R. H., 2000. Electrochemistry of copper activation of sphalerite. Inter. J. Miner. Process, 58: 57 – 66

Index of Terms

A

B

C

D

F

G

H

I

O

P

X

Y

Z

Index of Scholars

A

B

E

F

G

H

I

J

K

L

M

N

O

P

Q

R

S

T

U

V

W

GPSR Compliance

The European Union's (EU) General Product Safety Regulation (GPSR) is a set of rules that requires consumer products to be safe and our obligations to ensure this.

If you have any concerns about our products, you can contact us on ProductSafety@springernature.com

In case Publisher is established outside the EU, the EU authorized representative is:

Springer Nature Customer Service Center GmbH
Europaplatz 3
69115 Heidelberg, Germany

Batch number: 10321657

Printed by Printforce, the Netherlands